FOURTH INTERNATIONAL CONFERENCE ON COLLECTIVE PHENOMENA

ANNALS OF THE NEW YORK ACADEMY OF SCIENCES
Volume 373

FOURTH INTERNATIONAL CONFERENCE ON COLLECTIVE PHENOMENA

Edited by Joel L. Lebowitz

The New York Academy of Sciences
New York, New York
1981

Copyright © 1981 by The New York Academy of Sciences. All rights reserved. Under the provisions of the United States Copyright Act of 1976, individual readers of the Annals are permitted to make fair use of the material in them for teaching or research. Permission is granted to quote from the Annals provided that the customary acknowledgment is made of the source. Material in the Annals may be republished only by permission of The Academy. Address inquiries to the Executive Editor at The New York Academy of Sciences

Copying fees: The code at the bottom of the first page of each article in this Annual states the fee for each copy of the article made beyond the free copying permitted under Section 107 or 108 of the 1976 Copyright Act. (If no code appears, there is no fee.) This fee should be paid through the Copyright Clearance Center, Inc., Box 765, Schenectady, N.Y. 12301. For articles published prior to 1978, the copying fee is $1.75 per article.

Library of Congress Cataloging in Publication Data

International Conference on Collective Phenomena
  (4th : 1981 : Moscow, R.S.F.S.R.)
  Fourth International Conference on Collective Phenomena.

  (Annals of the New York Academy of Sciences; v. 373)
  1. Science—Congresses. I. Lebowitz, Joel Louis, 1930– . II. New York Academy of Sciences. III. Title. IV. Series.
Q11.N5 vol. 373 [Q1O1] 500s [500] 81-16792
                                                    AACR2

SP
Printed in the United States of America
ISBN 0–89766–135–4
ISBN 0–89766–136–2

ANNALS OF THE NEW YORK ACADEMY OF SCIENCES
VOLUME 373
October 30, 1981

FOURTH INTERNATIONAL CONFERENCE ON COLLECTIVE PHENOMENA*

Editor

JOEL L. LEBOWITZ

CONTENTS

Foreword. *By* JOEL L. LEBOWITZ	vii
The Arbitrariness in the Perturbation Series in Non-Abelian Gauge Theories. *By* H. EPSTEIN AND J. ILIOPOULOS	1
Possible Tests of Quantum Chromodynamics in Large Transverse Momentum Hadron Photoproduction. *By* D. SCHIFF	8
Photon-Photon Collisions. *By* PAUL KESSLER	15
The KMS Condition for *-Algebras. *By* J. ALCÁNTARA AND D. A. DUBIN	22
Lattice and Molecular Dynamics of Some Amine Intercalates of FeOCl. *By* R. H. HERBER	28
Quantum Chemical Aspects of Some Problems in Bioinorganic Chemistry. III. Some Ligand Properties of Metal Complexes: Population Analysis. *By* I. FISCHER-HJALMARS AND A. HENRIKSSON-ENFLO	37
Phospholipid Methylation and Membrane Function. *By* JULIUS AXELROD AND FUSAO HIRATA	51
Inverse Scattering, Ordinary Differential Equations of Painlevé-Type, and Hirota's Bilinear Formalism. *By* A. RAMANI	54
Explosively Heated Gaussian Objects. *By* F. J. MAYER AND D. J. TANNER	68
Surface Studies of Fusion Reactor Wall Materials at AFI. *By* T. FRIED, B. EMMOTH, AND M. BRAUN	77
A Quantum Model of Doubt. *By* YURI F. ORLOV	84
Agreement Through Fair Play. *By* ALEKSANDR YA. LERNER	93
Nonsmooth Analysis and the Theory of Fans. *By* ALEKSANDR D. IOFFE	101
On Multiple Regression for the Case with Error in Both Dependent and Independent Variables. *By* VIKTOR BRAILOVSKY	113
On Multivariate Linear Regression with Missing Data among the Independent Variables. *By* VIKTOR BRAILOVSKY	128
On Some Important Features of Extra Low Frequency and Low Frequency Electromagnetic Waves $0 \lesssim \omega \lesssim \omega_L$ in a Magnetoplasma Connected with the Influence of Ions. *By* YAKOV L. AL'PERT	138

*This volume is the result of the Fourth International Conference on Collective Phenomena, held on April 12, 13, and 14 in Moscow, USSR, sponsored by the New York Academy of Sciences.

The Effect of the Earth's Rotation on the Propagation of Weak Nonlinear Surface and Internal Long Oceanic Waves. *By* A. I. LEONOV 150

Recent Developments in Contour Dynamics for the Euler Equations. *By* NORMAN J. ZABUSKY 160

Confinement and Phase Transitions in Gauge Theories. *By* LAURENCE JACOBS .. 171

Some Considerations of Stability in Solidification of Lamellar Eutectics. *By* J. S. LANGER 179

On Two-Element Subsets in Groups. *By* L. V. BRAILOVSKY AND G. A. FREIMAN 183

Integer Programming and Number Theory. *By* P. L. BUZYTSKY AND G. A. FREIMAN 191

Trends in the Development of Computer Applications. *By* DANIEL D. MCCRACKEN 202

Doob-Meyer Decompositions for Two-Parameter Stochastic Processes. *By* ELY MERZBACH 205

Magnetic Properties of Relativistic Fermi Gas. *By* E. M. CHUDNOVSKY 208

Entropy and Irreversibility. *By* OLIVER PENROSE 211

Microscopic Dynamics and Macroscopic Laws. *By* JOEL L. LEBOWITZ 220

The New York Academy of Sciences believes that it has a responsibility to provide an open forum for discussion of scientific questions. The positions taken by the scientists whose papers are presented here are their own and not necessarily those of The Academy. The Academy has no intent to influence legislation by providing such forums.

FOREWORD

Joel L. Lebowitz*

*Departments of Mathematics and Physics
Rutgers University
New Brunswick, New Jersey 08903*

The Fourth International Conference on Collective Phenomena took place, like the preceding ones, not in a well-appointed academic lecture hall but in a cramped living room in a Moscow apartment under the discomforting surveillance of the KGB. Yet the room, indeed, the whole apartment, was full to overflowing with both local and foreign scientists, and the papers presented, as can be seen in this volume, were of high quality and interest.

Why should scientists with comfortable positions in Western universities travel to Moscow to attend an unauthorized conference in a crowded apartment? The answer to this question raises another, broader, issue, for, simply by their presence there, all who attended the conference were expressing their involvement with the cause of human rights and dignity. What does science have to do with human rights? Do scientists have a special duty to human rights?

Yes, for science is a humanistic enterprise—that is, it is, above all, a distinctly human adventure. The fact that our minds can actually comprehend something about the structure and nature of the universe is not something to be taken for granted—it is cause for great surprise, great excitement, and even greater awe. Given their vision and understanding of the universe, scientists are, or should be, particularly aware of the preciousness, uniqueness, and inherent dignity of human beings. Intellectual honesty and respect for truth should also make us recognize the indivisibility of this human dignity. Denying this dignity to anyone, in any country, diminishes all of us, in every country.

As scientists, we also realize how precarious human existence and human civilization are. The only long-term safeguard of our civilization, our values, and our lives is a completely peaceful world. Such a world is possible only if human rights and human dignity become a permanent and universally respected part of society.

It is, then, for these very selfish, very practical reasons that scientists, as intellectual leaders, should be concerned with human rights, both here and abroad. And it is particularly appropriate for us to start with the human rights of our scientific colleagues, which are cruelly violated in many parts of the world today. During the past few years, the New York Academy of Sciences has been deeply involved with this problem—mainly, though not exclusively, with regard to scientists in the Soviet Union.

Why single out the Soviet Union? Indeed, the Soviet Union is not the worst offender in this respect at present. One of our so-called friendly neighbors, such as Argentina, Chile, or Uruguay, probably bears that ignoble distinction. However, the Soviet Union is a country of especial concern to us because it has, like us, the capability to destroy the world. Also, the achievements of Soviet scientists are, in

---

*Chairman, Human Rights Committee of the New York Academy of Sciences.

many areas, equal to or superior to our own. It is, therefore, a country that we would, for purely pragmatic reasons, very much like to see pursuing a peaceful course. This requires, as I have said, that it respect the dignity and rights of all people.

This is not the case at the present time. All the local participants in the fourth of these conferences had applied for and been refused permission to emigrate to Israel (thus, "refusniks"). They had consequently lost their jobs or been demoted and, in all cases, had been shut out completely from the scientific life of the Soviet Union. Other "dissident" scientists—those who had spoken out for human rights—were unable to attend this meeting because they were in labor camps (Orlov, Shcharansky, Kovalev, etc.) or in exile (Sakharov).

These conferences are an extension of the Moscow Sunday Seminars begun in 1972 by a group of refusnik scientists so that they could keep up their professional work despite their exclusion from all official scientific activities. The Seminar soon gained a reputation for its high standards and for the devotion of its participants to science. It has taken place in the Brailovskys' living room since 1977 and has been visited by scores of Western scientists.

The Seminar's first attempt to organize an international conference in 1974 was foiled by the Soviet authorities—the organizers of the conference were arrested and held for fifteen days. Meetings did take place in 1977, 1978, and 1980, though under the surveillance of the authorities, with some intimidation of the local participants.

Unfortunately, the situation has worsened since the April 1980 conference. Viktor Brailovsky, the organizer and conference chairman, was arrested in November of that year and subsequently convicted of "defaming the Soviet Union." He was sentenced to five years of exile. The charges against him were based on his role in the publication of a short-lived unofficial magazine *Jews in the USSR*. The seminar itself, being "legal" by Soviet law, was not officially cited against him. There are, however, strong reasons to believe that the Seminar was, indeed, a cause of Dr. Brailovsky's problems. The Seminar has since been forcibly prevented from taking place on many occasions and its future is uncertain. There have also been other arrests of many dissident and some refusnik scientists.

Given this bleak situation, what can scientists do to effect changes for the better? This is a difficult question, to which I can only give partial and personal answers. First and foremost, it is very important that we not remain silent in the face of the continuing mistreatment of our colleagues—no matter how repetitious our protests appear to be. I believe that the Soviet authorities care about their image in the outside world and scientists are an influential group in the West and in some third world countries. What scientists think about the Soviet Union can affect arms agreements, technology transfers, and other matters that go beyond purely scientific exchanges. We should, therefore, persist in making our concern known to the Soviets in all kinds of ways: by protest actions and by keeping in touch with dissident and refusnik scientists.

These scientists greatly value contacts with their Western colleagues. They wish to have correspondence with and visits from us. We must not fail them. If you wish more information on how you can help scientists in trouble, in the USSR and elsewhere, please contact me or Mrs. Dorothy Hirsch, Committee of Concerned Scientists, 9 East Fortieth Street, New York, New York 10016.

I would like to thank Frederick Bartlett, Bill Boland, and India Trinley of the New York Academy of Sciences, Simon Levin of Cornell University, Valentin Turchin of the City College of New York, Lee Segal of Rensselaer Polytechnic Institute, and Martin Kruskal and Thomas Stix of Princeton University for their aid in the publication of this Annal.

THE ARBITRARINESS IN THE PERTURBATION SERIES IN NON-ABELIAN GAUGE THEORIES

H. Epstein

Institut des Hautes Etudes Scientifiques
91440 Bures-sur-Yvette, France

J. Iliopoulos

Laboratoire de Physique Théorique
de l'Ecole Normale Superieure
75231 Paris, cedex 05, France

INTRODUCTION

The greatest achievement of high energy physics in the last decade has been the realization that gauge theories describe all interactions among elementary particles. All phenomena observed today are compatible with a model based on the group $U(1) \times SU(2) \times SU(3)$. The first two factors correspond to the electromagnetic and weak interactions, while the third gives rise to the strong ones. There is ample experimental evidence to support these assertions, but the striking feature is that it is all derived from perturbative calculations. This was expected for the electroweak interactions because we were accustomed to thinking of them as weak coupling theories. Surprisingly, it also turned out to be true for the strong interactions in certain kinematic regions. It is only where perturbation theory is applicable that reliable numerical results can be obtained using the available field theory technology. So, in this sense, we can say that this recent progress was the triumph of renormalizable perturbation theory. Its domain of validity, starting with the interactions of photons with charged leptons, has been extended to cover most observed phenomena.

In spite of this success, it is today that the limitations of perturbation theory are most painfully felt. All gauge theories, which we mentioned above, exhibit essential nonperturbative behavior. For the $U(1) \times SU(2)$ electroweak theory, the nonperturbative phenomena are associated with spontaneous symmetry breaking. The problem is even more crucial for quantum chromodynamics, the $SU(3)$ gauge theory of strong interactions.

At large distances, we enter the strong coupling regime where perturbation theory is useless. This region, if the theory is any good for physics, must possess the property of confinement, i.e., it must forbid the appearance of asymptotic states corresponding to the fundamental fields (colored quarks as well as non-Abelian gluons). Numerical results suggest that the transition from the weak to the strong coupling regimes is rather abrupt,[1] but they are based on a truncated form of the theory in which space-time is replaced by a small finite lattice. There exist no direct computational methods in a strong coupling theory in the space-time continuum.

Because of all these problems, it is interesting to investigate how far the "weak coupling regime" extends. By this we mean the range of coupling constants for which perturbation theory can give meaningful results. Even expressed this way the problem is rather ill-defined, except, of course, for the case of a convergent series, a rather unlikely state of affairs. It is clear, however, that all these questions are connected with the behavior of the perturbation series and it is this last topic that we shall try to discuss.[2]

The general statement of the problem is the following: Let A be some physical quantity. Perturbation theory gives A as a power series of the form

$$A(g) = \sum_{n=0}^{\infty} g^n A_n, \qquad (1)$$

where g is a coupling constant whose role is to identify the different orders of the perturbation series. It will be defined more precisely in the explicit cases considered below. We are interested in the behavior of this series.

Two questions are relevant here:

1. Let $\hat{A}(g)$ be the corresponding quantity computed by the exact solution of the problem. What is its relation with $A(g)$ of (1), if any? As could have been anticipated, the answer to this question depends on the particular case considered. It is worth noticing that, even when (1) represents an absolutely convergent series, it does not necessarily converge to $\hat{A}(g)$.

2. Even if we forget about $\hat{A}(g)$, which, in any case, is unknown for almost every interesting problem, does the series (equation 1) define a function $A(g)$? In other words, is the perturbation series convergent in any generalized sense? This question has been extensively investigated and there are several results for quantum mechanical problems and superrenormalizable field theories.[2] For renormalizable field theories, however, the question is still ill-defined. In fact, in these theories, the coupling constant g is not given a priori, but, rather, is defined as part of the renormalization program. Usually, it is a value assigned to a certain Green function at prescribed values of the external momenta. However, alternative definitions are also possible and, if g_1 is a coupling constant defined by one prescription, we can define a new one, g_2, by a series of the form:

$$g_2 = g_1 + \sum_{n=2}^{\infty} a_n g_1^n, \qquad (2)$$

where the set of numbers a_n is completely arbitrary. Furthermore, (2) exhausts all possible perturbative definitions.

We now see why question 2 is ill-defined. Perturbation series in terms of g_1 or g_2 may have completely different convergence properties. Therefore, for renormalizable theories, the interesting question is a third one:

3. Is there an "optimal" definition of the coupling constant, i.e., one that makes the series (equation 1) for the quantity of interest $A(g)$ as good as possible?

We shall argue below that the answer to this question is negative for all physically interesting quantities in quantum chromodynamics.

The logic of the argument is the following: First, we present the main result, due to Gerard 't Hooft,[3] which shows, with a specifically chosen definition of the coupling constant, that Green functions of gauge-invariant operators have a domain of analyticity near the origin in the complex g^2 plane of the form

$$|\text{Im } g^2| < C (\text{Re } g^2)^2, \qquad (3)$$

where C is a constant. This domain contains the beginning of the positive real axis, but with a vanishing opening angle around it. Second, we show that this situation cannot be improved by a finite coupling constant renormalization of the form of (2).

'T HOOFT'S RESULT

Let us consider a two-point, one-particle irreducible Green function of a gauge-invariant operator. For example,

$$A(x) = \bar{q}(x) \, \Gamma \, q(x), \tag{4}$$

where $q(x)$ is the quark field operator, which has color, flavor, and Dirac indices and Γ is some matrix such that $A(x)$ is a color singlet. We define

$$\Gamma(k^2, \mu^2, g^2) = \int \langle 0 | T(A(x), A(x)) | 0 \rangle_c \, e^{ikx} dx \tag{5}$$

The Green function in the r.h.s. is the sum of all connected diagrams and μ^2 is a subtraction point used to renormalize the theory. g is the corresponding renormalized coupling constant. We do not need to specify any particular form of renormalization conditions. For dimensional reasons, Γ, apart from a trivial constant factor, is a function of two variables: $\Gamma(k^2/\mu^2, g^2)$. We are interested in the analytic properties of Γ, for fixed k^2/μ^2, near the origin in the complex g^2 plane. From general principles, we know that, for fixed real g, it is analytic in the cut $t = k^2/\mu^2$ plane. The cut is along the positive real axis. We shall use the renormalization group in order to translate this information into analyticity properties in g^2.

Γ satisfies the following equation:

$$\left[\mu^2 \frac{\partial}{\partial \mu^2} + \beta(g^2) \frac{\partial}{\partial g^2} + \gamma(g^2) \right] \Gamma = 0, \tag{6}$$

where β and γ are functions with a perturbation expansion in powers of g^2:

$$\beta(g^2) = b_0 g^4 + b_1 g^6 + b_2 g^8 + \ldots \tag{7}$$

$$\gamma(g^2) = \gamma_0 + \gamma_1 g^2 + \gamma_2 g^4 + \ldots \tag{8}$$

The β-function is the usual renormalization group function, but $\gamma(g^2)$, given by (8), is determined by the renormalization of the operator $A(x)$ and the Green function Γ. This is the origin of the constant term in the expansion (equation 8).

If $\beta(g^2)$ and $\gamma(g^2)$ were exactly known, then the solution of (6) would yield

$$\Gamma\left(\frac{k^2}{\mu^2}, g^2\right) = \Gamma\left(\left(1, \bar{g}^2\left(\frac{k^2}{\mu^2}, g^2\right)\right)\right) e^{-\int_{g^2}^{\bar{g}^2} \gamma(x)/\beta(x) \, dx}, \tag{9}$$

with $\bar{g}^2 \, (k^2/\mu^2, g^2)$ satisfying

$$\mu^2 \frac{\partial}{\partial \mu^2} \bar{g}^2 \left(\frac{k^2}{\mu^2}, g^2\right) = \beta(\bar{g}^2) \qquad \bar{g}^2 \, (1, g^2) = g^2. \tag{10}$$

The problem, of course, is that, in general, $\beta(g^2)$ and $\gamma(g^2)$ are only given by their first few terms in the expansions of equations 7 and 8. This is the crucial observation:

There exists a certain implicit renormalization scheme such that $\beta(g^2)$ and $\gamma(g^2)$ become known to all orders; in fact, they are polynomials in g^2. Let us perform the finite renormalization[3,4]

$$g^2 = \hat{g}^2 + \alpha_1 \hat{g}^4 + \alpha_2 \hat{g}^6 + \ldots . \tag{11}$$

Correspondingly, we define a new function, $\hat{\beta}(\hat{g}^2)$, with a perturbation expansion

$$\hat{\beta}(\hat{g}^2) = \hat{b}_0 \hat{g}^4 + \hat{b}_1 \hat{g}^6 + \hat{b}_2 \hat{g}^8 + \ldots . \tag{12}$$

It is well known that consistency of the two expansions imposes the conditions

$$b_0 = \hat{b}_0 \qquad b_1 = \hat{b}_1, \tag{13}$$

with all other $\hat{b}_i$s ($i \geq 2$) becoming functions of the b_is and the α_is. In particular, it is easy to verify that one can choose the α_is such that

$$\hat{b}_i = 0 \qquad i \geq 2, \tag{14}$$

i.e., in the $\hat{g}$ renormalization scheme, the β-function is given to all orders by its first two terms.

With a similar argument, by appropriately modifying the renormalization conditions of the operator $A(x)$ and the Green function Γ, we can choose $\hat{\gamma}(\hat{g}^2)$ to be given, exactly, by

$$\hat{\gamma}(\hat{g}^2) = \gamma_0 + \gamma_1 \hat{g}^2 \tag{15}$$

and all higher $\hat{\gamma}_i$ ($i \geq 2$) coefficients vanishing. Needless to say, we cannot write down the renormalization conditions that define the $\hat{g}^2$ scheme in closed form. Rather, we must first renormalize in some standard way and then, order by order, define the finite renormalizations so that (14) and (15) are satisfied. The important thing, however, is that such a scheme exists. The numerical values of the coefficients b_0, b_1, γ_0, and γ_1, which, as we saw, are scheme independent, are known for QCD:

$$b_0 = -\frac{1}{8\pi^2}\left(11 - \frac{2}{3}N_f\right)$$

$$b_1 = \frac{1}{(8\pi^2)^2}\left(\frac{19}{3}N_f - 51\right) \tag{16}$$

$$\gamma_0 = 2$$

$$\gamma_1 = -\frac{1}{\pi^2},$$

where N_f is the number of quark flavors and the last two values are given for the scalar operator $A(x) = \bar{q}(x)\,q(x)$. Now we can solve (9) and (10) explicitly, compute $\bar{g}^2(k^2/\mu^2, \hat{g}^2)$, and find the analytic properties of $\Gamma(k^2/\mu^2, \hat{g}^2)$ as a function of $\hat{g}^2$. Since the theory is asymptotically free (we assume $N_f \leq 16$ in (16)), the region of small $\hat{g}^2$ is connected with that of very large k^2. There, we have an accumulation of cuts corresponding to the different physical thresholds (mesons, baryon-antibaryon systems, etc.). After a straightforward computation, the result is that, for fixed $k^2 < 0$,

Epstein & Iliopoulos: Non-Abelian Gauge Theories

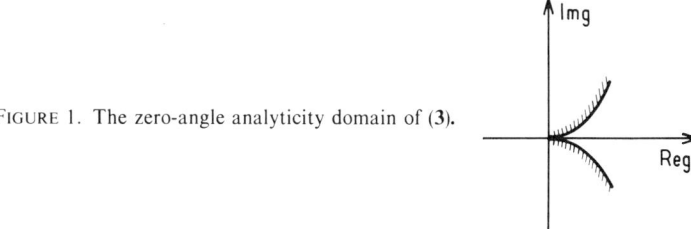

FIGURE 1. The zero-angle analyticity domain of (3).

the analyticity domain in $\hat{g}^2$ is given by (3) and shown in FIGURE 1. The vanishingly small angle around the real axis makes any resummation procedure doubtful. A continuation through the cut is problematic, unless one can evaluate the discontinuities across all inelastic cuts up to infinite energy. This is the conclusion of 't Hooft's argument.

CAN WE IMPROVE THIS RESULT?

As we said in the Introduction, one expects the analytic properties of the Green functions in the g^2-plane to depend on the definition of g^2. The bad behavior of Γ found in the last section was obtained for a particular scheme, the one that made the functions $\beta(g^2)$ and $\gamma(g^2)$ simple polynomials. Noticing that these functions are just certain combinations of Green functions of the theory at particular momenta, one may think that the price for giving them such good analytic behavior was precisely the disaster found in all the other Green functions. Is it possible to improve the behavior of Γ by relaxing the condition that β and γ are polynomials? We shall show here that the answer is negative.

In order to simplify the argument, let us assume that the boundaries of FIGURE 1 are circles. The problem is to find a transformation that maps the analyticity domain of FIGURE 1 to the one of FIGURE 2 in which we choose, as an example, a domain with a finite opening angle around the real axis. However, we must be careful. Not all transformations are acceptable. We want the new variable g_1^2 to be accessible to perturbation, also; i.e., we want g_1 and $\hat{g}$ to be related by an expansion of the form of (2) or (11). But the mapping from FIGURE 1 to FIGURE 2 is given by

$$g_1^2 = e^{-(1/\hat{g}^2)} \qquad (17)$$

or a power of it and it has no perturbation expansion.

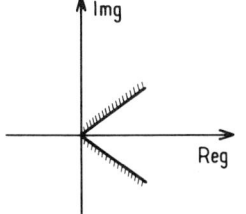

FIGURE 2. A finite-angle analyticity domain.

Let us recapitulate: We have assumed that there exists a function, $\Gamma(k^2, \hat{g}^2)$, whose successive derivatives with respect to $\hat{g}^2$ at the origin are given by the terms of the perturbation expansion. We saw that the known analyticity properties in k^2 can be translated into the zero angle domain in $\hat{g}^2$ shown in FIGURE 1. This implies a divergent perturbation series. We then found that a change of variables, $g_1^2 = f(\hat{g}^2)$, which admits a perturbation expansion, cannot improve the situation. By a similar argument, we can see immediately that no such change of variables can affect the opening angle of an analyticity domain of a Green function. In fact, we can make this statement stronger. We can consider any coupling constant renormalization of the form of (11), even if it does not correspond to the asymptotic expansion of some function $f(\hat{g}^2)$.

In this case, we can always determine the coefficients α_i of (11) so that the convergence properties of the expansion of $\Gamma(k^2/\mu^2, \hat{g}^2)$, at some particular values of k^2, are improved. This, however, will not change the situation for other values of k^2. Let us be more precise: We assume that the two-point function Γ is given by its perturbation expansion,

$$\Gamma = \sum_n \Gamma_n(z) g^{2n}, \tag{18}$$

where $z = \ln k^2/\mu^2$. A finite angle analyticity domain of the form of FIGURE 2 can be translated into a bound for the Γ_ns, at fixed z, by a power of $n!$. In order to simplify the argument, let us assume that

$$\Gamma_n(z) = z^n (n!)^p. \tag{19}$$

(This is not realistic because the leading logarithmic term in perturbation theory does not grow like $n!$.)

A transformation of the form

$$\hat{g}^2 = \sum_{n=1}^{\infty} \alpha_n g^{2n} \qquad \alpha_1 = 1 \tag{20}$$

will yield a new series for Γ,

$$\Gamma = \sum_n \hat{\Gamma}_n(z) \hat{g}^{2n}, \tag{21}$$

where the $\hat{\Gamma}_n$s will be polynomials in z of degree n. The coefficient of the last term, z^n, is precisely $(n!)^p$. All the other coefficients depend on the values of α_k of (20) and are, therefore, arbitrary. We shall show that, given $C > 0$, there exists an n_0 such that, for all $n \geq n_0$, $|\hat{\Gamma}_n(z)| > C(n!)^{p-1}$ for at least one value of z in the interval $-1 \leq z \leq 1$. The proof is simple. Let us rewrite $\hat{\Gamma}_n(z)$ as an expansion in terms of Legendre polynomials $P_k(z)$,

$$\hat{\Gamma}_n(z) = \sum_{k=1}^{n} \gamma_k P_k(z), \tag{22}$$

with $\gamma_n = (n!)^p$. We have

$$\gamma_n = (n!)^p = \int_{-1}^{1} P_n(z)\hat{\Gamma}_n(z)\, dz \leq 2\,|P_n(z)|_{\max}\,|\hat{\Gamma}_n(z)|_{\max}, \tag{23}$$

where $|P_n(z)|_{max}$ and $|\hat{\Gamma}_n(z)|_{max}$ are the maximum values of $P_n(z)$ and $\hat{\Gamma}_n(z)$ in the inverval $-1 \leq z \leq 1$. We thus obtain

$$|\hat{\Gamma}_n(z)|_{max} \geq \frac{(n!)^p}{2|P_n(z)|_{max}} \geq C(n!)^{p-1} \tag{24}$$

for n sufficiently large. This simple argument shows that we cannot enlarge the angle of analyticity by a change of variables and, therefore, we cannot avoid 't Hooft's result. It is clear that the argument given above can be adapted to a great many cases where the dependence of Γ_n on z can be quite different from the power behavior we assumed here.

ACKNOWLEDGMENTS

One of us (J.I.) is indebted to Prof. G. Tiktopoulos for his kind hospitality as well as for numerous suggestions and enlightening discussions.

REFERENCES

1. CREUTZ, M., L. JACOBS & C. REBBI. 1979. Phys. Rev. Lett. **42:** 1390–93; Phys. Rev. D **20:** 1915–22.
2. For a recent review of this problem see, for example, PARISI, G. 1980. *In* Hadron Structure and Lepton-Hadron Interactions. M. Lévy *et al.*, ed. Plenum Press. New York; ZINN-JUSTIN, J. 1980. *Ibid*.
3. 'T HOOFT, G. 1977. *In* Orbis Scientiae. A. Perlmutter *et al.*, Eds.: 699. Plenum Press. New York; 1977. *In* The Whys of Subnuclear Physics A. Zichichi, Ed. Plenum Press. New York.
4. KHURI, N. N. & O. A. MCBRYAN. 1979. Phys. Rev. D **20:** 881–86.

POSSIBLE TESTS OF QUANTUM CHROMODYNAMICS IN LARGE TRANSVERSE MOMENTUM HADRON PHOTOPRODUCTION

D. Schiff

Laboratoire de Physique Théorique et Hautes Energies
Associé au Centre National de la Recherche Scientifique
Université Paris-Sud
91405 Orsay, France

Introduction

Contrary to the situation in the 1960's, it is now possible to contemplate the possibility of constructing a fundamental theory of strong interactions. The success of QED for electromagnetic interactions encourages one to believe that a local gauge theory such as Quantum Chromodynamics (QCD)[1] might explain hadronic phenomena. This theory is based on a Yang-Mills Lagrangian with the SU(3) color group as an internal symmetry group.[2] Analogously with QED, it describes the interaction between quarks as a result of the exchange of spin 1 particles (instead of one photon, there are eight colored gluons that interact with one another).

It is remarkable that the general structure of QCD[3] matches a large number of features of the hadronic world: spectroscopy (quark model), current algebra, parton model structure of hard processes such as deep inelastic lepton hadron scattering, logarithmic scale violations, etc.

The theory of QCD may actually be divided into two branches: one concerned with long range behavior and the possibility of proving quark confinement,[4] and one that deals with the small distances relevant to hard processes.[5] A remarkable feature of the theory is that the coupling constant decreases at small distance.[6] This phenomenon, referred to as asymptotic freedom,[7] allows one to treat the interactions between quarks and gluons at work in hard processes by perturbation theory.

Perturbative QCD has achieved tremendous progress in explaining deep inelastic lepton-hadron scattering, e^+e^- annihilation, and lepton pair production.[8]

In this short review,[9] we shall deal with large p_T photoproduction, which presents many interesting features, among which is the fact that, unlike the above listed processes, it involves gluons at the Born term level. We shall show that, in the present generation of accelerator experiments, this reaction[10] does, indeed, yield very clear tests of perturbative QCD and a direct determination of the gluon distribution in the proton and of the gluon fragmentation function. In this sense, large p_T hadron photoproduction looks like a beautiful experiment with which to study the gluon.

General Framework

Let us first recall how the photon interacts with the target in the QCD framework.

Direct Coupling Contribution

Because of its pointlike nature, it may couple directly to one parton of the target, as shown by the "direct coupling" graphs of FIGURE 1: (a) the QCD Compton graph and (b) the fusion graph.

These simple subprocesses have some very nice features:

(1) Differences of cross sections $d\sigma(\gamma p \rightarrow h^+ X) - d\sigma(\gamma p \rightarrow h^- X)$ isolate the Compton graph of FIGURE 1a while remaining of the same order of magnitude as individual cross sections because of the electric charge factor in the coupling of the photon with the parton.

(2) The internal Fermi motion of the constituents of the hadron has a negligible effect on the cross section, which is a much more favorable situation than that of hadron-hadron collisions.

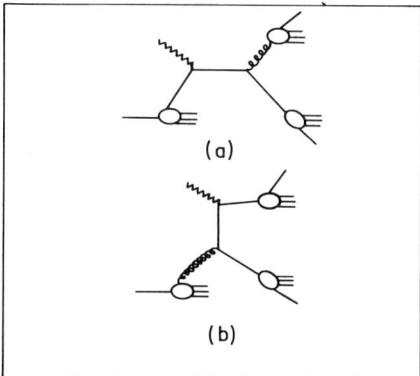

FIGURE 1. Direct coupling graphs for photoproduction. (a) Compton graph; (b) fusion graph.

(3) We shall see below that, since they present a specific 3-jet topology, they are characterized by kinematical constraints in correlation distributions; this property may be used to isolate these terms.

Anomalous Photon Distribution

The photon may convert into an almost colinear quark-antiquark pair; this mechanism yields a structure function proportional to log Q^2, which is renormalized by the QCD emission of colinear gluons; this is what is called the anomalous photon structure function.[11] One of the partons inside the photon then interacts, through a hard subprocess, with a parton from the target.

This contribution is interesting because the anomalous photon structure function is exactly calculable in the framework of perturbative QCD (unlike hadron structure functions where only the Q^2 evolution is calculable).

The Hadronized Photon Contribution

The photon may hadronize in a quasi-bound state of quarks and gluons. This yields a contribution to the structure function that is not determined by perturbative QCD calculations. It may be estimated in the framework of the Vector Dominance Model (VDM). We shall look for situations where this is suppressed relative to the two contributions mentioned above and may be viewed as a small background.

TESTS OF QCD

We are able to separate these different contributions and thus devise clear tests of QCD, as shown in Reference 9. First, the QCD Compton effect (FIGURE 1a) may be isolated by studying the quantity $\Delta^h(\mathbf{p})$, defined as

$$\Delta^h(\mathbf{p}) = E\frac{d\sigma}{d\mathbf{p}}(\gamma p \to h^+ X) - E\frac{d\sigma}{d\mathbf{p}}(\gamma p \to h^- X). \qquad (1)$$

We find, indeed, that subtracting cross sections, as in (1), allows us to get rid of most of the anomalous and hadronized photon contributions, the latter being not well controlled. Moreover, the theoretical predictions for $\Delta^h(\mathbf{p})$ are not obscured by the uncertainties encountered in large p_T hadronic collisions, namely those due to the treatment of the internal Fermi motion of partons inside hadrons and those implied by

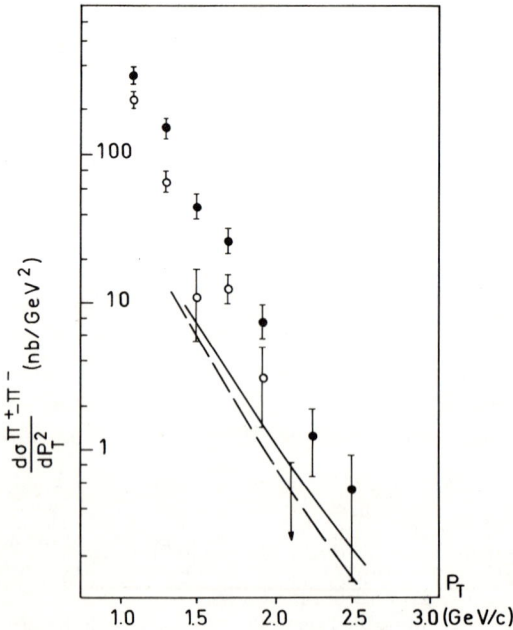

FIGURE 2. $d\sigma^{\pi^+ - \pi^-}/dp_T^2$ for $\theta_{\text{lab}} < 15°$ (full curve and full dots) and $y > 0.5$ (dashed curve and open dots) for $45 < E^\gamma < 70$ GeV. Data are from Reference 12.

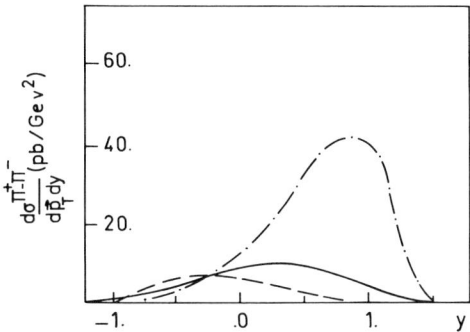

FIGURE 3. $d\sigma^{\pi^+-\pi^-}/d\mathbf{p}_T dy$ at $p_T = 3$ GeV/c and $s = 200$ GeV^2. Compton term (dash-dotted curve), anomalous (dashed curve), and VDM (full curve) contributions.

our current ignorance of gluon structure and fragmentation functions. Let us restrict our study to pion-inclusive production, for simplicity. $\Delta^\pi(\mathbf{p})$ may be written as

$$\Delta^\pi(\mathbf{p}) = d\sigma^{\pi^+-\pi^-}/dy d\mathbf{p}_T$$
$$= \int dx\, dz\, [4G_u^{val}(x, Q^2) - G_d^{val}(x, Q^2)] \frac{D_u^{\pi^+}(z, Q^2) - D_u^{\pi^-}(z, Q^2)}{z} \quad (2)$$
$$\times \frac{8}{3} \frac{\alpha \alpha_s(Q^2)}{\hat{s}^2} \left(-\frac{\hat{t}}{\hat{s}} - \frac{\hat{s}}{\hat{t}}\right) \delta(\hat{s} + \hat{t} + \hat{u}),$$

where the Gs and Ds are the well-determined valence structure and fragmentation functions and $\alpha_s(Q^2) = (12\pi/25) \log (Q^2/0.25)$. In FIGURE 2, we compare the y integrated cross section to already available data taken at energies between 45 and 70 GeV.[12] The anomalous photon and VDM contributions to $\Delta^\pi(\mathbf{p})$ are discussed in detail in Reference 9: The direct coupling term dominates, so it is meaningful to compare $\Delta^\pi(\mathbf{p})$ with the data. The difference between theory and experiment in the region $p_T > 2$ GeV/c is consistent with what is expected for the hadron-like background. Future experiments planned at CERN-SPS and FNAL[13] with p_T up to 5 GeV/c will thus provide meaningful measurements of the QCD Compton process. The prediction for the rapidity distribution at $p_T = 3$ GeV/c and $s = 200$ GeV^2 is shown in FIGURE 3.

GLUON FRAGMENTATION AND DISTRIBUTION FUNCTION DETERMINATIONS

A beautiful feature of large p_T photoproduction experiments is that they provide information about the gluon through the study of two-particle (toward-away) inclusive distributions. Since the direct coupling term is the dominant contribution to cross section differences, one may concentrate on it when discussing the following quantities:

$$\Delta^\pm(\mathbf{p}_1\mathbf{p}_2) = \frac{d\sigma(\gamma p \to \pi^+\pi^\pm X)}{d\mathbf{p}_{T_1} dy_1 d p_{2x} dy_2} - \frac{d\sigma(\gamma p \to \pi^-\pi^\pm X)}{d\mathbf{p}_{T_1} dy_1 d p_{2x} dy_2}. \quad (3)$$

The study of $\Delta^+(\mathbf{p}_1\mathbf{p}_2) + \Delta^-(\mathbf{p}_1\mathbf{p}_2)$ (respectively, $\Delta^+(\mathbf{p}_1\mathbf{p}_2) - \Delta^-(\mathbf{p}_1\mathbf{p}_2)$) turns out to give direct access to the gluon fragmentation (distribution in the proton) function, since only the Compton (fusion) graph contributes.

Gluon Fragmentation Function

It is useful to make the following definition (p_T^{jet} is the outgoing jet transverse momentum):

$$\frac{d\sigma^c}{dz_2\,dp_T^{jet}} = \int_{p_{T_1}^{min}} d\mathbf{p}_{T_1} \int d\mathbf{p}_{2x}\, \delta\left(z_2 - \frac{|p_{2x}|}{p_T^{jet}}\right)$$
$$\times \int_{y_1^{min}}^{y_1^{max}} dy_1 dy_2 \delta\left(p_T^{jet} - \frac{\sqrt{s}}{e^{y_1} + e^{y_2}}\right) [\Delta^+(\mathbf{p}_1, \mathbf{p}_2) + \Delta^-(\mathbf{p}_1, \mathbf{p}_2)], \quad (4)$$

which can be simply written as

$$\frac{d\sigma^c}{dz_2 dp_T^{jet}} = D_g^{h_2}(z_2, \overline{Q}^2)\, d(p_{T_1}^{min}, p_T^{jet}, s, y_1^{min}, y_1^{max}), \quad (5)$$

with $\overline{Q}^2 \simeq (p_T^{jet})^2$ and $D_g^{h_2}$ being the gluon fragmentation function; d is straightforwardly calculated from valence quark distribution and fragmentation functions.[9] As shown in FIGURE 4, $d\sigma^c/dz_2 dp_T^{jet}$ is fairly large and, thus, should allow a precise measurement of $D_g(z, Q^2)$. The scaling violation may also be investigated.[9]

Gluon Distribution Function in the Proton

A similar study may be performed using $\Delta^-(\mathbf{p}_1\mathbf{p}_2) - \Delta^+(\mathbf{p}_1\mathbf{p}_2)$. We may define, with obvious notations,

$$\frac{d\sigma^F}{dx} = \int dy_1 d\mathbf{p}_{T_1} dp_{2x} \frac{d\sigma^{\Delta^- - \Delta^+}}{dy d\mathbf{p}_{T_1} dp_{2x} dx}, \quad (6)$$

which is proportional to the gluon distribution function $G_g(x, \overline{Q}^2)/x$, where $\overline{Q}^2$ is the

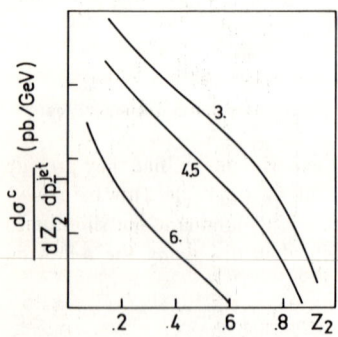

FIGURE 4. $d\sigma^c/dz_2 dp_T^{jet}$. Curves are labeled with the value of p_T^{jet}. ($s = 200$ GeV^2).

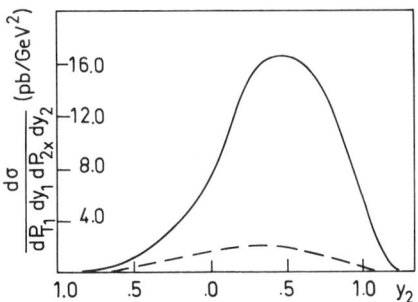

FIGURE 5. $d\sigma/dp_{T_1}dy_1dp_{2_x}dy_2$ for $y_1 = 0$, $p_{T_1} = p_{2_x} = 3$ GeV/c and $s = 200$ GeV². The full (dashed) curve is the anomalous (hadronized) contribution.

mean value of Q^2 in the integration range. Estimates of this quantity indicate the feasibility of the experimental measurement.[9]

OBSERVATION OF THE ANOMALOUS PHOTON STRUCTURE FUNCTION

Since the anomalous photon structure function is a very interesting QCD object, it is of great interest to measure it. This should, in principle, be possible through γ-γ collisions in e^+e^- annihilation reactions. The rate seems to be too small for the present generation of e^+e^- machines. However, photoproduction offers us a way to undertake this task in the near future. The strategy we propose is to take advantage of the fact that final jets produced in direct coupling reactions and those produced in processes involving the photon structure function have very different topologies. The direct coupling results in two large p_T final jets and a backward low p_T jet made of proton fragments; in the second case, a fourth jet made of low p_T fragments of the photon is observed in the forward direction.

Thus, for selecting the anomalous photon contribution we, first, keep only 4-jet events, taking advantage of the kinematical constraint satisfied by direct coupling 3-jet events at the level of double-inclusive jet cross sections:

$$d\sigma/dp_T^{jet}dy_1dy_2 \propto \delta(\sqrt{s} - p_T(e^{y_1} + e^{y_2})).$$

Then, we select symmetric large p_T pairs. This allows us to get rid of the Fermi motion enhancement effect[14]; thus, theoretical calculations are reliable—in particular, we discuss in Reference 9 how this implies that the VDM background is very small, the anomalous photon component then being the dominant contribution. In FIGURE 5, we show the resulting double inclusive cross section, which is measurable with the expected integrated luminosity (~10 events/picobarn) of the SPS tagged photon experiment.[13]

CONCLUSION

In conclusion, let us repeat that large p_T photoproduction will make it possible in the near future to test perturbative QCD through isolation of the QCD Compton effect and extraction of the anomalous photon structure function. Moreover, it will yield very important information on some of the gluon properties upon which theorists have been speculating for a long time.

REFERENCES

1. FRITZSCH H., & M. GELL-MANN. 1972. 16th International Conference on High Energy Physics (Batavia). J. D. Jackson and A. Roberts, Eds.; FRITZSCH, H., M. GELL-MANN & H. LEUTWYLER. 1973. Phys. Lett. B **47**: 365.
2. YANG, C. N. & P. MILLS. 1954. Phys. Rev. **96**: 191.
3. See, for instance, ELLIS, J. 1979. 9th International Symposium on Lepton and Photon Interactions FNAL.
4. Some of the relevant questions are discussed in MARCIANO, W. & H. PAGELS. 1978. Phys. Rep. **36C**: 139.
5. See, for instance, DOKSHITZER, YU. L., D. I. DYAKONOV & S. I. TROYAN. 1980. Phys. Rep. **58**: 271.
6. POLITZER, H. D. 1973. Phys. Rev. Lett. **30**: 1346; GROSS, D. I. & F. W. WILCZEK. 1973. Phys. Rev. Lett. **30**: 1343; KHRIPLOVICH, I. B. 1969. Yad. Fiz. (Sov. J. Nucl. Phys.) **10**: 409.
7. POLITZER, H. D. 1974. Phys. Rep. **14C**: 129.
8. See, for instance, FIELD, R. 1978. La Jolla Institute Summer Workshop.
9. This is a summary of a work in collaboration with M. FONTANNAZ, A. MANTRACH, & B. PIRE. 1980. Phys. Lett. B **89**: 263; **94**: 509; 1980. Z. Physik. C **6**: 241, 347.
10. For a complete list of references on the subject, see Reference 9.
11. WITTEN, E. 1977. Nucl. Phys. **B120**: 189; LEWELLYN SMITH, C. H. 1978. Phys. Lett. B **79**: 83.
12. WA4 collaboration preliminary data quoted in Reference 9.
13. NA14 CERN-London-Orsay-Saclay-Southampton collaboration. Wide band beam and tagged photon beam FNAL facilities.
14. BAIER, R., J. ENGELS & B. PETERSSON. 1979. Z. Physik C **2**: 265.

PHOTON-PHOTON COLLISIONS

Paul Kessler

Laboratoire de Physique Corpusculaire
Collège de France
75231 Paris, France

HISTORICAL BACKGROUND

Processes of the type shown in FIGURE 1 were studied by some of the pioneers of quantum electrodynamics, in particular by Landau and Lifshitz,[1] even in the early thirties. However, their application to electron colliding-beam physics started with two papers published in the *Physical Review* in 1960, by F. Low[2] and by Calogero and Zemach.[3]

Low suggested measuring the lifetime of the π^0 through the reaction $ee \to ee\pi^0$ (FIGURE 2). For the analysis of that process, he proposed a formula of extreme simplicity and elegance, in which both electron beams were replaced by equivalent-photon spectra, namely,

$$\sigma_{ee \to ee\pi^0} = \int N(k_1) N(k_2) \sigma_{\gamma\gamma \to \pi^0}(k_1, k_2) \, dk_1 \, dk_2,$$

with the equivalent-photon spectra given by the so-called Williams-Weizsäcker approximation, i.e.,

$$N(k_i) = \frac{\alpha}{\pi} \frac{1}{k_i} \frac{E^2 + (E - k_i)^2}{E^2} \ln \frac{E}{m_e} \quad (E = \text{beam energy}, k_i = \text{photon energy})$$

and the cross section $\sigma_{\gamma\gamma \to \pi^0}$, related to τ, the lifetime of the π^0, given by

$$\sigma_{\gamma\gamma \to \pi^0}(k_1, k_2) = \frac{8\pi^2}{m_\pi \tau} \delta(4k_1 k_2 - m^2).$$

Calogero and Zemach, on the other hand, suggested studying pair production in ee colliding beam machines, in particular $ee \to ee\pi^+\pi^-$. Their paper contained a number of prophetic ideas, as well: (1) That their process provides the possibility of studying the pion-pion intereaction in the absence of any spectator hadrons. (2) Forward-scattered electrons may be "tagged" as soon as they have lost a fraction of their energy, since their curvature in the magnetic field will be modified. (3) Although those processes are of fourth order in α, they are not too small, since they are enhanced by log E/m_e factors. (4) In FIGURE 3, diagram 1 gives rise to even angular momentum states of the pion pair, whereas diagram 2 produces a pure $J = 1$ state. It is possible to eliminate diagram 2 by an appropriate choice of the experimental configuration.

After those fundamental papers appeared, they were apparently forgotten for almost ten years. The reason was, obviously, that electron storage rings of the first generation (the first Stanford storage ring, VEPP-2 at Novosibirsk, ACO at Orsay) would not have allowed one to investigate those processes seriously, since neither their energies nor their luminosities were sufficient.

16   Annals New York Academy of Sciences

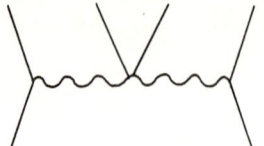

FIGURE 1. General Feynman diagram for photon-photon collision processes.

In 1969, De Celles and Goehl published a study on the production of the S-wave pion-pion resonance σ in $e\,e$ collisions.[4]

In 1968–69, our group at the Collège de France started a theoretical investigation of photon-photon collisions in electron storage rings. We soon came to the following conclusions:[5]

1. When the electrons are tagged at angles very close to 0° (a few milliradians or so), the theoretical background (i.e., the contribution of other diagrams in α^4) becomes entirely negligible.

2. Under such conditions, the equivalent-photon approximation (see above) may be applied at either electron vertex with high accuracy; thus, the external vertices in the diagram of FIGURE 1 may be factored out and, so to speak, "forgotten"; since both virtual photons are kinematically extremely close to real photons, one is, indeed, studying photon-photon collisions.

3. With the new generation of storage rings then planned or under construction (ADONE at Frascati; COPPELIA at Orsay, later replaced by DCI; DORIS at DESY; SPEAR at SLAC) of beam energy $\simeq$ 1.5–3 GeV and expected luminosity $\simeq$ 10^{32} cm^{-2} s^{-1} (actually, that expectation was later found to be over-optimistic), counting rates would be high enough, even taking realistic cuts in acceptance into account, to allow the study of such reactions as $\gamma\gamma \to e^+\,e^-,\,\mu^+\,\mu^-,\,\pi^+\,\pi^-,\,\pi^0,\,\eta$.

Photon-photon collisions became popular after the Kiev Conference in 1970, with the reports of Brodsky et al., a theoretical group from SLAC and Cornell,[6] and Balakin et al. from Novosibirsk.[7] It should be mentioned that the "discovery" of photon-photon collisions at Novosibirsk followed from an experiment in which a large number of coplanar but noncollinear particle pairs were observed; those pairs were finally identified with $e^+\,e^-$ pairs produced in the process $e^+\,e^- \to e^+\,e^-\,e^+\,e^-$.[8]

In the years 1970–73, photon-photon collisions became a very fashionable field of theoretical activity (see below), and experiments were performed at Frascati on the $e^+\,e^-$ storage ring ADONE.[9,10] Those experiments, however, were only checks of quantum electrodynamics (just as the pioneering experiment at Novosibirsk), i.e., only the reactions $e^+\,e^- \to e^+\,e^-\,e^+\,e^-$ and $e^+\,e^- \to e^+\,e^-\,\mu^+\,\mu^-$ were studied. Nevertheless, the experimental efforts of the early seventies were of great importance,

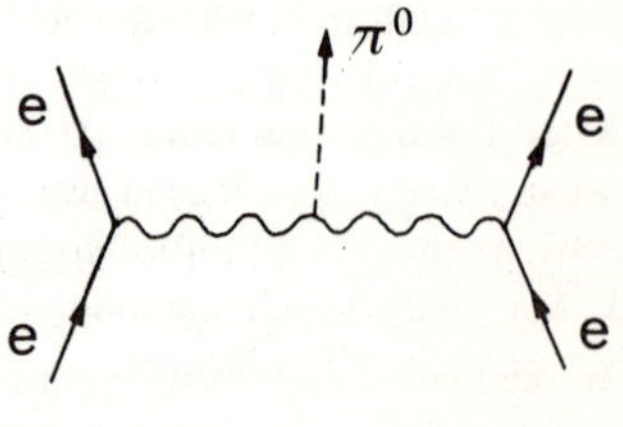

FIGURE 2. Feynman diagram for $e\,e \to e\,e\,\pi^0$.

since it was thus shown that photon-photon collision processes may be detected, properly identified, and correctly analyzed.

THEORETICAL IMPLICATIONS

As soon as it became obvious, in the summer of 1970, that photon-photon collisions would become a new area of investigation in high-energy physics, many theorists started speculating on theoretical implications and performing all kinds of computations in that field. Several hundreds of theoretical papers were published in the period 1970-73. The corresponding references can be found in the extensive reports by Terazawa[11] and Budnev et al.,[12] and also in the Proceedings of the International Colloquium on Photon-Photon Collisions, which took place at the Collège de France in 1973.[13]

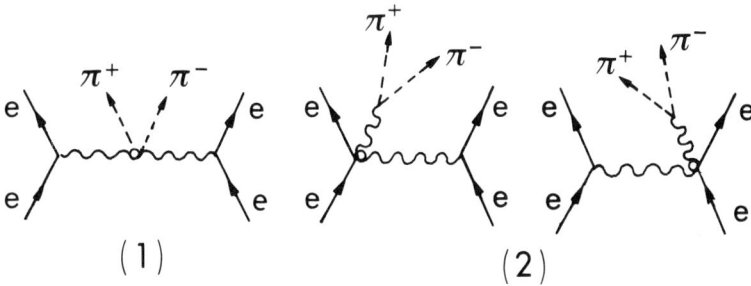

FIGURE 3. Feynman diagrams for $e\,e \to e\,e\,\pi^+\,\pi^-$. (1) Diagram involving the photon-photon collision mechanism. (2) Other diagrams.

Theoretical ideas discussed in those papers include:
1. Applications of low-energy theorems of current algebra to soft-pion production.
2. Study of meson-pair production, applying dispersion relations, Regge-pole theory, and the quark model.
3. Study of multihadron production, using vector dominance, Regge-pole theory, and the quark model.
4. Study of resonance production, applying quark and other models.
5. Study of the deep-inelastic configuration (Q^2 and/or Q'^2 large, calling Q^2, Q'^2 either photon's four-momentum squared).

Whereas, in the years 1974-77, little interest was shown by high-energy physicists in photon-photon collisions (after the discovery of the J/ψ, their attention was focused on other processes and problems), a new stimulus was provided in that field from 1978 on, when it was shown by some authors[14] that photon-photon processes at high energy may be an ideal place for checking quantum chromodynamics. Indeed, in those processes, due to the absence of any spectator hadrons and to the extent that hadronic jets in the final state may be identified with quarks or gluons, one has only to deal with

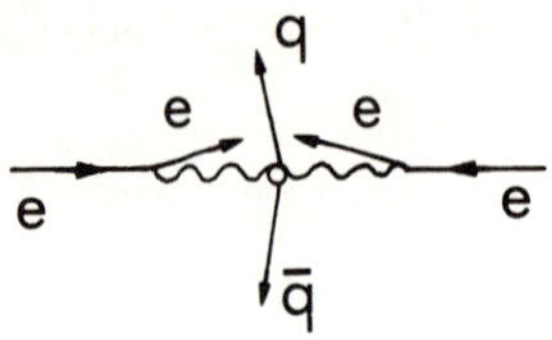

FIGURE 4. Diagram for $ee \to eeq\bar{q}$, in the configuration where both electrons are emitted at small angles and both quarks at large angles.

electrons, photons, quarks, and gluons. Therefore, in so far as photons at high energy are assumed to have a pointlike (rather than hadron-like) behavior, everything is calculable from QCD combined with QED. Several crucial checks of QCD may thus be performed, such as (considering only the simplest processes): (1) testing the quark propagator in the process $ee \to eeq\bar{q}$, with both electrons scattered at small angles and both quark jets emitted with large transverse momenta (FIGURE 4); (2) testing the photon's or electron's structure function in $ee \to eeq\bar{q}$ with one electron scattered at a small angle and the other one at a large angle, in which case only one quark jet would be emitted at large transverse momentum, while the other would go almost along the beam axis (FIGURE 5).

RECENT EXPERIMENTAL RESULTS

During the last several years, a number of interesting experiments bearing on hadron production in photon-photon collisions were performed in various laboratories. In particular, now that the new high energy electron-positron storage rings (PETRA at Hamburg and PEP at Stanford, both with 15–20 GeV per beam) are coming into operation, those processes are becoming—as we predicted ten years ago—an important area of experimental investigation.

Experimental results were presented at the very recent Workshop on $\gamma\gamma$ Collisions (Amiens, France, April 8–12, 1980). They included the following:

1. At Orsay (France), where the low energy electron-positron storage ring DCI (about 1 GeV per beam) is operating, several measurements of photon-photon processes have been performed. They included checks of quantum electrodynamics (i.e., lepton-pair production); also, quite recently, pion-pair production was measured in the threshold region and found to be quite abundant (about twice as many pion-pairs as predicted by the Born-term model).[15]

2. At Stanford, where the storage ring SPEAR (about 2–4 GeV per beam) is operating, a measurement of the process $ee \to ee\eta'$ led to the determination of the partial decay width of the resonance η' (958 MeV) into two photons. The value found happens to be in good agreement with predictions based on the quark model. Upper

FIGURE 5. Diagram for $ee \to eeq\bar{q}$, in the configuration where one electron and one quark are emitted at small angles, the other electron and the other quark at large angles.

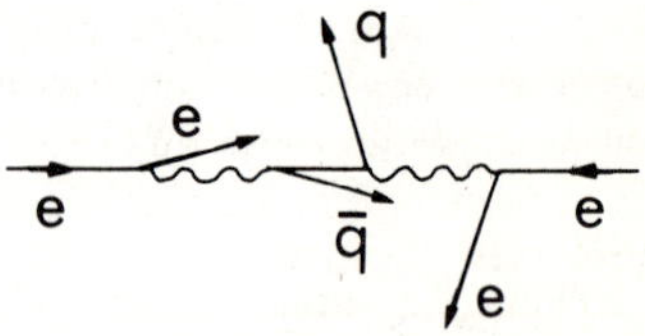

limits were also determined for the two-photon decay widths of the resonance f^0 (1270 MeV), A_2 (1310 MeV), and f' (1515 MeV).[16]

3. At Hamburg, where the new storage ring PETRA has been operating since 1979, several groups started studying photon-photon processes. Apart from checks of QED (lepton-pair production), two groups (PLUTO and TASSO)[17,18] have already shown some results on hadron production. In particular, both groups measured the two-photon decay width of the f^0; the values determined by them differ slightly (but theoretical predictions of that decay width also diverge to a large extent). In addition, both groups measured the cross section σ ($\gamma\gamma \rightarrow$ hadrons) at $\gamma\gamma$ center-of-mass energies varying from about 1 to 8 GeV. According to both measurements, that cross section appears to decrease in that range from about 1000 to 300 nanobarns; it thus seems to lie somewhat higher, especially at low energy, than is predicted on the basis of Regge-pole models. Finally, the most spectacular experimental result so far obtained is the detection of several jet pairs, apparently due to $\gamma\gamma \rightarrow q\bar{q}$, by the PLUTO group.

The Future

Interesting as they are, the above-mentioned experimental measurements are probably only the beginning of a systematic study of high energy photon-photon collision processes at PETRA and PEP. Still more promising for photon-photon physics is LEP, the new European project for an electron-positron storage ring of about 70 GeV per beam, which should be built in the next few years.[19]

However, it must be emphasized that photon-photon collision experiments are not at all easy ones. Tagging of the outgoing electrons at (almost) 0°, as we had suggested, and as was indeed carried out at some storage rings (ADONE at Frascati, DCI at Orsay), appears to be increasingly difficult to perform with higher energy rings for technical reasons (the 0° tagging counters are overwhelmed with electrons with degraded energy originating from bremsstrahlung processes). Therefore, PETRA and PEP have tagging systems at finite angles only (>1°), and the same will be true, more or less, of LEP. It follows that, on the one hand, counting rates are drastically reduced by using such finite-angle tagging systems (since most of the electrons come out at angles very close to 0°), and that, on the other hand, analysis of experiments becomes more difficult (to some extent, the photons exchanged are no longer "almost real"). Therefore, the experimentalists may prefer (as they actually did in some cases) to perform experiments without tagging the electrons at all. But, in that case, other difficulties appear: Background suppression may become difficult (background is provided mainly by beam-gas collisions and by e^+e^- processes where not all of the outgoing particles are observed); also, the reconstitution of photon-photon events may become hazardous, since too many parameters are missing (the electrons remain unseen, and not all particles produced in the $\gamma\gamma$ collision are observed in the central detector, since it doesn't have a full 4π acceptance). A possible compromise is to use "single tagging," i.e., to detect and measure only one of both outgoing electrons (as was done in some of the experiments at PETRA); but single tagging also has its shortcomings.[20]

Thus, the future of photon-photon collisions is very much dependent upon technical efforts to improve the detection apparatus. On the one hand, electron tagging systems should be allowed to operate as close as possible to 0°; on the other hand, central detectors to be used for studying those processes should have an acceptance as close as possible to 4π. It is certainly worthwhile to undertake such efforts for the sake of a new, fascinating, and probably very fertile field in elementary particle physics.

SUMMARY

Photon-photon collisions in electron-positron storage rings are a relatively new field in high-energy physics. The historical development of that field was described. Theoretical implications were briefly discussed, and some details were given of recent experimental results obtained in various laboratories (Orsay, Stanford, and DESY-Hamburg). Future developments in that area, in connection with technical problems that remain to be solved, were also discussed.

REFERENCES

1. LANDAU, L. D. & E. M. LIFSHITZ. 1934. Phys. Z. Sowjetunion. **6:** 244–57.
2. LOW, F. E. 1960. Phys. Rev. **120:** 582–83.
3. CALOGERO, F. & CH. ZEMACH. 1960. Phys. Rev. **120:** 1860–66.
4. DE CELLES, P. C. & J. E. GOEHL, JR. 1969. Phys. Rev. **184:** 1617–21.
5. ARTEAGA-ROMERO, N., A. JACCARINI & P. KESSLER. 1969. C. R. Acad. Sci. Ser. B **269:** 153–56 & 1129–33.
6. BRODSKY, S. J., T. KINOSHITA & H. TERAZAWA. 1970. Phys. Rev. Lett. **25:** 972–75.
7. BALAKIN, V. E., V. M. BUDNEV & I. F. GINZBURG. 1970. Zh. Eksp. Teor. Fiz. Pis'ma Red. **11:** 559–62. (JETP Lett. **11:** 388–90.)
8. BALAKIN, V. E., A. D. BUKIN, E. V. PAKHTUSOVA, V. A. SIDOROV & A. G. KHAKHPASHEV. 1971. Phys. Lett. B **34:** 663–64.
9. BACCI, C., G. PENSO, G. SALVINI, R. BALDINI-CELIO, G. CAPON, C. MENCUCCINI, G. P. MURTAS, A. REALE, M. SPINETTI & B. STELLA. 1972. Lett. Nuovo Cimento **3:** 709–14.
10. BARBIELLINI, G., S. ORITO, T. TSURU, R. VISENTIN, F. CERADINI, M. CONVERSI, S. D'ANGELO, M. L. FERRER, L. PAOLUZI & R. SANTONICO. 1974. Phys. Rev. Lett. **32:** 385–88.
11. TERAZAWA, H. 1973. Rev. Mod. Phys. **45:** 615–62.
12. BUDNEV, V. M., I. F. GINZBURG, G. V. MEDELIN & V. G. SERBO. 1975. Phys. Rep. **15C:** 181–282.
13. Proceedings of the International Colloquium on Photon-Photon Collisions in Electron-Positron Storage Rings. 1974. J. Phys. (Paris) **35** (Colloque C-2, supplement to nr. 3): 1–126.
14. BRODSKY, S. J., T. DE GRAND, J. GUNION & J. WEIS. 1979. Phys. Rev. D **19:** 1418–43.
15. BROSSARD, M., A. FALVARD, J. JOUSSET, B. MICHEL, G. MONTAROU, J. C. MONTRET, P. REICHSTADT, A. COURAU & J. HAISSINSKI. 1980. *In* Proceedings of the International Workshop on $\gamma\gamma$ Collisions, Lecture Notes in Physics. Vol. 134. G. Cochard & P. Kessler, Eds.: 212–37. Springer-Verlag. Berlin, Heidelberg, New York.
16. JENNI, P. 1980. *In* Proceedings of the International Workshop on $\gamma\gamma$ Collisions, Lecture Notes in Physics, Vol. 134. G. Cochard & P. Kessler, Eds.: 47–71. Springer-Verlag. Berlin, Heidelberg, New York.
17. BERGER, CH. 1980. *In* Proceedings of the International Workshop on $\gamma\gamma$ Collisions, Lecture Notes in Physics, Vol. 134. G. Cochard & P. Kessler, Eds.: 82–107. Springer-Verlag. Berlin, Heidelberg, New York.

18. HILGER, E. 1980. In Proceedings of the International Workshop on $\gamma\gamma$ Collisions, Lecture Notes in Physics, Vol. 134. G. Cochard & P. Kessler, Eds.: 108–36. Springer-Verlag. Berlin, Heidelberg, New York.
19. FIELD, J. H. 1980. In Proceedings of the International Workshop on $\gamma\gamma$ Collisions, Lecture Notes in Physics, Vol. 134. G. Cochard & P. Kessler, Eds.: 248–97. Springer-Verlag. Berlin, Heidelberg, New York.
20. See discussion in: ARTEAGA-ROMERO N., C. CARIMALO, G. COCHARD, A. COURAU, P. KESSLER, A. NICOLAIDIS, S. ONG & J. PARISI. Update on Photon-Photon Collisions. II. Problems of Analysis. Laboratoire de Physique Corpusculaire Report LPC 80/06. Unpublished.

THE KMS CONDITION FOR *-ALGEBRAS*

J. Alcántara[†] and D. A. Dubin

Faculty of Mathematics
The Open University
Milton Keynes, MK7 6AA England

Introduction

In References 1 and 2, we were successful in interpreting the formal continuum Heisenberg Hamiltonian as a continuous derivation on a certain topological *-algebra. This derivation is not an infinitesimal generator and the representations in which it is implemented can be completely characterized. To define KMS states for this model (the continuum Heisenberg ferromagnet), we need an extension of the Tomita-Takesaki theory to topological *-algebras. We have found that the geometrical approach to the Tomita-Takesaki theory due to Rieffel and Van Daele is suitable for this purpose. We partially justify our approach by showing that KMS states obey Sewell's and Bogolubov's inequalities.

In what follows, $\mathcal{A}$ will denote a Hausdorff unital topological *-algebra, and $i\delta$ a distinguished *-derivation on $\mathcal{A}$, assumed continuous for simplicity. The pair, $(\mathcal{A}, \delta)$, will be referred to as a dynamical system.

The Rieffel–Van Daele Theory

Let us start by describing the Rieffel–Van Daele results in a form convenient for us.

PROPOSITION 1. Let $\mathcal{H}$ be a complex separable Hilbert space and $\mathcal{K}$ a closed real subspace such that $\mathcal{K} \cap i\mathcal{K} = \{0\}$ and $\mathcal{K} + i\mathcal{K}$ is dense in, but not equal to, $\mathcal{H}$. Then there exists a strongly continuous one-parameter unitary group $\{\Delta^{it} : t \in \mathbf{R}\}$ on $\mathcal{H}$ such that:

a. $\Delta^{it}\mathcal{K} = \mathcal{K}$ for all $t \in \mathbf{R}$
b. To every pair $\Phi, \Psi \in \mathcal{K}$, there is a function $F: \mathbf{C} \to \mathbf{C}$, which is analytic in the strip $S(-1,0)$, bounded and continuous on its closure, and takes boundary values

$$F(t) = \langle\Phi, \Delta^{it}\Psi\rangle, \qquad F(t - i) = \langle\Delta^{it}\Psi, \Phi\rangle. \tag{1}$$

The unitary group $\{\Delta^{it} : t \in \mathbf{R}\}$ is the unique unitary group satisfying the above two conditions.

Let $P: \mathcal{H} \to \mathcal{K}, Q: \mathcal{H} \to i\mathcal{K}$ be the indicated real orthogonal projections, and set $R = P + Q$. Then the modular group is related to the subspace projections by

---

*Part of the material presented by one of us (D.A.D.) at the Fourth International Conference on Collective Phenomena.
[†]Submitted in partial fulfillment of the requirements for the Ph.D degree, The Open University. Supported by the British Council and U.P.C.H. (Lima, Peru).

$$\Delta^{it} = (2 - R)^{it} R^{-it} \qquad (t \in \mathbf{R}). \qquad (2)$$

We have introduced the notation

$$S(a,b) = \{z \in \mathbf{C} : a < \mathrm{Im}\, z < b\}. \qquad (3)$$

By $An(a,b)$ we shall mean analyticity in $S(a,b)$, boundedness and continuity in its closure, and boundary values explicitly stated. Although this makes for extra notation, these analyticity conditions occur frequently enough to make the abbreviation worthwhile.

It is shown in Reference 3 that if M is a von Neumann algebra with a cyclic and separating vector ω, acting on $\mathcal{H}$, and M_s denotes the collection of self-adjoint elements of M, then the closure of $[M_s\omega]$ ($=\mathcal{K}$) is a closed real subspace of $\mathcal{H}$ such that $\mathcal{K} \cap i\mathcal{K} = \{0\}$ and $\mathcal{K} + i\mathcal{K}$ is dense in $\mathcal{H}$. In Reference 1, another procedure is given for the construction of a closed real subspace with the required properties, independently of von Neumann algebra considerations.

The KMS Condition

The following lemma concerning the implementability of derivations is standard.

LEMMA 1. A state of ϕ on the dynamical system $(\mathcal{A},\delta)$ is said to be stationary if $\phi \circ \delta = 0$. In the GNS representation of a stationary state ϕ, there exists a densely defined symmetric operator H_ϕ for which

$$H_\phi \Omega_\phi = 0, \qquad \pi_\phi(\delta(f))\Phi = [H_\phi, \pi_\phi(f)]\Phi, \qquad (4)$$

for all $f \in \mathcal{A}$, $\Phi \in \mathcal{D}_\phi$.‡

We shall now define KMS states. It will be noted that the fact that δ is not required to be an infinitesimal generator causes us to introduce the suppositions that H_ϕ is essentially self-adjoint and that the real subspace $\mathcal{K}_\phi$ is temporally stable.

DEFINITION 1. A state ϕ on the dynamical system $(\mathcal{A},\delta)$ is said to be β-KMS if:

a. ϕ is stationary
b. The implementing operator H_ϕ is essentially self-adjoint
c. Let $\mathcal{K}_\phi$ be the closed real subspace§

$$\mathcal{K}_\phi = cl - \{[f]_\phi : f \in \mathcal{A}_h\} \qquad (5)$$

and let $\{\tau_t^\phi : t \in \mathbf{R}\}$ be the unitary group generated by H_ϕ^{**}.
Then $\tau_t^\phi \mathcal{K}_\phi = \mathcal{K}_\phi$ for all $t \in \mathbf{R}$.

d. For every pair $f, g \in \mathcal{A}$, there is a function F_{fg} satisfying $An(0,\beta)$, with boundary values

---

‡If L is the closed left ideal $\{f : \phi(f^*f) = 0\}$ in $\mathcal{A}$, then $\mathcal{D}_\phi = \mathcal{A}/L$, with sesquilinear form $\langle [f]_\phi, [g]_\phi \rangle = \phi(f^*g)$.
§$\mathcal{A}_h$ is the real vector space of Hermitian elements of $\mathcal{A}$, $\{f : f^* = f\}$.

$$F_{fg}(t) = <[f^*]_\phi, \tau_t^\phi[g]_\phi>, \qquad F_{fg}(t + i\beta) = <\tau_t^\phi[g^*]_\phi, [f]_\phi>. \qquad (6)$$

The following technical lemma is necessary in order to apply the Rieffel–Van Daele theory. For a proof, see Theorem 3.9 of Reference 3.

LEMMA 2. Let $\{W_n\}_{n \in \mathbb{N}}$ be a sequence of functions satisfying An(0,β). EbSuppose that, in the sup-norm, $W_n(t) \to G(t)$ and $W_n(t + i\beta) \to H(t)$. Then there exists a function W satisfying An(0,β) with boundary values $W(t) = G(t)$, $W(t + i\beta) = H(t)$.

COROLLARY 1. Let ϕ be a β-KMS state on $(\mathcal{A},\delta)$. For every pair $\Phi, \Psi \in \mathcal{H}_\phi$, there exists a function F satisfying An(0,β) with boundary values

$$F(t) = <\Phi, \tau_t^\phi \Psi>, \qquad F(t + i\beta) = <\tau_t^\phi \Psi, \Phi>. \qquad (7)$$

LEMMA 3. For a β-KMS state ϕ on $(\mathcal{A},\delta)$, the subspace $\mathcal{H}_\phi$ satisfies $\mathcal{H}_\phi \cap i\mathcal{H}_\phi = \{0\}$.

Proof. If $\Psi \in \mathcal{H}_\phi \cap i\mathcal{H}_\phi$, then $\Psi, i\Psi \in \mathcal{H}_\phi$. By Corollary 1, for all $\Phi \in \mathcal{H}_\phi$ there exist functions F,G satisfying An(0,β) with boundary values

$$F(t) = <\Phi, \tau_t^\phi \Psi>, \qquad F(t + i\beta) = <\tau_t^\phi \Psi, \Phi>,$$

$$G(t) = <\Phi, \tau_t^\phi i\Psi>, \qquad G(t + i\beta) = <\tau_t^\phi i\Psi, \Phi>.$$

The function $H = F + iG$ satisfies An(0,β) with boundary values

$$H(t) = 0, \qquad H(t + i\beta) = 2F(t + i\beta).$$

By Theorem 12.8 of Reference 4, $H = 0$. Then $F(t + i\beta) = <\tau_t^\phi \Psi, \Phi>$ and, setting $t = 0$ and $\Phi = \Psi$, the desired conclusion follows. □

PROPOSITION 2. For a β-KMS state ϕ on $(\mathcal{A},\delta)$, Ω_ϕ is separating for $\pi_\phi(\mathcal{A})$.

Proof. First, we prove that if $\phi(f^*f) = 0$, then $\phi(ff^*) = 0$. Assuming $\phi(f^*f) = 0$, $f = f_1 + if_2, f_1, f_2 \in \mathcal{A}_h$, we get $[f_1]_\phi + i[f_2]_\phi = 0$.
By LEMMA 3, $[f_1]_\phi = [f_2]_\phi = 0$, so $\phi(ff^*) = 0$.
If $\pi_\phi(f)\Omega_\phi = 0$, then $\pi_\phi(gf)\Omega_\phi = \pi_\phi(g)\pi_\phi(f)\Omega_\phi = 0$, or, equivalently, $\phi(f^*g^*gf) = 0$, for all $g \in \mathcal{A}$. Therefore, $\phi(gff^*g^*) = 0$, or $\pi_\phi(f^*)\pi_\phi(g^*)\Omega_\phi = 0$, for all $g \in \mathcal{A}$. Since Ω_ϕ is cyclic and $\pi_\phi(f^*)$ is closeable, we get that $\pi_\phi(f^*) = 0$. Taking adjoints gives $\pi_\phi(f^*)^* = 0$, which implies $\pi_\phi(f) = 0$, since $\pi_\phi(f) \subset \pi_\phi(f^*)^*$. □

The modular structure arises from a conjugation on $\mathcal{H}_\phi + i\mathcal{H}_\phi$. We now prove that the required conjugation is closed.

LEMMA 4. For a β-KMS state ϕ on $(\mathcal{A},\delta)$, let S_ϕ be an operator with domain $\mathcal{H}_\phi + i\mathcal{H}_\phi$ and

$$S_\phi(\Phi + i\Psi) = \Phi - i\psi \qquad (\Phi, \Psi \in \mathcal{H}_\phi). \qquad (8)$$

Then S_ϕ is densely defined and closed.

Proof. Assume that $\Phi_n + i\Psi_n \to 0$ and $\Phi_n - i\Psi_n$ is convergent, where $\Phi_n, \Psi_n \in \mathcal{H}_\phi$. Since $\mathcal{H}_\phi$ is closed and Φ_n, Ψ_n are convergent, there exist $\Phi, \Psi \in \mathcal{H}_\phi$ such that $\Phi_n \to \Phi$

and $\Psi_n \to \Psi$. Now, $\Phi_n + i\Psi_n \to 0$ gives $\Phi, \Psi \in \mathcal{H}_\phi \cap i\mathcal{H}_\phi$, which implies $\Phi = \Psi = 0$. Consequently, $\Phi_n - i\Psi_n \to 0$, so S_ϕ is closed. □

Definition 2. With the notation as above, the modular operator associated with the β-KMS state ϕ is given by

$$\Delta_\phi = S_\phi^* S_\phi. \tag{9}$$

From this, Δ_ϕ is strictly positive and self-adjoint. The polar decomposition of S_ϕ,

$$S_\phi = J_\phi \Delta_\phi^{1/2}, \tag{10}$$

then serves to define the modular conjugation J_ϕ, an antiunitary involution.

PROPOSITION 3. Let ϕ be a β-KMS state on $(\mathcal{A}, \delta)$, Δ_ϕ the modular operator, and $\{\tau_t^\phi : t \in \mathbf{R}\}$ the unitary group generated by the implementing Hamiltonian H_ϕ^{**}. Then

$$\tau_{\beta t}^\phi = \Delta_\phi^{it} \tag{11a}$$

or

$$H_\phi^{**} = -\beta^{-1} \ln \Delta_\phi. \tag{11b}$$

Proof. Recall that $\tau_{\beta t}^\phi$ leaves $\mathcal{H}_\phi$ invariant for all $t \in \mathbf{R}$. For every pair $\Phi, \Psi \in \mathcal{H}_\phi$, there is a function F satisfying An($-1,0$) with boundary values

$$F(t) = \langle \Phi, \tau_{\beta t}^\phi \Psi \rangle, \qquad F(t - i) = \langle \tau_{\beta t}^\phi \Psi, \Phi \rangle.$$

By the uniqueness results mentioned in PROPOSITION 1, equation 11a, and, hence, equation 11b, follows. □

CORRELATION INEQUALITIES

We now turn our attention to the inequalities of Sewell[5] and Bogolubov.[6]

LEMMA 5. Let H be a self-adjoint operator on a separable Hilbert space, and $v \in \text{dom}(H) \cap \text{dom}(\rho_\beta)$, where $\rho_t = \exp(-tH/2)$ for $t \in \mathbf{R}$. Then

$$\beta \langle v, Hv \rangle \geq \|v\|^2 \ln(\|v\|^2 \|\rho_\beta v\|^{-2}). \tag{12}$$

Proof. Recall that Jensen's inequality (Theorem 3.3 in Reference 4) states if μ is a probability measure, $f \in L^1(\mu)$ is real, with range in $[a,b]$, and ϕ is convex on (a,b), then

$$\phi\left(\int f \, d\mu\right) \leq \int (\phi \circ f) \, d\mu. \tag{13}$$

Now if $H = \int_{-\infty}^\infty \lambda \, dE(\lambda)$ is the spectral resolution of H, then

$$\langle v, \phi(H) v \rangle = \int_{-\infty}^\infty \phi(\lambda) \, d\langle v, E(\lambda) v \rangle.$$

With $f = 1$ and $d\mu = \|v\|^{-2} d<v, E(\lambda)v>$, Jensen's inequality gives

$$\phi(\|v\|^{-2}<v, Hv>) \leq \|v\|^{-2}<v, \phi(H)v>.$$

Taking $\phi(t) = \exp(-\beta t/2)$, squaring, using the Cauchy-Schwarz inequality, taking logarithms, and multiplying by -1 gives the result. □

PROPOSITION 4. A β-KMS state ϕ on $(\mathcal{A},\delta)$ obeys Sewell's inequality

$$-\beta\phi(\delta(f^*)f) \geq \phi(f^*f)\ln[\phi(f^*f)\phi(ff^*)^{-1}]. \tag{14}$$

Conversely, if a state ϕ satisfies Sewell's inequality, then it is stationary. If, in addition, the self-adjointness and stability conditions b and c of Definition 1 are satisfied, ϕ is β-KMS.

Proof. Take $H = H_\phi^{**}$ in the previous lemma, and $v = [f]_\phi$. Then $\|v\|^2 = \phi(f^*f)$, $<v, Hv> = -\phi(\delta(f^*)f)$, and

$$\|\rho_\beta v\|^2 = \|\Delta_\phi^{1/2} v\|^2 = \|J_\phi \Delta_\phi^{1/2} v\|^2 = \|[f^*]_\phi\|^2 = \phi(ff^*).$$

Inequality 12 gives (14) with these values substituted.

For the converse, we start by showing stationarity. From Sewell's inequality, it follows that

$$\text{Im}\phi(\delta(f^*)f) = 0.$$

For $f = g^* - 1$, we get $\text{Im}\phi(\delta(g)) = 0$. Replacing g by ig gives $\text{Re}\phi(\delta(g)) = 0$, so $\phi(\delta(g)) = 0$. Now assume, in addition, b and c of Definition 1. For $k \in C_c(\mathbf{R}))$, consider

$$\nu(k) = \int k(-\lambda) d<[f^*]_\phi, E(\lambda)[f^*]_\phi>, \qquad \mu(k) = \int k(\lambda) d<[f]_\phi, E(\lambda)[f]_\phi>.$$

Sewell (Theorem 3 of Reference 5) has shown that $\nu < \mu$ and that the Radon-Nikodym derivative is $d\nu/d\mu(\lambda) = \exp(-\beta\lambda)$. Here $\{E(\lambda) : \lambda \in \mathbf{R}\}$ is the spectral family for H_ϕ^{**}. The remainder of Sewell's proof that ϕ is β-KMS now goes through (Lemma 12 of Reference 5). □

PROPOSITION 5. A β-KMS state ϕ on $(\mathcal{A},\delta)$ obeys Bogolubov's inequality,

$$|\phi([f,g])|^2 \leq (\beta/2)\phi(f^*f + ff^*)\phi([g^*,\delta(g)]).$$

Proof. For a proof, see Theorem 1 of Reference 6. □

In our work so far, we have not considered the Green's functions. It seems to us that their existence must be considered separately for different models, e.g., the conditions to ensure existence for the ideal Bose gas will be rather different from the conditions for the continuum Heisenberg ferromagnet. They should lead to the same conclusion, however—that, for a state ϕ in which δ is implemented and generates a $\tau_t^\phi = \exp(it H_\phi^{**})$,

$$\tau_t^\phi(\mathcal{D}_\phi^{**}) = \mathcal{D}_\phi^{**}.$$

For then we can define

$$G^\phi(f_1,t_1;\ldots f_n,t_n) = <\Omega_\phi, \tau_{t_1}^\phi \pi_\phi^{**}(f_1)\tau_{t_2-t_1}^\phi \ldots \tau_{t_n-t_{n-1}}^\gamma \pi_\phi^{**}(f_n)\tau_{t_n}^\phi \Omega_\phi>.$$

We leave this question open.

Summary

We define KMS states for any topological *-algebra equipped with a distinguished *-derivation. In the representation associated with a KMS state, the GNS cyclic vector is separating. The modular operator bears the usual relation to the self-adjoint extension of the Hamiltonian implementing the derivation. A KMS state obeys Sewell's and Bogolubov's inequalities. Under an additional conditional domain condition, Sewell's inequality implies the KMS condition.

References

1. ALCÁNTARA, J. 1979. Ph.D. Thesis. The Open University.
2. ALCÁNTARA, J. & D. A. DUBIN. 1981. States on the Current Algebra. Reports on Mathematical Physics. In press.
3. Rieffel, M. A. & A. VAN DAELE. 1977. Pacific J. Math. **69:** 187.
4. RUDIN, W. 1974. Real and Complex Analysis. McGraw-Hill. New York.
5. SEWELL, G. L. 1977. Commun. Math. Phys. **55:** 53.
6. POWERS, R. T. 1976. Commun. Math. Phys. **51:** 151.

LATTICE AND MOLECULAR DYNAMICS OF SOME AMINE INTERCALATES OF FeOCl*

R. H. Herber

Department of Chemistry
Rutgers University
New Brunswick, New Jersey 08903

Introduction

Because of their interesting electrical, magnetic, electrochemical, and chemical properties, intercalation compounds formed when organic molecules are inserted into a layer structure inorganic matrix have received considerable attention in recent years.[1-3] Of particular interest in this context are the amine intercalates of FeOCl, which have been studied in considerable detail in our laboratories.[4,5] Iron(III)oxychloride is a layered compound belonging to the orthorhombic space group $P_{mnm}(D_{2h}^{13})$, with two formula units per unit cell. The original crystal structure data of Goldsztaub[6] has more recently been refined by Lind[7] and shows that the structure consists of a stack of double-layer sheets of *cis* $FeCl_2O_4$ octahedra linked together with shared edges. Of particular significance is the fact that the chlorine atoms lie on a plane that defines the edge of each layer, and the bonding across these layers is assumed to be of the van der Waals type. Since this interlayer interaction is quite weak, the layers are readily pried apart and it is possible to insert a wide variety of Lewis base molecules into this structure.[2,3,8] The unit cell dimensions of the unintercalated matrix are $a = 3.780$, $b = 7.917$, and $c = 3.302$ Å. On intercalation, only the b-axis unit cell dimension undergoes a large increase; the magnitude of this increase has been used to infer the ordering of the intercalant "guest" molecule within the solid.[5] Numerous speculations have been suggested concerning the orientation of the "guest" molecules within the FeOCl lattice; recent detailed studies have suggested that each case must be considered on its own merits, since the steric requirements of the particular intercalant, as well as the Lewis basicity of the lone-pair electrons (which appear to be a necessary, but not sufficient, condition for the formation of an intercalate with FeOCl) need to be taken into consideration.

In the present study, the intercalates formed with 1,4-pyridiazine (DIAZ), *N,N*-dimethyl-4-aminopyridine (DMAP), and diazabicyclo(2,2,2)octane, triethylenediamine (TED) are discussed in detail in terms of both the lattice dynamics of the intercalate and the molecular dynamics of the intercalant.

Experiment

The FeOCl used in these studies was synthesized by methods described in the literature.[9] DIAZ, DMAP, and TED were purchased from Aldrich Chem. Co. and

---

*This research was supported, in part, by a grant from the Division of Materials Research, National Science Foundation, no. DMR 7808615A01, and by the Center for Computer and Information Services, Rutgers University.

used without further purification. Ethanol used in the intercalation procedure was thoroughly dried prior to use. All manipulations were carried out in an inert atmosphere under the rigorous exclusion of moisture. The composition of the final (equilibrium) product was established by elemental analysis for C, H, N, and Cl and by a colorimetric procedure for Fe. FeOCl $(1,4-C_4H_4N_2)_{1/3.5}$ was prepared by maintaining FeOCl in contact with an ethanol solution of DIAZ at 24 °C for three months in a sealed glass ampul. Analysis: Found (calc'd) C: 10.58 (10.54), H: 1.33 (0.89), N: 5.81 (6.15), Cl: 27.36 (27.23), Fe: 40.58 (42.90). FeOCl $(C_6H_{10}N_2)_{1/6}$ was obtained by the sealed tube method after three weeks at 50 °C. Analysis: Found (calc'd) C: 11.62 (9.56), H: 1.72 (1.34), N: 3.51 (3.72). Cl: 26.90 (28.21), Fe 44.04 (44.44). FeOCl $(C_6H_{12}N_2)_{1/6}(H_2O)_{1/3}$ was prepared analogously, the product being isolated from the ethanol solution of the base after 17 d at 50 °C. The water is presumed to have arisen from the base (which was 97% pure as purchased), but this point is not definitively established. Analysis: Found (calc'd) C: 8.48 (9.12), H: 2.05 (1.79), N: 3.32 (3.55), Cl: 27.14 (26.93), Fe 41.02 (42.42).

All the products were characterized by their infrared spectra in KBr (~0.5%) and their powder pattern x-ray diffraction. The latter data, obtained using Cu K_α radiation and a Phillips-Norelco model 596238 diffractometer, also permitted the calculation of the unit cell dimensions of the intercalates. Infrared data were obtained using a Perkin-Elmer model 283 grating spectrometer, with the sample at room temperature.

[57]Fe Mössbauer data were accumulated using the constant acceleration spectrometer, velocity calibrations, and data reduction procedures described earlier.[10] The sample temperatures were maintained to better than ±0.5 K in the temperature range $78 \leq T \leq 320$ K, using an Air Products Model APD-E proportional temperature controller with two Au(Co)/constantan thermocouples to monitor the experimental data. All isomer shifts in this paper are reported with respect to the centroid of a room temperature spectrum of NBS SRM metallic iron.

Results

The unit cell b-axis lattice parameters were calculated from the 010 and 020 reflections of the powder pattern x-ray diffraction data and are included in the data summarized in TABLE 1. These powder patterns also showed the absence of the systematic FeOCl b-axis reflections in the intercalates, from which it is concluded that these solids represent effectively "stage 1" compounds; that is, *every* van der Waals layer has been expanded to accommodate the Lewis base "guest" molecules in the structure. The most obvious location for the intercalant within the van der Waals layer is a cavity in which there are three iron atom nearest neighbors. From the observed stoichiometry, it is clear that, in the DIAZ intercalate, essentially every cavity is filled, as is true for the intercalate formed with isopropylamine, 1,2-pyridazine, and pyridine (one modification).[11] In the TED and DMAP intercalates, on the other hand, only every other cavity contains an intercalant moiety, as has been observed for other large guest molecular species, including trimethyl phosphite, trimethyl phosphine, 2,6-lutidine, dodecylamine, and quinuclidine.[5,11]

Despite the considerable difference in the length of the major molecular axes of

TABLE 1
SUMMARY OF DATA FOR FeOCl AND THE THREE INTERCALATES DISCUSSED IN THE TEXT

	FeOCl	FeOCl (DIAZ)$_{1/6}$	FeOCl (DMAP)$_{1/6}$	FeOCl (TED)$_{1/6}$ (H$_2$O)$_{1/3}$
IS(300 K), mm s^{-1}	0.358 ± 0.008	0.45 ± 0.02	0.456 ± 0.008	0.465 ± 0.008
IS(75 K), mm s^{-1}	0.558 ± 0.008*	0.50 ± 0.04	0.500 ± 0.008	0.509 ± 0.010
QS(300 K), mm s^{-1}	0.940 ± 0.008	0.70 ± 0.05	0.709 ± 0.008	0.700 ± 0.008
QS(75 K), mm s^{-1}	0.963 ± 0.008*	1.20, 1.11†	1.125 ± 0.015	1.010 ± 0.008
d(IS)/dT × 10^4, mm s^{-1} K^{-1}	5.291	—	5.17 (T ≥ 200 K)	4.9 (T ≥ 220 K)
d(ln A)/dT × 10^3, K^{-1}	1.170	2.83	2.94	2.30
θ_M, K	341	219	215	243
b, Å	7.91	13.33	13.41	13.06
H_i(4.2 K), KOe	432 ± 1	—	429 ± 1	—
V_{zz}	(−)		(−)	

*Extrapolated from data above T_N.
†The second figure refers to the partially deintercalated sample—see text.

DMAP and TED or DIAZ, the *b*-axis unit cell dimensions of the FeOCl intercalates are very similar, being 5.50, 5.16, and 5.42 Å, respectively. This result lends strong support to the hypothesis that, in these solids, the long molecular axis (which contains the nitrogen atom lone pair) is oriented parallel or nearly parallel to the van der Waals layer plane; i.e., parallel to the *ac* plane.

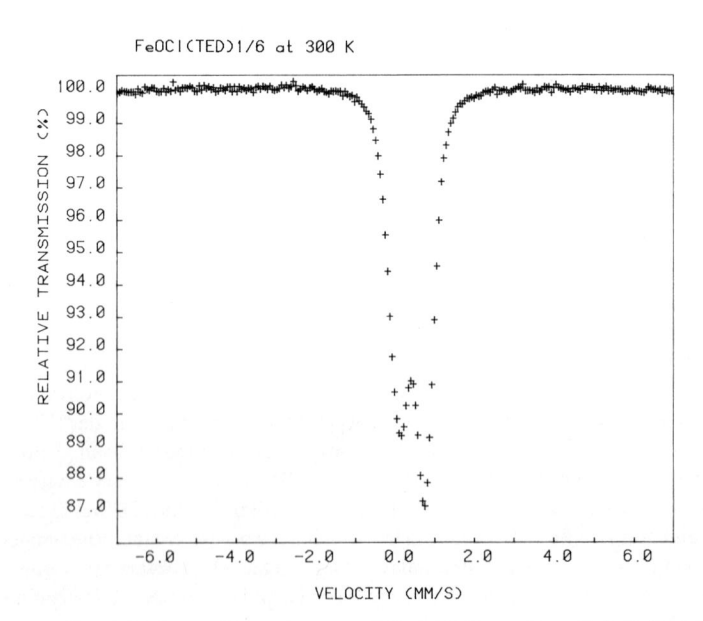

FIGURE 1. ^{57}Fe Mössbauer effect spectrum of the TED intercalate, FeOCl [1,4 diazabicyclo(2,2,2)octane]$_{1/6}$ (H$_2$O)$_{1/3}$, at 300 K.

Herber: Lattice and Molecular Dynamics

The ^{57}Fe Mössbauer effect parameters, and their temperature dependence, are summarized in TABLE 1. At temperatures well above the magnetic ordering temperature, the Mössbauer spectra consist of well-resolved doublets characterized in terms of the isomer shift (IS), which is related to the electron density at the ^{57}Fe nucleus; the quadrupole splitting (QS), which is related to the shape of the charge distribution around the iron atom; and the area under the resonance curve (A), which is related to the mean square amplitude of the Mössbauer-active atom and, hence, to the details of the thermal excitation of the lattice. A representative room temperature Mössbauer spectrum is shown in FIGURE 1.

The IS of unintercalated FeOCl, ~0.35 mm s^{-1} at 300 K, is characteristic of Fe(III) high spin[12] and is not significantly different in the several intercalates, as summarized in TABLE 1. In contrast to the behavior evidenced by unintercalated FeOCl (above the magnetic ordering temperature), for which the temperature dependence of the isomer shift is well fitted by a linear correlation function, IS(T) for the three intercalates is not a smooth function of temperature. For the DMAP

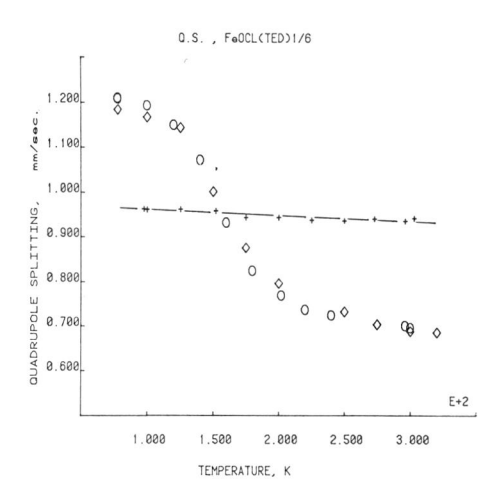

FIGURE 2. Temperature dependence of the isomer shift parameter for the TED intercalate over the temperature range 78 ≤ T ≤ 320 K. The isomer shift is reported relative to the center of an α-Fe spectrum at 296 ± 1 K.

intercalate, for example, the data below ~200 K show a significant scatter from which no unique temperature dependence can be extracted. Above 200 K, d(IS)/dT is approximately 5.17 × 10^{-4} mm s^{-1} K^{-1}, which is very similar to the value observed for FeOCl over the range 95 ≤ T ≤ 320 K.[4]

The temperature dependence of the isomer shift for the TED intercalate is summarized graphically in FIGURE 2, from which it is noted that the normal negative slope second order Doppler shift behavior is not followed over the whole temperature range, but, rather, IS appears to go through a maximum at ~220 K. This behavior is both reproducible among different samples and independent of whether adjacent data points are obtained on sample heating or sample cooling cycles. The broadness of the maximum in FIGURE 2 indicates that the change in the isomer shift is not due to a first order phase transition (which would be more abrupt in the IS(T) data and observable in the temperature dependence of the recoil-free fraction—see below), but, rather, may be presumed to arise out of the temperature dependence of the binding across the van der Waals gap. A qualitatively similar behavior is observed in all three intercalates discussed in this study, and clearly arises from the consequences of the b-axis expansion in these materials relative to unintercalated FeOCl. This point will be more fully discussed below.

The temperature dependence of the quadrupole splitting parameter for the three intercalates shows an even more striking behavior. As noted earlier,[4] the temperature coefficient of the QS parameter of FeOCl is very small. QS varies from ~0.962 mm s^{-1} at 100 K to ~0.940 mm s^{-1} at 300 K. In contrast, the QS parameter for the three intercalates shows a smooth sigmoid temperature dependence. The data for the DMAP intercalate are shown in FIGURE 3 and for the TED intercalate in FIGURE 4. In both figures, the data for unintercalated FeOCl are also shown; it is clear that the unusual temperature dependence of the quadrupole hyperfine interaction is a property of the intercalate, rather than of the neat lattice. Finally, it should be noted in this context that the QS of the intercalate is larger than that of FeOCl at low temperatures and smaller at high temperatures, over the range $78 \leq T \leq 300$ K. For the DMAP intercalate, the inflection point in QS(T) occurs at ~140 K; this temperature corresponds approximately to the onset of the increase in the isomer-shift with increasing temperature, as noted above. Completely analogous observations are made in the case of the TED intercalate, for which the maximum slope of the temperature dependence of QS occurs at ~153 K, and, again, this corresponds to the temperature at which the temperature dependence of the isomer shift changes sign (i.e., becomes positive on increasing T). From these data, there can be little doubt that the two temperature dependencies—IS(T) and QS(T)— for the intercalates reflect the energetics of the bonding forces across the van der Waals layer. Indeed, the present data suggest that, in these intercalates, this bonding interaction is on the order of 0.3–0.5 kcal mol^{-1}. At low temperatures, when the mean thermal excitation energy ($\sim kT$) is less than this energy, the matrix behaves as a three-dimensional solid, while, when kT is significantly larger than this value, the bonding energy across the van der Waals gap is small compared to kT, and the intercalate demonstrates a quasi two-dimensional behavior, the intralayer bonding forces predominating over the interlayer bonding forces.

Similar behavior is observed for the DIAZ intercalate, for which the temperature dependence of the quadrupole splitting parameter is given in FIGURE 5. For this sample, an additional complication is observed, since QS(T) appears to depend on the sample history, especially its thermal treatment. In examining this material via Mössbauer spectroscopy, the sample was mounted at room temperature and measured at ~280 K. The sample was then cooled to ~80 K while under vacuum, and a sequence of measurements was made at progressively higher temperatures, until 273 K was reached (sample still under vacuum). Reaching 273 K required about four hours of measurement time. Subsequently, the temperature was lowered and data accumulated until liquid nitrogen temperature was reached. As indicated in FIGURE 5, in which the increasing temperature regime and the decreasing temperature regime are indicated by appropriate symbols, the two data sets show what appears to be hysteresis. In fact,

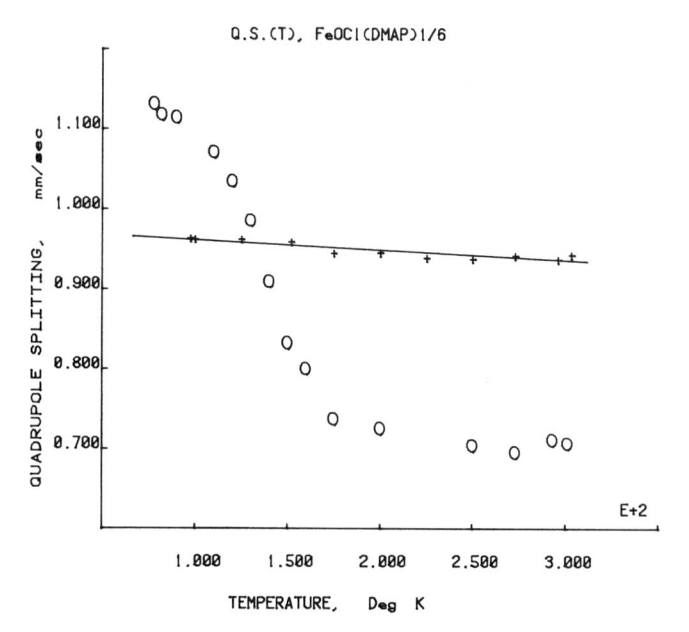

FIGURE 3. Temperature dependence of the quadrupole splitting parameter for the DMAP intercalate (circles) and the unintercalated matrix (crosses).

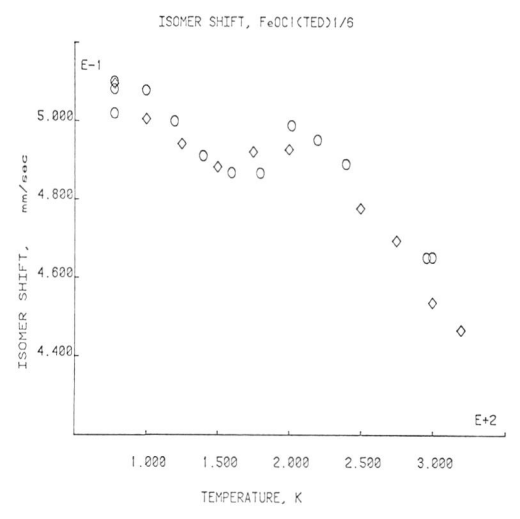

FIGURE 4. Temperature dependence of the quadrupole splitting parameter for the TED intercalate (circles and diamonds) and the unintercalated matrix (crosses).

it is assumed that, while accumulating data at the highest temperature point (273 K) under vacuum, a partial deintercalation occurred, giving rise to the results summarized in the figure.

It is interesting to note, in this context, that a similar deintercalation does not appear to occur in the case of the DMAP or TED intercalate, although the latter is smaller (in the axial molecular dimension) by about 10% than the DIAZ molecule. Both of these intercalates appear to be stable under vacuum at room temperatures for periods from several hours to days, and no evidence for deintercalation at 300 K is apparent from the Mössbauer effect data.

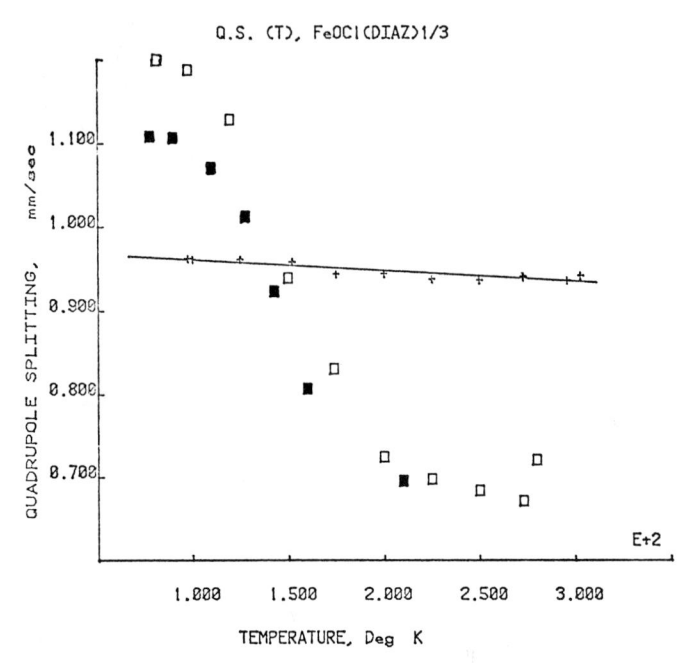

FIGURE 5. Temperature dependence of the quadrupole splitting parameter for the DIAZ intercalate (squares) and the unintercalated matrix (crosses). For the intercalate, the data acquired for temperatures increasing from 78 K are indicated by open squares and those acquired for temperatures decreasing from 280 K are indicated by filled squares (see text).

A further parameter that can be extracted from the temperature-dependent Mössbauer effect data is a characteristic temperature, θ_M, which is related to the strength of the lattice energy and is effectively an average over the phonon spectrum of the solid. Starting with the Debye theory of solids, it can be shown that the temperature dependence of the area under the resonance curve is given by[13]

$$\frac{d(\ln A)}{dT} = -\frac{3E_\gamma^2}{M_{\text{eff}} c^2 k_B \theta_m^2}, \qquad (1)$$

where A is the normalized area, E_γ is the Mössbauer gamma ray energy, k_B is the Boltzmann constant, M_{eff} is an effective mass parameter, and θ_M is the lattice temperature of the solid as probed by the Mössbauer active atom. For a true Debye solid, all the quantities except θ_M are known, so θ_M can be readily calculated from the experimental data.

One weakness of this approach is that, for a real solid, the phonon spectrum, $g(\omega)$, does not obey a Debye-like behavior; hence, the calculated lattice temperature represents only one possible averaging over all the phonon states. In addition, in a solid of the type considered here, the effective mass, M_{eff}, does not necessarily correspond to the atomic mass of the probe atom (56.95 amu for ^{57}Fe), but, rather, must take into account the covalency of the bonding of the probe atom to its nearest neighbor environment.

As has been discussed earlier,[4,5] when the temperature dependence of the isomer shift parameter is well fitted by a linear regression, it is possible to obtain a value of M_{eff}. However, due to the complex behavior of IS(T) noted above, this analysis is not possible for the three intercalates under discussion. In the present case, an atomic mass has been assumed for M_{eff} in all four cases (the three intercalates plus the neat lattice), and a value of θ_M has been calculated from (1), which, on substitution of numerical values, becomes

$$\theta_M = 11.659 \, (-d \, (\ln A)/dT)^{-1/2}. \tag{2}$$

These θ_M values, which are summarized in TABLE 1, can only be considered relative values because of the above approximation, and are useful only for providing an internal comparison between the four solids.

With this limitation in mind, it is interesting to note that θ_M for the three intercalates is 100 to 120 degrees lower than it is for the unintercalated matrix, in agreement with the conclusions extracted from the temperature dependence of the hyperfine interaction parameters discussed above. Thus, these data demonstrate a significant softening of the lattice on intercalation, with the insertion of Lewis base "guest" molecules into the van der Waals layer resulting in a weakening of the bonding across this gap. Since the probe atom in these experiments is the iron atom of the FeOCl matrix, the clear fact that emerges from these results is that the mean square amplitude of vibration of the probe atom in the intercalates is significantly larger than it is in the unintercalated lattice. Although there is a resolvable quadrupole hyperfine interaction extractable from the Mössbauer data (i.e., the two components of the doublet are well resolved), it has, so far, not been possible to extract a meaningful value for the vibrational anisotropy from the Mössbauer data, although the intensity asymmetry of the doublet shows a significant temperature dependence. Further analysis of these data in terms of a Gol'danskii-Karyagin effect[14] is currently being undertaken.

Finally, in the context of the present lattice dynamical discussion, it is appropriate to note that, at temperatures well above the magnetic ordering temperature ($T \geq$ ~120 K), the recoil-free fraction data do not reflect any significant lattice softening at the characteristic temperatures discussed above in connection with the isomer shift and quadrupole splitting parameters. The data both above and below ~200 K are well fitted by a linear regression for ln A versus T. It appears from this observation that the recoil-free fraction is not sufficiently sensitive to the transition from a three-dimensional solid at low temperatures to a quasi two-dimensional solid at high temperatures to be reflected in the data concerning the area under the resonance curve. Since the recoil-free fraction reflects an averaging over all phonon states and since the iron atom is well bound within the two-dimensional layer structure in the *ac* plane, even at high temperatures, the mean amplitude of vibration averaged over all angles is apparently dominated by the bonding forces in the *ac* plane. Further studies of this interesting observation for intercalates, with even larger *b*-axis expansions than those here considered, are currently in progress.

ACKNOWLEDGMENTS

The samples used in this study, as well as some of the initial data from the Mössbauer and x-ray studies, were obtained by Dr. Yonezo Maeda of Kyushu University while a visiting scientist in our laboratories, and his work is hereby gratefully acknowledged. The computer programs used in data reduction were largely written by Mr. T. K. McGuire, to whom thanks are due for his helpful collaboration.

REFERENCES

1. VAN BRUGGEN, C. F., C. HAAS & H. W. MYRON, Eds. 1980. Proc. Int. Conf. Layered Materials and Intercalates. North Holland Pub. Co. Amsterdam.
2. KANAMARU, F. & M. KOIZUMI. 1974. Jpn. J. Appl. Phys. **13:** 1319.
3. KIKKAWA, S., F. KANAMARU & M. KOIZUMI. 1979. Bull. Chem. Soc. Jpn. **52:** 963.
4. HERBER, R. H. & Y. MAEDA. 1980. Phys. Status Solidi B **99:** 352–56.
5. HERBER, R. H. & Y. MAEDA. 1980. Inorg. Chem. **19:** 3411.
6. GOLDSZTAUB, S. 1934. C.R. Acad. Sci. **198:** 667; 1935. Bull. Soc. Franc. Miner. Cryst. **58:** 6.
7. LIND, M. D. 1970. Acta Crystallgr. Sect. B **26:** 1058.
8. HAGENMULLER, P., J. PORTIER, B. BARBE & P. BOUCHER. 1967. Z. Anorg. Allg. Chem. **355:** 209; PALVADEAU, P., L. COIC, J. ROUSEL & J. PORTIER. 1978. Mater. Res. Bull. **13:** 221.
9. HALBERT, T. R. & J. SCANLON. 1979. Mater. Res. Bull. **14:** 415.
10. REIN, A. J. & R. H. HERBER. 1975. J. Chem. Phys. **63:** 1021 and references therein.
11. HERBER, R. H. & Y. MAEDA. 1981. Inorg. Chem. **20:** 1409.
12. GREENWOOD, N. N. & T. C. GIBB. 1971. Mössbauer Spectroscopy. Chapman Hall. London.
13. GIBB, T. C. 1976. Principles of Mössbauer Spectroscopy. Chapter 6 and references therein. Chapman Hall. London.
14. Reference 13: 143–4, and references therein; HAZONY, Y. & R. H. HERBER. 1974. J. Phys. (Paris) **66:** 131.

QUANTUM CHEMICAL ASPECTS OF SOME PROBLEMS IN BIOINORGANIC CHEMISTRY. III. SOME LIGAND PROPERTIES OF METAL COMPLEXES: POPULATION ANALYSIS*

I. Fischer-Hjalmars and A. Henriksson-Enflo

Institute of Theoretical Physics
University of Stockholm
Vanadisvägen 9, S-113 46 Stockholm, Sweden

INTRODUCTION

In order to throw light upon the influence of metals on biochemical reactions, a reseach project was started some years ago.[1] Our main goal is to study the binding of metals in general, and transition metals in particular, to constituents of the living cell, especially to amino acids and simple polypeptides. For this purpose we are using quantum chemical calculations on suitably chosen model compounds. Since it is known that metals often form chelates with organic compounds, we have chosen a simple planar chelate as our model, $[(C_2H_2X_2)_2 M]^n$, shown in FIGURE 1.

In previous articles (I,[2] II,[3] IV,[4] and V[5]) of the present series, we have reported results from ab initio calculations on such models. The ligating atoms, X, chosen were those most common in a biological context, i.e., X = NH, O, or S. The metals were Be, Mg, Ni, and Zn, with charges of the whole complex n = +2, 0, −2, and Li with n = +1, 0, −1.

The analysis of our computational results for the complexes shows that certain properties, as could be expected, are dependent on the kind of metal involved. Examples of such properties are binding energies, total charge on the metal, and the population of s, p, and d valence orbitals of the metal. The variation of these properties is of paramount importance to the possible biological reactions of the different metals.

On the other hand, we have also found that many properties are typical for a certain kind of ligand and a certain total charge n of the complex, almost independent of the metal in question. As an example, TABLE 1 summarizes calculated orbital energies of the following species: the free ligand $C_2H_2O_2$, the dimer $(C_2H_2O_2)_2$ and the complexes $[(C_2H_2O_2)_2 M]^0$, with M = Be, Ni, Cu, and Zn. Metal orbitals and ligand orbitals, although mixed together, can easily be distinguished from each other. As expected, the metal orbitals have different energies in different complexes. The ligand orbital energies are, however, rather constant. As an example, the deepest $p\pi$ (b_{1u}) orbital has an energy varying between −15.0 and −16.2 eV. Even with a considerable mixing of a metal orbital into a ligand orbital, the energy is kept almost constant (see, e.g., the a_g, C,O orbital −19.5 to −20.4 eV).

The orbital pattern shown by TABLE 1 is in harmony with measurements of

---

*This research was supported by the Swedish Work Environment Fund and the Swedish Natural Science Research Council.

ionization potentials. Furlani and Cauletti have reviewed ionization measurements on different series of transition metal complexes with ligands not very different from our model compounds.[6] From their tables, it is seen that the ionization potential of the uppermost ligand π-orbital in acetyl acetonate complexes is found between the limits of 8.06 and 8.49 eV in a series with nine different metals as central atom. They have found similar results for other ligand orbitals and other complexes.

In the present article, we shall mainly concentrate upon the distribution of electronic density within the ligands. As is well known, the density in a bond region is a measure of the strength of the bond and the density in an atomic region can be used as an index of the reactivity of that part of the molecule.

At present, we shall limit ourselves to a description of the electronic density as a function of the ligating atom X and the total charge of the complex, n. We shall use the Mulliken population analysis to describe the density.[7] As is well known, the Mulliken scheme is not completely perfect, neither as a measure of the actual electronic density nor as a reactivity index. Nevertheless, the scheme is very useful for a study of trends in shifts of densities in a series of similar molecules and as an indicator of situations where the possible reactivity index deserves a closer analysis.

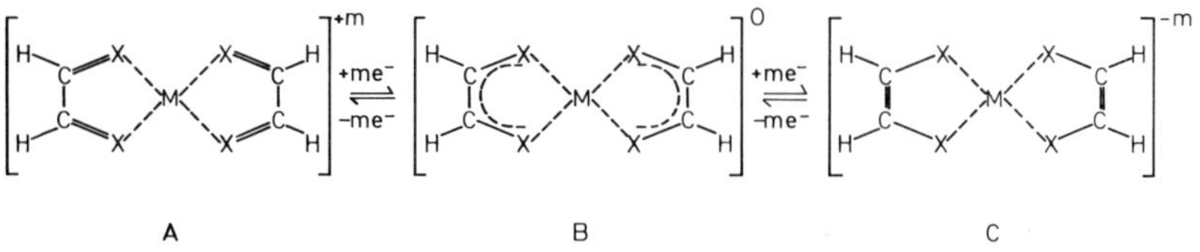

FIGURE 1. Model complexes studied. Formulas A and C can be represented by simple valence bond structures: Formula A as built up from a positively charged metal ion and two neutral ligands with clear double bond character of the C=X bonds, formula C as built up from a positively charged metal ion and two ligands with two negative charges each, the C=C bond having a clear double bond character. For formula B, no definite valence bond structure can be written. In the present study, the metal M is Be ($m = 2$) or Li ($m = 1$). The ligands X are O or S.

The Mulliken population analysis of complexes with different central atoms shows a uniformity, similar to the orbital energy uniformity of TABLE 1. As an example, charges on C and H atoms derived from gross atomic populations are collected in TABLE 2. Both the total charge and its σ- and π-parts are seen to be almost identical for metals as different as Be and Mg on one hand and transition metals on the other. TABLE 2 also includes overlap populations of bonds between the metal and O or S atoms of the ligand. These populations show larger variations, particularly when X = S. They are part of the properties that are typical for the specific metal and will not be discussed in detail here.

In the following, we shall concentrate upon the ligand part of the complexes. Then it is sufficient to analyze results obtained for one divalent and one monovalent metal. We have chosen the Be and Li complexes for this purpose.

In Results, we first discuss the results obtained for the free ligands $(C_2H_2X_2)^n$, X = O or S and $n = 0, -1$, or -2. Then we continue to analyze the results for the

TABLE 1
CALCULATED ORBITAL ENERGIES (IN EV) FOR $[M(C_2H_2O_2)_2]^0$

Orbital	type	Mono	Di	Be	Ni	Cu	Zn
a_g	C,O	−27.0	−26.5	−26.3	−26.4	−26.3	−26.2
b_{3u}	C,O	—	−26.5	−26.0	−26.2	−26.1	−26.1
b_{2u}	C,O	−21.9	−21.5	−21.9	−21.7	−21.6	−21.6
b_{1g}	C,H,O	—	−21.3	−21.1	−21.5	−21.5	−20.1(0.8)
b_{1g}	M	—	—	—	—	−19.8(1.0)	−22.7(1.1)
a_g	M	—	—	—	−19.1(0.9)	−20.7(0.9)	−21.5(1.5)
b_{2g}	M	—	—	—	−17.3(1.3)	−19.0(2.0)	−21.2(2.0)
b_{3g}	M	—	—	—	−16.8(1.8)	−18.8(2.0)	−21.0(2.0)
a_g	M	—	—	—	−15.4(1.8)	−18.4(1.0)	−21.0(1.8)
a_g	C,O	−20.3	−19.8	−20.1	−20.4(0.4)	−19.8(1.2)	−19.5(0.5)
b_{3u}	C,O	—	−19.7	−19.9	−19.8	−19.8	−19.8
a_g	O,C	−17.8	−17.4	−18.6	−17.8(1.0)	−17.5(1.1)	−18.0(0.2)
b_{3u}	O,C	—	−16.9	−16.9	−17.4	−17.4	−17.3
b_{2u}	O,C	−17.1	−16.6	−17.0	−17.0	−16.9	−16.8
b_{1g}	O,C	—	−16.3	−16.1	−16.6	−16.5	−16.4
b_{1u}	O,C	−15.5	−15.0	−16.2	−16.1	−16.0	−16.0
b_{2g}	O,C	—	−15.0	−15.7	−15.1(0.7)	−15.4	−15.6(0.1)
b_{2u}	O	−13.8	−13.3	−14.1	−13.9	−13.9	−13.9
b_{1g}	O	—	−13.0	−13.0	−14.2(0.1)	−12.5(0.2)	−14.3(0.1)
b_{3g}	O	−13.3	−12.7	−14.1	−13.8(0.2)	−13.9	−13.9
a_u	O	—	−12.6	−13.6	−13.9	−13.8	−13.7
a_g	O	−12.0	−11.5	−12.4	−12.1	−12.2	−12.2
b_{3u}	O	—	−11.3	−11.9	−11.9	−11.8	−11.9
b_{1u}	C,O	—	—	−5.0	−4.9	−4.8	−4.8

NOTE: M = Be, Ni, Cu, Zn. Di means dimer, i.e., no metal. Mono represents the free glyoxal monomer in its *cis* structure with the same geometry as in the complex. Most of the orbitals are ligand orbitals with clear ligand character, very stable from molecule to molecule. The five d-orbitals are clumped together in a special group, even though the orbital energies vary considerably. Values in parenthesis are d-orbital populations of the orbital (populations less than 0.1 are not given).

TABLE 2
CHARGES DERIVED FROM GROSS ATOMIC POPULATIONS ON C AND H ATOMS AND MX OVERLAP POPULATIONS OF COMPLEXES $[(C_2H_2X_2)_2M]^0$

M:	Be	Mg	Ni	Zn
X = O				
Cσ	0.08	0.08	0.08	0.08
Cπ	0.08	0.07	0.08	0.08
H	0.21	0.21	0.22	0.21
MOσ	0.30	0.27	0.34	0.30
MOπ	0.04	0.07	0.07	0.07
X = S				
Cσ	−0.51	−0.49	−0.47	−0.49
Cπ	0.03	0.04	0.01	0.03
H	0.27	0.27	0.26	0.26
MSσ	0.39	0.30	0.52	0.48
MSπ	0.04	0.10	0.09	0.10

complexes $[(C_2H_2X_2)_2 M]^n$, X = O or S, M = Be with n = 2, 0, −2 and M = Li with n = 1, −1. A general discussion is presented in Discussion and Conclusions.

METHOD OF COMPUTATION

Single configuration ab initio self-consistent-field (SCF) calculations within the molecular orbital–linear combination of atomic orbitals (MO-LCAO) framework were performed. The joint MOLECULE-ALCHEMY program system[8] was used. The basis set was of Gaussian type and of double zeta quality, except for a minimal basis for the K and L shells of third row elements and for the K shell of sulfur. Polarization functions of *p*-type were added to Li and Be and of *p*- and *d*-type to Ni, Cu, and Zn. Mulliken population analysis was carried out with a computer program, POPUL, kindly supplied by Johansen.[9] The chosen geometry was intermediate between the geometries optimal for positively and negatively charged complexes.

Some computational details are given in References 2 and 3 and others can be obtained upon request.

RESULTS

Specification of Populations Considered

The Mulliken population analysis[7] is a tool for describing the localization of the electrons to different regions of ordinary space. The analysis has many levels of refinement. Most frequently, only gross atomic populations are discussed in the literature. Here, we shall use the next level of refinement, dividing space into atomic and bonding regions. The electron densities in these two regions are represented by the net atomic populations and the overlap populations, respectively.

We define the *extra net population* as the difference between the net atomic population of the atom in the molecule and the number of electrons of the neutral atom (H, C, O, S) or metal ion (Li$^+$, Be^{2+}). For the calculation of these differences, we assume the atomic valence shell configurations to be $(s)^1 (p\sigma)^2 (p\pi)^1$ for C, $(s)^2 (p\sigma)^3 (p\pi)^1$ for O and S, and $(1s)^2$ for Li$^+$ and Be^{2+}. Both overlap and net populations are further partitioned into their σ and π parts.

The results of the calculations are given as diagrams in FIGURES 2–7. The upper row shows overlap populations of the bonds, CH, CC, CX, and MX. The π- and σ-overlaps are shown separately. Their sum, the total overlap, is shown in FIGURES 2 and 5. For each bond, there are three (or two) columns, corresponding to the different overall charges of the whole complex. The second row shows corresponding diagrams of the extra net population of the atoms, H, C, X, and M. Columns below the baseline (lack of electrons, i.e., net positive charge on the atom) show that the atom has donated part of its electrons to bonding regions or other atoms. Columns above the baseline show that the atom has attracted even more electrons than the free atom.

The addition of π-electrons to the molecule lowers the value of n. The sum of all π-contributions for a fixed value of n, both overlap and extra net population, must be equal to the number of added π-electrons. In the σ-part of the system, the total number of electrons is the same for all values of n. Increase of overlap or extra net population in one place means, therefore, decrease in another. The sum of all σ-contributions belonging to a fixed value of n must vanish.

Effect of Total Charge n on Free Glyoxal $(C_2H_2O_2)^n$

Energy calculations[3] show, in harmony with chemical experience, that free $[C_2H_2O_2]^n$ is neutral. However, in order to evaluate the effect of a metal ion on the ligand in complexes with a different total charge n, it is essential to begin with a study

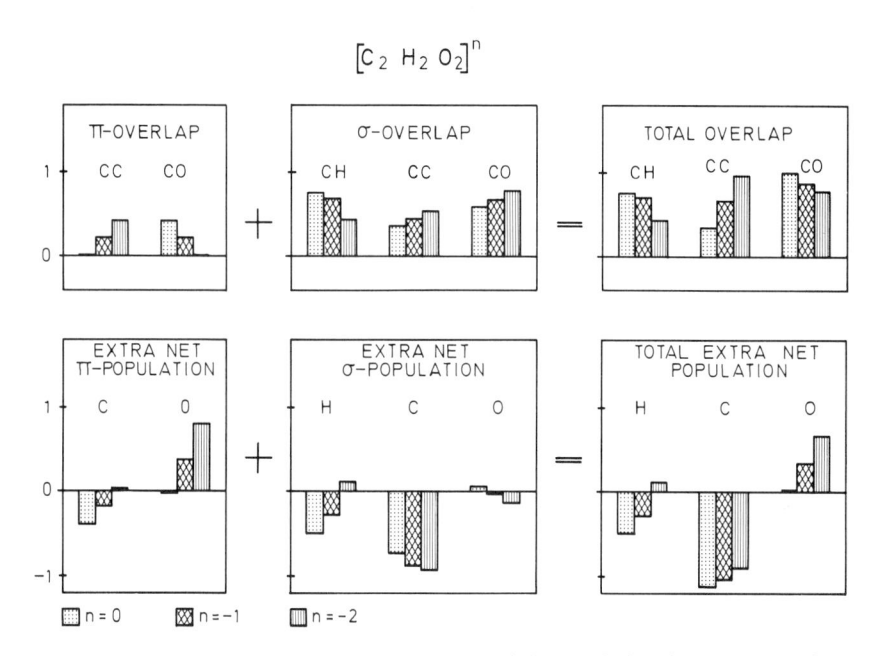

FIGURE 2. Diagrams obtained from Mulliken population analysis. The upper row shows overlap populations (π, σ, and total) for the different bonds in the molecule. The lower row shows extra net populations (π, σ, and total) on the different atoms in the molecule. For definition of extra net population, see text. This figure represents results for $[C_2H_2O_2]^n$. The first column in each group of three corresponds to the case of total charge n of the whole complex equal to 0, the next to $n = -1$, and the last to $n = -2$.

of the changes caused by addition of electrons to the free ligand, even if the ions formed may be merely artifacts. FIGURE 2 shows the population of such ions.

In the neutral $C_2H_2O_2$, $n = 0$, represented by the first column in each group of three, there is no π-overlap in the CC-bond, only in the CO-bond. When $n = -1$ (middle column) the π-overlap of CC and CO are equal, and when $n = -2$ (last column) π-overlap is only found in the CC-bond. The diagram of extra net π-population shows that both C and O populations increase when electrons are added, but that the larger part goes to the O atoms.

Since the number of σ-electrons is constant, one might expect the σ-population to be rather independent of the n-value. FIGURE 2 shows, however, sizable changes in the σ-system. The CH overlap is considerably decreased when the molecule is negatively charged. When $n = -2$, the CH overlap is approximately halved compared to $n = 0$. But both CC and CO σ-overlaps increase with addition of π-electrons. Net populations are also changed. In particular, the H atom picks up electron density, partly from the CH bond and partly from the C atom.

The two diagrams to the far right in FIGURE 2 show that the π- and σ-parts together build up a CC bond in the state $n = -2$, with close to three times as much overlap as in the single bond of $n = 0$. The strong CO bond of $n = 0$ will only loose 20% of its overlap when the molecule is transformed into a dianion. As a result, with overlap as a measure, the bonding in the whole O—C—C—O region is stronger in the dianion than in the neutral molecule. Nevertheless, the total bonding overlap is diminished. In fact, the decrease of the CH overlap more than counterbalances the increase in the rest of the molecule. The decrease in the sum of bonding overlap is in harmony with energy calculations. As shown in Reference 3, the binding energy of the dianion is 9 eV smaller than that of the neutral molecule. But it should be mentioned that the responsibility for the destabilization of the dianion is not due only to the CH bonds but also to the nonbonded interaction.

As can be inferred from FIGURE 1c, addition of two π-electrons to neutral glyoxal will lead to a structure with a CC double bond and $p\pi$ lone pairs on each O atom, i.e., negative charges on those two atoms only. However, the total extra net population of

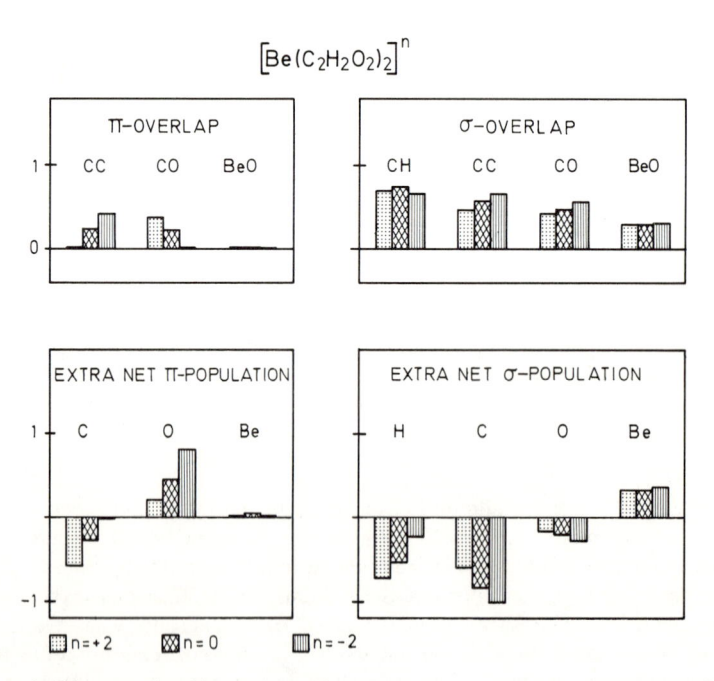

FIGURE 3. Diagrams showing overlap and extra net populations (π and σ) for the complex $[Be(C_2H_2O_2)_2]^n$, $n = +2, 0, -2$.

$[Li(C_2H_2O_2)_2]^n$

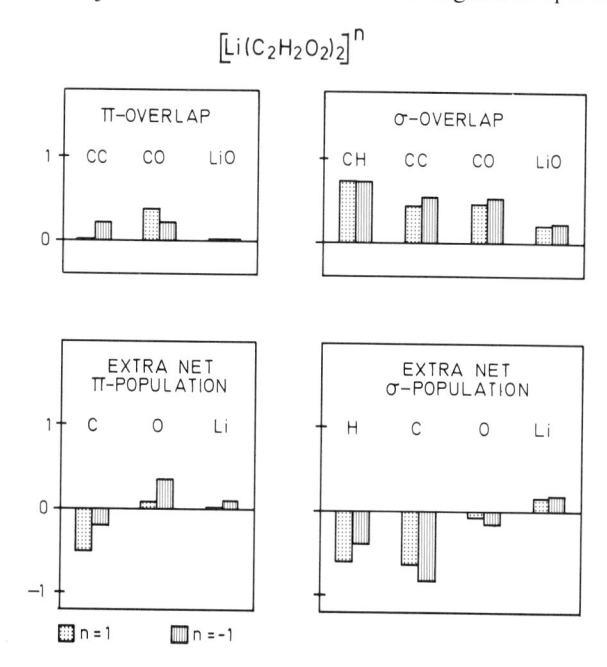

FIGURE 4. Diagrams showing overlap and extra net populations (π and σ) for the complex $[Li(C_2H_2O_2)_2]^n$, $n = +1, -1$.

FIGURE 2 shows that all the atoms will get their share of the additional charge, though the amounts will vary. The share of the C atom is approximately one third that of the O atom, but the H atom has almost as much as the O atom. Thus, the additional charge is pushed towards the outskirts of the molecule, the H and O atoms, and only a minor part is left at the backbone of C atoms.

In summary, the changes brought about by the addition of two π-electrons to a $C_2H_2O_2$ molecule are: (1) increased net electron populations on O, H, and C atoms, in total, three units, (2) increased populations in the sum of CC and CO bonds but decrease in CH bonds, in total, reduced bonding overlap by half a unit, (3) increase of negative overlap (repulsion) in nonbonded regions, in total, half a unit.

Effect of the Metal Ion on the Electron Population of the Ligands in Complexes $[(C_2H_2O_2)_2M]^n$

When a complex between a positive metal ion and neutral ligands, $C_2H_2X_2$, is formed, bonds between the M ion and the X atoms are established. The covalent part of these bonds is essentially σ-type (see TABLE 2). The metal ion also acquires some extra net atomic population, mainly σ-type. We shall now discuss the details of this electron transfer.

Overlap populations and extra net populations in the complex $[(C_2H_2O_2)_2 Be]^n$ are shown in FIGURE 3 for $n = 2, 0, -2$. These n-values correspond to the values $n = 0$, -1, -2 of free $C_2H_2O_2$. Similarly, FIGURE 4 shows the populations of $[(C_2H_2O_2)_2 \text{Li}]^n$, $n = 1, -1$, corresponding to $n = 0, -1$ of free $C_2H_2O_2$.

The first thing to observe is that both MO overlap and M net population are almost constant when n is varied. Since similar trends are found in FIGURES 3 and 4, we shall first discuss the Be complex, where the trends are most obvious. We shall begin with a comparison between $[(C_2H_2O_2)_2 \text{Be}]^{2+}$ and $C_2H_2O_2$. Afterwards, we shall comment on the effect of electron addition.

The total amount of σ-population accepted by the Be atom and the four BeO bonds is around one and a half units. This population is donated mainly by the O atoms and the CO bonds. But the electronegative O atoms try to compensate for their loss by attraction of π-electrons, mainly from the C atoms, which, in their turn, attract σ-electrons from the H atoms. The final population of the complex is such that the total net population of the O atom is almost unchanged. The CO bond has lost, mainly in σ but also in π. The C atom has lost more in π than it gained in σ. The CC and CH bonds are almost unchanged, but the H atom has lost a substantial amount. The main losers are the CO bond and the H atom, which have donated almost equal parts to the metal region.

A very similar pattern is shown by the Li-complex of FIGURE 4, the only difference being that the amount of population transferred is smaller than that in the Be-complex.

The next step is to see what happens when π-electrons are added to the Be-complex. This picture is in many respects similar to the picture found in the last section. The increase of π-overlap in the CC bond and the decrease of π in the CO bond is obvious. All increases in net population in $[(C_2H_2O_2)_2 \text{Be}]^n$ are similar to, but somewhat smaller than, those in $(C_2H_2O_2)^n$.

The main difference between the population shifts with n of free ligand (FIGURE 2) and metal complex (FIGURE 3) is a damping effect from the metal ion on the repulsion between the π- and σ-populations of the ligand. The σ-population is still pushed away from the C atom towards the H atom, but the CH overlap is not reduced. Another effect, not shown in FIGURE 3, is that the metal ion will reduce the nonbonded repulsion in the complex compared to the repulsion in the free ligand.

The consequence of these two differences is as follows. Since addition of electrons slightly increases the total bonding overlap of the complex, and since nonbonded repulsion has a slower rate of increase with the number of added electrons, the electron affinity of the system increases. In fact, energy calculations show that three electrons can be added to the original system, i.e., that the complex with $n = -1$ is the most stable one (see References 3 and 5).

The details of the shifts in populations by addition of electrons to $[(C_2H_2O_2)_2 \text{Li}]^+$ are the same as those described for the Be-complex, the only difference being that all effects are smaller.

These calculated shifts can be compared with the reactivity of the methine (CH) hydrogen of amino acids.[10] Experimentally, it is found that this H atom is very stable in the free amino acid but, in the presence of a metal ion, Cu(II) or Co(III), it easily undergoes substitution reactions: For example, addition of acetaldehyde (CH_3CHO) to glycine (NH_2CH_2COOH) in the presence of Cu^{2+} yields threonine ($CH_3CH(OH)CH(NH_2)COOH$). This effect could be compared with the extra net population of the H atom in our model complex. In the free monomer, the extra net population is larger than that in the metal complex (see FIGURES 2 and 3). When the H atom has a low electronic population, i.e., a positive charge, it is obviously easier for a proton to leave the molecule. Experiments also indicate that the reactivity of this methine hydrogen decreases as the total charge of the complex becomes more negative. This, also, conforms to our calculations, showing that the extra net populations on the hydrogen increase as the total charge n becomes more negative.

Effect of Total Charge n on Free Thiogyoxal $(C_2H_2S_2)^n$

The overlap and net populations of $(C_2H_2S_2)^n$ are presented in FIGURE 5 for $n = 0$, -1, and -2. Before studying the effect of electron addition, we shall comment on the electron distribution of the neutral molecule, comparing it to $C_2H_2O_2$ of FIGURE 2. In particular, we want to compare the polarity of the CO and CS bonds.

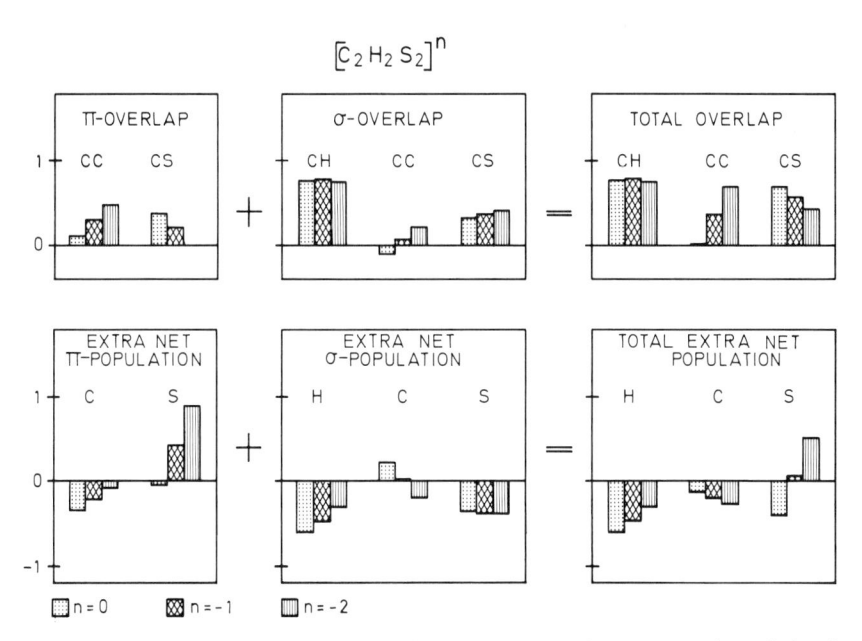

FIGURE 5. Diagrams showing overlap and extra net populations (π, σ, and total) for the complex $[C_2H_2S_2]^n$, $n = 0, -1, -2$.

Within the Mulliken scheme, we define the polarity of a CX bond as the difference between the extra net populations of the two atoms. Since populations are partitioned into σ- and π-parts, we can define σ-polarity and π-polarity by this means.

The polarity of the CO bond in the neutral molecule is obvious from FIGURE 2 and conforms to the usual chemical picture. The total extra net population on the O atom is larger than that on the C atom by more than one unit. The diagrams of π- and

σ-populations show that one third of this difference comes from the π-population and two thirds from the σ-population. After the addition of two extra π-electrons, the σ-polarity is almost the same, but the π-polarity increases to twice the original value, thus attaining the same value as the σ-polarity. The total polarity is about 40 percent larger in the dianion than in the neutral molecule.

The polarity of the CS bond in neutral thioglyoxal is more unclear (see FIGURE 5). In fact, the σ-population shows that the C atom is more negative than the S atom by more than half a unit. On the other hand, the π-population shows the opposite polarity, the π-population on S being one third unit larger than that on C. Since σ dominates, the total population indicates that C is more negative than S. Addition of two extra π-electrons to the molecule reduces the σ-polarity to one third. Simultaneously, the π-polarity increases by a factor of three. Since the π-polarity dominates in the dianion, S is clearly more negative than C. In the monoanion, S is also more negative than C, but to a smaller degree. This complicated pattern mirrors the well-known difficulty of finding an acceptable electronegativity value of S.[11]

Comparing $C_2H_2S_2$ (FIGURE 5) to $C_2H_2O_2$ (FIGURE 2), we see that the π-populations of the two molecules are very similar. This is to be expected from the finding that the CO and CS bonds have the same π-polarity. On the other hand, the fact that C is more electronegative than S with respect to σ-electrons has substantial consequences. It is seen that, in $C_2H_2S_2$, σ-electrons are donated to the C atom both from the S atom and the CS bond. The piling up of negative σ-charge on the C atom results in repulsive, negative σ-overlap between the two C atoms, thus preventing the formation of a CC σ-bond. The formation of the CH bond is no problem, since the H atom has a net positive charge. The lack of overlap population in the CC region, obvious from the total overlap diagram, should imply that $C_2H_2S_2$ is not a stable molecule. In fact, Hoyer et al. remark that, as far as they are aware, the neutral $C_2H_2S_2$ does not exist, but the dianion is well known.[12]

Addition of π-electrons to $C_2H_2S_2$ leads to changes of the π-population similar to those found for $(C_2H_2O_2)^{2-}$ above. The response of the σ-population to the increased number of π-electrons is, however, somewhat different. The reason may be that the π-charges on the C and S atoms are more distant than those on the C and O atoms. The repulsion between the π- and σ-systems is, therefore, less pronounced in $C_2H_2S_2$. Part of the σ-population on the C atom is transferred to the H atom, but the CH overlap is not changed. When the excess σ-charge of the C atom is removed, a weak CC σ-bond is formed. It is remarkable that the π-overlap of the CC bond is more than twice the σ-overlap. The total bonding overlap is larger in the anions than in the neutral molecule. The sum of bonding and nonbonding overlap is found to be largest in the monoanion. One should not expect overlap populations to give accurate information about stability, but it is gratifying that the largest sum of overlap population coincides with energy calculations in predicting the monoanion $(C_2H_2S_{2na\%})^-$ to be the most stable.[3] Similarly, in the case of glyoxal, the neutral $C_2H_2O_2$ has the largest overlap and is the most stable, according to energy values.[3]

The final electron distribution in the dianion is such that the additional charge is mainly found on S atoms where the net population has increased by almost one unit. The increase on the H atoms is only one third and the C atoms have lost population. There is a small increase in bonding overlap and some increase in nonbonded repulsion, though only half as much as in $(C_2H_2O_2)^{2-}$.

Effect of the Metal Ion on the Electronic Population of the Ligands in Complexes $[(C_2H_2S_2)_2M]^n$

Overlap populations and net populations of the complex $[(C_2H_2S_2)_2 Be]^n$, with $n = 2, 0, -2$, are shown in FIGURE 6 and $[(C_2H_2S_2)_2 Li]^n$ with $n = 1, -1$ in FIGURE 7. As in the corresponding oxygen compounds, the overlap population of the MS bond is seen to be almost constant as n is varied. The net atomic population on the M ion only shows minor variations with the charge n. Much larger population shifts are found in the ligand part. We shall now discuss them in the same way as above.

The total amount of σ-population accepted by the Be atom and the BeS bonds is more than two units. This population is donated mainly by the S atoms and the CS

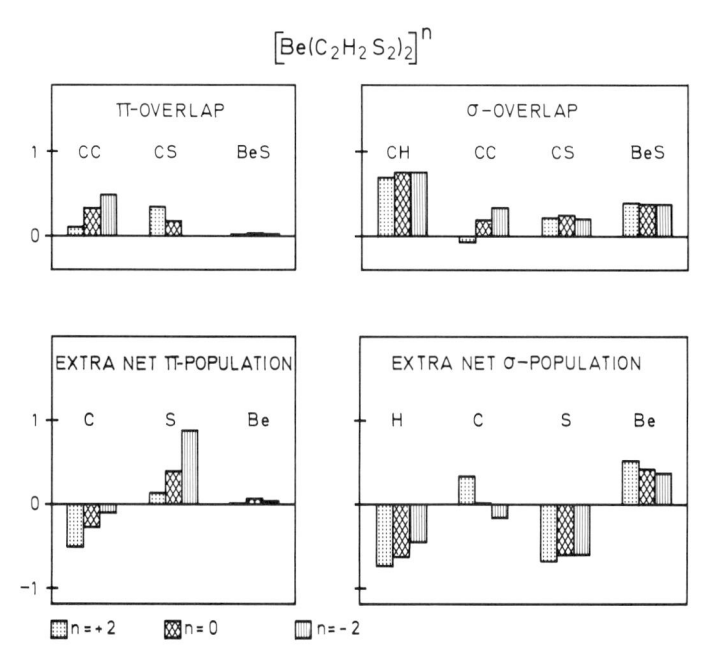

FIGURE 6. Diagrams showing overlap and extra net populations (π and σ) for the complex $[Be(C_2H_2S_2)_2]^n$, $n = +2, 0, -2$.

bonds. The loss of σ-population on S is only partly compensated for by the attraction of π-population from the C atoms, followed by transfer of σ-population from H to C. The final picture is not as clear as for the oxygen compound. All atoms and bonds (except the nonexisting CC bond) have lost part of their population to the metal region.

When extra π-electrons are added to the Be complex, the π-populations for $n = 0$ and $n = -2$ are found to be almost the same as those for the free ligands with $n = -1$ and $n = -2$, respectively. As for the σ-population, the loss of electrons from the S atom and the CS bond remains at the same level as for $n = 2$. The CC σ-overlap increases somewhat more than in the free ligand.

The final population for $n = -2$ is very similar to that of $(C_2H_2S_2)^{2-}$. The net population has increased on S by almost one unit and on H by one third. The C atoms have lost a little. The nonbonded repulsion has increased by the same amount as in the free ligands. The main difference is found in the bonding overlap. The CH is the same, but CS has decreased and CC has increased compared to $(C_2H_2S_2)^{2-}$.

The populations of $[Li(C_2H_2S_2)_2]^n$ of FIGURE 7 show the same trends as the Be-complex, though all effects are smaller. The total transfer of σ-population to the metal region is one and a half unit at the expense of S and H populations. At the same time, the CC bond is strengthened. Added π-electrons cause increased populations, especially on S, but also on H and in the CC bond, in both π- and σ-parts.

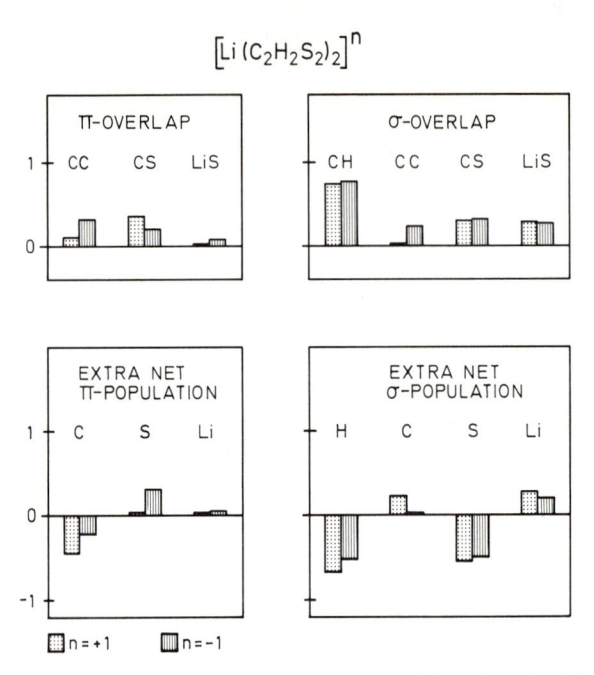

FIGURE 7. Diagrams showing overlap and extra net populations (π and σ) for the complex $[Li(C_2H_2S_2)_2]^n$, $n = +1, -1$.

DISCUSSION AND CONCLUSIONS

In our study of metal complexes $[(C_2H_2X_2)_2 M]^n$ as models of metals in biology, we have found that several properties are typical of the kind of ligating atom X and the charge n on the whole complex, but almost independent of the kind of metal ion M. The electron density distribution within the ligand is such a property. Since Mulliken population analysis[7] is an acceptable tool for this kind of study, we have applied such an analysis to our model compound. We want more details than are provided by gross atomic populations, e.g., information about the strength of bonds. The present discussion is therefore devoted to overlap populations of bond regions and "extra net populations" of atomic regions, defined above. Both kinds of population are partitioned into their σ- and π-parts.

In the study of the charge distribution within the whole complex, it is informative to build up the compound from hypothetically free ligands, $(C_2H_2X_2)^n$, and positively charged metal ions. For this purpose, our analysis begins with free ligands $(C_2H_2X_2)^n$ with varying total charge n. It is found that the addition of two π-electrons to $C_2H_2X_2$ will substantially change not only the π-population, but also the σ-population. The extra charge of $[C_2H_2O_2]^n$ is found to be shared between O and H atoms and, to a minor extent, by the C atoms. The σ-part of CC and CO overlaps increase and the CH overlap decreases. There is also a substantial increase of nonbonded repulsion. The extra charge of $[C_2H_2S_2]^n$ goes mainly to the S atoms, but a smaller part is found on the H atoms. The C atoms, on the other hand, lose population in this molecule, but the CC overlap increases somewhat.

It could be that the difference between $(C_2H_2O_2)^{2-}$ and $(C_2H_2S_2)^{2-}$ depends, to a certain extent, upon the different length of the CX bonds. The space available for the extra π-charge of $(C_2H_2S_2)^{2-}$ is larger and, as a consequence, the inductive effect on the σ-system is weaker. The difference in size may also be important with respect to the influence of the metal ion on the ligand population in the complexes $[(C_2H_2X_2)_2M]^n$.

When a metal complex is formed, the covalent part of the MX bonds is found to be almost pure σ. The extra net population on M is also σ-type. The population donated to the M and MX regions causes rearrangements of the electronic population within the whole ligand. In compounds with X = O, the main donors are the CO bonds and the H atoms. In compounds with X = S, the main donors are the S and H atoms and the CS bonds. The finding that the H atoms lose electrons to the metal is in harmony with experiments showing that metal ions activate the reactivity of amino acids.[10]

To obtain a better understanding of the difference between O complexes and S complexes, the polarity of the CX bond in $(C_2H_2X_2)^n$ was studied within the Mulliken scheme. We find, as expected, that O is more negative than C for the CO bond, both in the σ- and the π-part. When π-electrons are added to $H_2C_2O_2$, the σ-polarity is almost constant, but the π-polarity increases.

For the neutral $H_2C_2S_2$, we find that the C atom is more negative than S in the σ-part, but that S is more negative than C in the π-part. The difference between the σ extra populations is more than half a unit and the corresponding π difference is one third unit. The sum of σ- and π-polarity indicates, therefore, that C is more negative than S. When one π-electron is added, the σ-polarity goes down and the π-polarity goes up. In the monoanion we find, therefore, that S is more negative than C. In the dianion, the electronegativity of S is still larger than C. Since we have also found that the neutral $C_2H_2S_2$ is less stable than $(C_2H_2S_2)^-$, we conclude that, in actually existing compounds, S seems to be more negative than C, but that the balance between the two atoms is very subtle indeed.

As to the reactivity of the different complexes, we can predict different reactivity for complexes with different ligating atoms X. It is thus intriguing to study mixed complexes with different ligands X, Y. Such studies are under way.[13] Our aim is to study the reactivity of metal complexes of amino acids. It is well known that, for example, substitution of methine hydrogen atoms is drastically enhanced when amino acids form complexes with metals.[10]

In this study, we have found that many properties are independent of the type of metal, but highly dependent on the total charge n of the complex. In metal–amino acid complexes in solution, this total charge can be changed by pH changes. Moreover, at a certain pH value, different metals will have different ligands, since they have different pK values.[14,15] As a consequence, they will have different reactivity.

Acknowledgments

We are indebted to J. Almlöf, University of Oslo, H. Johansen, Technical University of Denmark, and U. Wahlgren, University of Stockholm, for kindly supplying computer programs.

References

1. FISCHER-HJALMARS, I., M. SUNDBOM & H. VOKAL. 1974. *In* Molecular and Quantum Pharmacology. E. Bergmann and B. Pullman, Eds.: 583–86. Reidel. Dordrecht, Holland.
2. DEMOULIN, D., I. FISCHER-HJALMARS, A. HENRIKSSON-ENFLO, J. A. PAPPAS & M. SUNDBOM. 1977. Int. J. Quantum Chem. **12:** 351–69.
3. BLOMBERG, M. R. A., I. FISCHER-HJALMARS & A. HENRIKSSON-ENFLO. 1980. Isr. J. Chem. **19:** 143–48.
4. BLOMBERG, M. R. A. & U. WAHLGREN. 1980. Chem. Phys. **49:** 117–26.
5. FISCHER-HJALMARS, I. & A. HENRIKSSON-ENFLO. 1980. Int. J. Quantum Chem. **18:** 409–20.
6. FURLANI, C. & C. CAULETTI. 1978. Structure and Bonding, Vol. 35. J. D. Dunitz, J. B. Goodenough, P. Hemmerich, J. A. Ibers, C. K. Jörgensen, J. B. Nielands, D. Reinen, and R. J. P. Williams, Eds.: 119–69. Springer-Verlag. Berlin.
7. MULLIKEN, R. S. 1955. J. Chem. Phys. **23:** 1833–40.
8. ALMLÖF, J. 1974. USIP Report 74-29. University of Stockholm (for the integral part and some of the closed shell SCF calculations); BAGUS, P. S., B. LIU, M. YOSHIMINE, A. D. MCLEAN & U. WAHLGREN are the authors of the ALCHEMY SCF program.
9. JOHANSEN, H. 1974. The author of the program POPUL.
10. PHIPPS, D. A. 1979. J. Mol. Catalysis. **5:** 81–107.
11. PAULING, L. 1960. The Nature of the Chemical Bond, 3rd edit. Cornell University Press. Ithaca, N.Y.
12. HOYER, E., W. DIETZSCH & W. SCHROTH. 1971. Z. Chem. **11:** 41–53.
13. DAVSTAD, K. Unpublished results. University of Stockholm.
14. SILLÉN, L. G. & A. E. MARTELL, Eds. 1971. Stability Constants of Metal-Ion Complexes (Suppl. 1). Special Publication No. 25. The Chemical Society. London.
15. PERRIN, D. D. & R. P. AGARWAL. 1973. *In* Metal Ions in Biological Systems, Vol. 2. H. Sigel, Ed.: 168–206. Marcel Dekker. New York.

PHOSPHOLIPID METHYLATION AND MEMBRANE FUNCTION

Julius Axelrod and Fusao Hirata

*National Institute of Mental Health
Bethesda, Maryland 20014*

Hormones, neurotransmitters, antigens, and many drugs initiate their effects by interacting with receptors on the surfaces of cells. How their specific messages are carried from the receptor through the membrane is a problem of major biological importance. In the past two years, our laboratory has found that methylation of membrane phospholipids plays an important role in the transduction of signals through biomembranes.

We have found two enzymes in plasma membranes that methylate phosphatidylethanolamine to phosphatidylcholine. S-adenosylmethionine serves as the methyl donor in these reactions. The first enzyme (methyltransferase I) methylates phosphotidylethanolamine to phosphatidyl N-monomethylethanolamine; this enzyme requires Mg^{2+} and has a low K_m for S-adenosylmethionine. The second enzyme (methyltransferase II) converts phosphatidyl N-monomethylethanolamine to phosphatidylcholine by two successive methylations. Methyltransferase II has a high K_m for S-adenosylmethionine and does not require Mg^{2+}.

The substrate and products of these enzymes are asymmetrically distributed in erythrocyte membranes. Phosphatidylethanolamine faces the cytoplasmic side of the membrane, while phosphatidylcholine is orientated towards the outer surface. Using normal and inside-out red cell ghosts, we found that the phospholipid transferases were asymmetrically localized in membranes. Like its substrate, methyltransferase I faces the cytoplasmic side and methyltransferase II, like its product, is located in the external surface of the membrane. As phosphatidylethanolamine is successively methylated on the inner surface of the red cell membrane, the metabolites are rapidly translocated across the membrane. The final product, phosphatydylcholine, is located in the outer surface of the membrane. This flip-flop of phospholipids requires Mg^{2+}—a cation necessary for the first methylation step. The product of methyltransferase I, phosphatidyl N-monomethylethanolamine, decreases membrane viscosity. This, in turn, increases membrane fluidity and lateral mobility of proteins.

The elucidation of the enzymatic processes involved in the methylation and translocation of phospholipids in membranes made it possible to study the effect of agonists interacting with surface receptors upon transduction of biological signals through the membrane. The initial studies were carried out on β-adrenergic receptors present on reticulocyte membranes. When rat reticulocyte ghosts were loaded with ^{3}H-S-adenosylmethionine and then stimulated with the β-adrenergic agonist isoproterenol, there was a considerable increase in the ^{3}H-methylated phospholipids. There was also a rapid translocation of these lipids from the inside to the outside surface of the membrane. The increase in phospholipid methylation by isoproterenol was stereospecific and blocked by the β-adrenergic agonist propanolol. The increase in phospholipid synthesis by a number of catecholamine agonists showed the same order of potency as the activation of adenylate cyclase. Direct stimulation of adenylate cyclase with cholera toxin or sodium fluoride showed no elevation in phospholipid methylating activity, indicating that the β-adrenergic receptor, but not the adenylate cyclase activity, activates the methyltransferase enzymes. Using resealed reticulocyte ghosts, we found that the activity of isoproterenol-sensitive adenylate cyclase was doubled at a concentration of S-adenosylmethionine that affects the synthesis of phosphatidyl-N-monomethylethanolamine. The latter compound has been shown to increase membrane fluidity. All these observations indicate that binding the β-adrenergic receptor with an agonist increases phospholipid methylation, which, in turn, decreases membrane viscosity and thus facilitates the lateral mobility of the receptors. These events provide a greater chance for the β-adrenergic receptor to collide and couple with the β-adenylate cyclase. HeLa cells, astrocytomas, and lymphocytes contain β-adrenergic receptors. Phospholipid methylation can be markedly stimulated in these cells by isoproterenol.

The effect of phospholipid methylation on the availability of β-adrenergic receptor sites was examined in rat reticulocytes. When S-adenosylmethionine was introduced into reticulocyte ghosts, there was a marked increase in the number of β-adrenergic receptor sites that can bind to the β-adrenergic ligand ^{3}H-dihydroalpranalol. The unmasking of these β-adrenergic receptors is dependent upon the synthesis of phosphatidylcholine, but not phosphatidyl-N-monomethylethanolamine. In HeLa cells, the number of β-adrenergic receptors decreased when phospholipid methylation was inhibited by 3-deazaadenosine. When the methyltransferase inhibitor was removed from the cell by washing, the receptors reappeared on the cell surface.

Other membrane phenomena are affected by phospholipid methylation. One of these is Ca^{2+} ATPase activity in erythrocytes. Erythrocyte ghosts incubated with S-adenosylmethionine increased Ca^{2+}-stimulated ATP hydrolyses and Ca^{2+} efflux. The increase in phospholipid methylation and Ca^{2+} ATPase activity was blocked by the methyltransferase inhibitor S-adenosylhomocysteine. The increase in Ca^{2+} ATPase activity closely paralleled the synthesis of phosphatidyl N-monomethylethanolamine, but not phosphatidylcholine.

The effect of phospholipid methylation was also examined in rat mast cells. These cells release histamine when stimulated by the lectin conconavalin A (ConA) in the presence of phosphatidylserine. When ^{3}H-methionine was introduced into rat mast cells, ConA caused a marked increase in the incorporation of ^{3}H methyl groups into lipids. This increase paralleled the release of histamine. The addition of ConA to these cells resulted in a rapid decarboxylation and methylation of ^{14}phosphatidylserine (previously introduced into cells) to phosphatidylcholine. These findings implicate phosphatidylserine decarboxylation and phospholipid methylation in the activation of membranes by ConA and in the subsequent exocytotic release of histamine from mast cells.

The binding of ConA to lymphocytes causes an increase in DNA synthesis and mitogenesis. Our studies with phospholipid methylation and ConA prompted experiments on lymphocyte mitogenesis. Incubation of rat lymphocytes with ConA and ^{3}H-methionine resulted in a transient doubling of the incorporation of ^{3}H methyl groups into membrane lipids; there was also a 2- to 3-fold increase in the accumulation of ^{3}H-lysophosphatidylcholine after ConA. The stimulation of lymphocytes with ConA also caused a release of ^{14}C-arachidonic from lymphocytes that were prelabeled with the fatty acid. Both arachidonic acid and lysophosphatidylcholine are products of the metabolism of phosphatidylcholine by phospholipase A_2. A paralleled activity with respect to a variety of lectins on mitogenesis and phospholipid methylation was found. Additional experiments with methyltransferase inhibitors clearly showed that phospholipid methylation is involved in lymphocyte mitogenesis.

REFERENCES

1. HIRATA, F., O. H. VIVEROS, E. J. DILIBERTO, JR. & J. AXELROD. 1978. Identification and properties of two methyltransferases in the conversion of phosphatidylethanolamine to phosphatidylcholine. Proc. Nat. Acad. Sci. USA **75**: 1718–21.
2. HIRATA, F. & J. AXELROD. 1978. Enzymatic synthesis and rapid translocation of phosphatidylcholine by two methyltransferases in erythrocyte membranes. Proc. Nat. Acad. Sci. USA **75**: 2348–52.
3. HIRATA, F. & J. AXELROD. 1978. Enzymatic methylation of phosphatidylethanolamine increases erythrocyte membrane fluidity. Nature **275**: 219–20.
4. HIRATA, F., W. J. STRITTMATTER & J. AXELROD. 1979. β-adrenergic receptor agonists increase phospholipid methylation, membrane fluidity and β-adrenergic receptor-adenylate cyclase coupling. Proc. Nat. Acad. Sci. USA **76**: 368–72.

INVERSE SCATTERING, ORDINARY DIFFERENTIAL EQUATIONS OF PAINLEVÉ-TYPE, AND HIROTA'S BILINEAR FORMALISM*

A. Ramani

Laboratoire de Physique Théorique et Hautes Energies
Université Paris-Sud
91405 Orsay, France

INTRODUCTION

Use of the inverse scattering transform (IST) method as a tool to solve partial differential equations (PDEs) has recently been the object of increasing interest.[1-5] The classical approach consists of first defining a formal "scattering problem," or eigenvalue problem, in which the solution u of the original PDE plays the role of a "potential." One then solves the eigenvalue problem at initial time, using the initial condition for u as potential. One thus gets the scattering data (reflection coefficients, bound states energies, and norming constants). The crucial point is that the time evolution of these scattering data is given by a linear equation when u evolves under the original PDE. It is, therefore, quite easy to compute these data at a later time. Finally, one reconstructs the potential at a later time from the time-evolved scattering data. This new potential is then just the time-evolved value of u. Only equations for which all these steps are possible may truly be called "solvable by IST or IST-type." This classical method, however, involves a heavy analytical apparatus.

Much work has been done on IST-type and similar equations along completely different lines of thought, and often in a more algebraical and less analytical way.[6-12] Among these lines of thought figure, on the one hand, the so-called "bilinear formalism" that was first described by Ryogo Hirota,[7-10] and, on the other hand, the connection with Painlevé-type ordinary differential equations (ODEs) first noted by M. J. Ablowitz and H. Segur.[11,12] The purpose of this paper is to show the remarkable convergence of some of the results of these two approaches, despite their entirely different spirits.

Ablowitz *et al.* have formally proposed a conjecture along the latter approach, namely that, whenever an ansatz reduces an IST-type PDE exactly into an ODE, the result is a Painlevé-type equation, i.e., its solution has no movable critical singularities (singularities other than poles).[12] This conjecture, which we will call the "Painlevé conjecture," if proved, would lead to a criterion for determining whether an equation is IST-type or not. This criterion would consist of trying ansätze to reduce a PDE to ODEs and then trying to find movable branch points in a systematic way. A necessary condition is that no such movable branch points are found. This is not a sufficient condition, of course, but, so far, it has proven restrictive enough to discriminate correctly between classical IST-type equations and very similar equations known not to be IST-type.

*This work was included in a thesis defended at Université Paris-Sud (Orsay) 6 April 1981.

On the other hand, the Hirota "bilinear method"[7-10] can be used to formalize a wide variety of equations without any reference to a scattering problem whatsoever. In many cases, it can also help find particular solutions to those equations, namely, traveling wave solutions and "two quasi-solitons" solutions.[8,13] The latter are solutions that separate asymptotically at large times, both positive and negative, into two traveling waves that nonlinearly "cross" each other at some finite time and then re-emerge unchanged, just like two solitons. The expression "quasi-soliton" has been used because, to be properly called a soliton, a traveling wave must be able to cross, not only a single other traveling wave, but anything, which is *not* true in the case of all these equations. Indeed, it is generally *not* possible to find solutions that separate asymptotically into three or more traveling waves at both large positive and large negative times.

It is only for a small number of these equations that one can find solutions where an arbitrarily large number of traveling waves cross each other. This is not yet quite sufficient to call these waves "solitons", since it has not yet been proved that they would also re-emerge after crossing something which is not made out of traveling waves (in terms of inverse scattering, this would be the contribution of the continuous spectrum). Such solutions are, however, highly suggestive of soliton-like behavior, and, thus, of a connection with IST. Equations for which "N-quasi-solitons" solutions exist for arbitrary N will be said to pass the Hirota test. Of course, all IST-type equations pass this test, since they have multisoliton solutions. But some new equations that have not yet been intensively studied also pass this test. This is a strong indication that they are also IST-type.

There is a remarkable convergence between the Hirota test and the criterion related to the Painlevé conjecture. Namely, whenever an equation obtained through the bilinear formalism failed the Hirota test, it also failed the Painlevé criterion—it was possible to find an ansatz that reduced it to an ODE that was not Painlevé-type, i.e., that admitted solutions with movable critical singularities. Conversely, all the ansätze the author could imagine for all equations that passed the Hirota test reduced them to ODEs such that no solutions with movable critical singularities could be found; thus, the Painlevé criterion also suggested that they were IST-type.

The conjecture relating IST with the Painlevé property is compatible with the Hirota test in all cases yet known, and this is the result the author wishes to present in this paper.

In the following section, the aspects of bilinear formalism that are relevant to this purpose are briefly developed. Similarly, the connection between IST and Painlevé-type equations is briefly described, after which the equations that were studied are listed, with particular emphasis on those for which both methods suggest a connection with IST, although no eigenvalue problem is known for them at this time.

Hirota's Bilinear Formalism

We shall here sketch some basic points of the bilinear formalism introduced by R. Hirota.[7-10,13] It is a very powerful method of simplifying algebraic calculations, but one would not do it justice by looking at it only as a purely formal tool for the manipulation of complicated expressions. It also gives a great deal of insight into many problems related to IST, with which it has a very deep and important connection.

The basic symbol D of the formalism is defined by

$$(D_x^a D_y^b \ldots) f(x, y, \ldots) \cdot g(x, y, \ldots) = \left(\frac{\partial}{\partial x} - \frac{\partial}{\partial x'}\right)^a \left(\frac{\partial}{\partial y} - \frac{\partial}{\partial y'}\right)^b \ldots$$

$$f(x, y, \ldots) g(x', y', \ldots) \Big|_{x'=x, y'=y, \ldots}.$$

With this symbol, one can write many results in a very compact and simple way. In particular, it quite conveniently represents some nonlinear partial differential equations. The simplest way to represent an equation in this formalism is

$$\mathcal{P}(D_x, D_t, \ldots) f \cdot f = 0,$$

where $\mathcal{P}$ is some polynomial that can always be taken to be even, as the action of any odd product of Ds on the dot product of a function with itself vanishes.

Conveniently choosing $\mathcal{P}$ can lead to equations that are known to be IST-type. For instance,

$$(D_x D_t + D_x^4) f \cdot f = 0$$

leads to the Korteweg–de Vries (KdV) equation

$$u_t + u_{xxx} + 6uu_x = 0$$

for $u = 2 \, (\partial^2/\partial x^2) \ln f$. In general, any polynomial would lead to an equation that, in terms of $u = \ln f$ or $u = (\partial/\partial x)^n \ln f$ for some integral n, looks strikingly similar to classical IST-type equations.

One can thus write a priori bilinear equations and generate new PDEs, then investigate whether they can be solved by the inverse scattering transform method. In some cases, one can even represent the eigenvalue problem itself in terms of the formalism. Indeed, the system

$$\left. \begin{array}{l} \psi_{xx} + (\lambda + u)\psi = 0 \\ \psi_t = u_x \psi + (4\lambda - 2u)\psi_x \end{array} \right\},$$

which is the eigenvalue problem for the KdV equation, can be written as

$$\left. \begin{array}{l} D_x^2 g \cdot f + \lambda g f = 0 \\ (D_t + D_x^3) g \cdot f = 3\lambda \, D_x g \cdot f \end{array} \right\},$$

with $g = \psi f$.

However, it is not possible to predict, by the mere consideration of the structure of the bilinear equation, whether there is an eigenvalue problem related to it or not, much less provide a systematic way to find it. For instance, while the KdV equation is related to a second-order eigenvalue problem, the equation

$$(D_x D_t + D_x^6) f \cdot f = 0 \quad \text{with} \quad u = 2 \frac{\partial}{\partial x^2} \ln f$$

leads to the Sawada-Kotera equation[14]

$$u_t + 45 \, u^2 u_x + 15(u_x u_{xx} + u u_{xxx}) + u_{xxxxx} = 0,$$

which can be solved through IST, but with a third order problem, while the equation

$$(D_x D_t + D_x^8) f \cdot f = 0$$

cannot be solved by IST at all. However, a whole hierarchy of equations of arbitrary high odd order is solved by the same second order problem as KdV, and they can be written in bilinear form, but not quite so simply.

There might be, however, some roundabout way, using the bilinear formalism, to show whether an eigenvalue problem that will solve an equation given in bilinear form exists or not, although there does not seem to be any direct way. The idea behind this is that, if an equation can be solved through IST, there must exist N-solitons solutions. Conversely, the existence of N-solitons solutions is such a strong property that one would like to believe that any equation that has such solutions for arbitrary N is solvable through IST.

The crucial point is that, when an equation is written in bilinear form, there exists a systematic way to look for solutions that resemble N-solitons solutions, at least for each given N. In a finite number of steps, one can either find such a solution, or show unambiguously that such a solution does not exist.[13] There are cases where one can prove by induction that solutions exist for all N, but it is not clear that this can always be done. At this point, we have found cases where such solutions are known only for some small values of N, the problem of finding a solution for larger values of N being too long to be solved in practice, although it is essentially trivial, as one has only to check that some formidable-looking quantity vanishes.

Actually, even if one can prove that these solutions exist for all N, one has not yet proven that they really describe solitons. Indeed, the proper definition of a soliton is a solitary wave that emerges and becomes isolated asymptotically when time goes to infinity by both positive and negative values, whatever the situation at any finite time is. In other words, a real soliton must be able to "cross" anything, whether this anything is made out of other solitons or not. If the equation can really be solved by IST, the contribution of the continuous spectrum is not made out of solitons, but solitons do "cross" it.

Thus, even the existence of solutions where N solitary waves cross each other, for arbitrary N, is not a proof that these waves are actually solitons. If such solutions do exist for all N, however, one is tempted to believe that they really describe solitons, and that the equation can be solved by IST.

This statement must be qualified to exclude the following situation. Given one IST-type equation, one can construct equations that would automatically be satisfied by any solution of the first equation, but in such a way that it need not be solvable by IST itself. Such an equation would have, among its solutions, all the N-solitons solutions of the first equation. However, one could presumably tell that something is amiss, because it could have other solitary wave solutions that do *not* exhibit soliton-like behavior (see Appendix A). One should thus require the existence of solutions where N solitary waves cross each other, not only for arbitrarily large N, but also for any combination of the different kinds of solitary waves that satisfy the equation, before one concludes that the equation can be solved by IST.

The method used to look for solutions with N solitary waves has been described in detail elsewhere[13] and we will not reproduce it here. The existence of such solutions for small N (2 or 3) usually gives very strong restrictions on the form of the equation and the exact values of the coefficients of each monomial. It is often true that, once these restrictions are met, one can prove by induction that solutions exist for all N. But there are some cases where it is not yet known whether solutions exist for N larger than some small integer (see Comparison of the Results of the Two Approaches). Even then, the similarity between these restrictions and those imposed by the Painlevé conjecture (see below) is striking.

THE PAINLEVÉ CONJECTURE

It has been known for some time that, when ansätze are used to reduce some of the classical IST-type equations, the modified Korteweg–de Vries equation, for instance, the resulting ODEs have the following remarkable property: Their solutions have no movable critical singularities, i.e., the location of all singularities except poles are independent of the integration constants.[11] We call this property the Painlevé property (or P property) because of Painlevé's classic work on this subject.[15,16] We will call these equations P-type equations.

A formal conjecture has been recently proposed: For any PDE of IST class, every ODE obtained by exact reduction of this PDE through some ansatz has the P property.[22]

This statement should be qualified by allowing for transformations in some cases. For the Sine-Gordon equation,[4,5] for instance, the solution of the ODE that is the reduced form of the Sine-Gordon equation does have movable logarithmic branch points, but its exponential has *pure poles,* which is not generally true for the exponential of a function with logarithmic singularities.

If proved, this conjecture would provide a test for determining whether a PDE can be solved by IST or not. The test would consist of taking all the ODEs into which the PDE can be reduced and checking that they have the P property. But this test can only give a necessary condition. We certainly do not claim that all PDEs that reduce to P-type ODEs can be solved by IST. There are known counterexamples of equations that are, in a sense, simpler than IST, since they can be shown to be equivalent to linear equations through some local transformation (see Appendix B). The whole IST method essentially consists of relating the original equation to some linear equation through some complicated nonlocal transformation.

If one tries to formulate a converse to this conjecture, one would claim that if a nonlinear PDE is such that all its reductions to ODEs have the Painlevé property, then it must be somehow related to a linear problem. But one must be ready to admit, at least, local transformations, IST-type nonlocal transformations, combinations of both, and maybe also entirely different kinds of nonlocal transformations. Recent work on monodromy-preserving deformations of linear ordinary differential equations indicates just such a possibility.[17] Thus, from the start, any test based on this conjecture can lead only to a necessary condition.

There is a second problem, namely, that there is, to our knowledge, no systematic way to obtain all the possible ansätze that reduce a PDE to an ODE, although we know of a couple of heuristic techniques to look for them. Since one can only find a necessary condition anyway, for the two reasons stated above, there is no point in looking for a proof that the ODEs are, indeed, P-type, which is very difficult; it is enough to find a necessary condition, namely, that no movable critical singularity is found, when properly looked for. The more efficient the search that fails to exhibit the movable singularity, the more confidence one has that this failure is significant.

An algorithm has been developed to study singularities with an algebraic leading order, to check whether these singularities are pure poles rather than algebraic or logarithmic branch points.[12] This algorithm has been described at length elsewhere,[12] and we will not describe it here. It will not, by its essence, detect equations that have solutions with essential movable singularities. Such equations certainly exist; an example is

$$w'' = w'^2 \frac{2w - 1}{w^2 + 1},$$

with general solution $w = \tan \ln (Az + B)$.

However, for many PDEs that have structures that look like classical IST equations, this algorithm has proven restrictive enough to discriminate correctly between those that are, indeed, IST-type and those that are known not to be IST-type, like the complex modified Korteweg–de Vries equation,

$$u_t + u_{xxx} + 2 \, (|u^2|u)_x = 0,$$

for instance. In fact, the failure to find branch points when using this algorithm is a very remarkable property of the equation, a property that is very easily lost if one modifies the coefficient of some monomial, in a way very similar to the loss of the N-solitons solutions, as explained above (not only similar but also strikingly parallel, as shown in the next section).

We propose the following criterion. Given a nonlinear PDE, try heuristically to reduce it to ODEs. For each ODE obtained, apply the algorithm to find movable critical singularities. If any is found, the conjecture says that the equation is not IST-type. If none is found, one would suspect that the equation is, indeed, IST-type, but only if there is an a priori reason to suspect that it is, for instance, if it closely resembles a classical IST-type equation. But this is precisely the case of the equations obtained through Hirota's bilinear formalism, as explained in the previous section. For the algorithm's failure to find movable critical singularities for these equations strongly suggests a connection with IST. The remarkable fact is that this property is always correlated with the existence of N-quasi-solitons solutions.

COMPARISON OF THE RESULTS OF THE TWO APPROACHES

This section lists a number of equations for which the existence of N-solitons solutions, on the one hand, and the Painlevé property, on the other hand, have been investigated. The equations are represented by Hirota's formalism since it is much more illuminating than the explicit form.

If the equation is equivalent to some classical IST-type equation,[1-5] this is mentioned and the exact connection is given. In these cases, N-solitons solutions obviously exist. One can check that they are, indeed, obtained through Hirota's formalism, as expected. All these equations satisfy the Painlevé test (given above); there is no known counterexample to date. Thus, whenever an equation is given a specific name below, it will mean that it is a classical IST-type equation, that there are N-solitons solutions, and that the Painlevé test is satisfied.

For the other equations, the expression "no movable critical singularities" means that all movable singularities with algebraic leading terms have been found to be pure poles using the algorithm[12] referred to above. No attempt has been made to find movable essential singularities.

$$D_x(D_t + D_x^{2n+1})f \cdot f = 0 \qquad (1)$$

$n = 1$ Korteweg–de Vries[1-5]

$$u_t + u_{xxx} + 6uu_x = 0 \qquad \text{for} \quad u = 2\frac{\partial^2}{\partial x^2}\ln f$$

$n = 2$ Sawada–Kotera[14]
$$u_t + u_{xxxxx} + 15(u_x u_{xx} + uu_{xxx}) + 45u^2 u_x = 0 \qquad \text{(same } u\text{)}$$
$n \geq 3$

- no N-solitons solutions for $N \geq 3$[18]
- movable critical singularities

Both criteria suggest it is *not* IST-type.

$$D_t(D_t + D_x^{2n+1})f \cdot f = 0 \qquad (2)$$

$n = 1$

$$u_{tt} + u_{txxx} + 3u_{tx}u_{xx} = 0 \qquad \text{for} \quad u = 2\ln f$$

- N-solitons solutions for all N
- no movable critical singularities for ansätze

$$u(x, t) = U[x - vt]$$
$$u(x, t) = U[x/(3t)^{1/3}]$$

Both criteria suggest a connection with IST.

$n \geq 2$

- no N-solitons solutions for $N \geq 3$[18]
- movable critical singularities

Both criteria suggest it is *not* IST-type.

$$(D_x^6 + D_x^3 D_t + \lambda D_t^2)f \cdot f = 0 \qquad (3)$$

$$u_{xxxxxx} + 15u_{xx}u_{xxxx} + 15u_{xx}^3 + u_{txxx} + 3u_{xt}u_{xx} + \lambda u_{tt} = 0 \qquad u = 2\ln f$$

- no 3-solitons solutions unless $\lambda = -1/5$ (not known whether there are N-solitons solutions for $N \geq 4$ if $\lambda = -1/5$)
- Painlevé property is satisfied
  —for $u(x, t) = U(x - vt)$, λ arbitrary
  —for $u(x, t) = U(x/(3t)^{1/3})$, if and only if $\lambda = -1/5$

Both criteria suggest a connection with IST if and only if $\lambda = -1/5$.

Ramani: Inverse Scattering Transform & Painlevé Conjecture   61

$$(D_x(D_x^{2n+1} + D_t) + D_y^2)f \cdot f = 0 \tag{4}$$

$n = 1$ Kadomtsev–Petviashvili[4]

$$\frac{\partial}{\partial x}(u_t + u_{xxx} + 6uu_x) + u_{yy} = 0 \quad \text{for} \quad u = 2\frac{\partial^2}{\partial x^2}\ln f$$

$n \geq 2$
- no N-solitons solutions for $N \geq 3$[18]
- movable critical singularities

$$(D_x(D_x^3 + D_t) + D_y^2 + D_z^2)f \cdot f = 0 \tag{5}$$

- no N-solitons solutions for $N \geq 3$[18]
- movable critical singularities

In these last two cases, both criteria suggest no connection with IST.

$$\left.\begin{array}{l} D_x^2 f \cdot g = 0 \\ (D_t + D_x^{2n+1})f \cdot g = 0 \end{array}\right\} \tag{6}$$

The first equation implies

$$\frac{\partial^2}{\partial x^2}u + v^2 = 0,$$

with

$$u = \ln fg, \quad v = \frac{\partial}{\partial x}(\ln f/g).$$

Expanding the second equation, one can substitute $-v^2$ for $(\partial^2/\partial x^2) u$. Taking one partial derivative with respect to x, one gets an equation for v only.

$n = 1$ modified Korteweg–de Vries equation (MKdV)[4-5]

$$v_t + v_{xxx} - 6v^2 v_x = 0$$

$n = 2$ and 3 second and third equations in the MKdV hierarchy[4]

$$v_t + v_{xxxxx} - 10u^2 u_{xxx} - 40uu_x u_{xx} - 10u_x^3 + 30u^4 u_x = 0$$

and

$$v_t + v_{xxxxxxx} - 14u^2 u_{xxxxx} - 84uu_x u_{xxxx} - 140u\, u_{xx} u_{xxx} - 126\, u_x^2\, u_{xxx}$$
$$- 182 u_x\, u_{xx}^2 + 420 u^2\, u_x^3 + 280\, u^3\, u_x u_{xx} - 140\, u^6\, u_x = 0$$

Note that, in the case of $D_x(D_t + D_x^{2n+1})$ for $n = 2$, the Sawada-Kotera equation is *not* the second equation of the Korteweg–de Vries hierarchy.[14]

$n \geq 4$ the equations are *not* higher equations of the MKdV hierarchy
- no N-solitons solutions for $N \geq 3$[18]
- movable critical singularities

Again, both criteria suggest that these equations are not IST-type for $n \geq 4$.

$$(D_x^3 + D_t)f \cdot g = 0$$
$$(D_x^4 + \lambda D_x D_t)f \cdot g = 0 \qquad (7)$$

The condition for the absence of movable critical singularities is $\lambda = 1$. It is precisely the same as the condition for the existence of 3-solitons solutions (it is not known whether N-solitons solutions exist for $N \geq 4$). A connection with IST is suggested by both criteria if and only if $\lambda = 1$.

$$(D_x^6 + D_x^3 D_t + \lambda D_t^2)f \cdot g = 0$$
$$(D_x^3 + \mu D_t)f \cdot g = 0 \qquad (8)$$

Here, there are movable critical singularities unless $\lambda = -1/5$ and $\mu = +1/5$. These conditions are just what is needed for the existence of 3-solitons solutions (again, one does not know whether N-solitons solutions exist for $N \geq 4$). Both criteria, again, agree in suggesting a connection with IST if and only if $\lambda = -1/5$ and $\mu = +1/5$.

$$(D_x^{2n+1} + D_t)f \cdot g = 0$$
$$(D_x D_t)f \cdot g = 0 \qquad (9)$$

Here, unlike case 6, one cannot get a single polynomial equation (although one can get an ugly-looking rational equation). It is convenient to write a system of equations for $u = \ln fg$ and $v = \ln (f/g)$.

$n = 1$

$$u_{xt} + v_x v_t = 0$$
$$v_t + v_{xxx} + v_x^3 + 3v_x u_{xx} = 0$$

- This system does not have solutions with movable critical singularities for either of the two ansätze,

$$u(x, t) = U(x - ct), \ v(x, t) = V(x - ct)$$

and

$$u(x, t) = U(x/(3t)^{1/3}), \ v(x, t) = \frac{1}{(3t)^{1/3}} V(x/(3t)^{1/3})$$

- N-solitons solutions can be found for all N.[18]

Thus, both criteria suggest that this system can be solved by IST.

$n \geq 2$

- movable critical singularities
- no N-solitons solutions for $N \geq 3$[18]

In all these cases, there is total agreement between the results of both approaches. In particular, for the following equations and systems,

$$D_t(D_t + D_x^3)f \cdot f = 0 \qquad (10)$$
$$(D_x^6 + D_x^3 D_t - 1/5 \ D_t^2)f \cdot f = 0 \qquad (11)$$

$$(D_x^3 + D_t) f \cdot g = 0$$
$$(D_x^4 + D_x D_t) f \cdot g = 0 \quad (12)$$

$$(D_x^6 + D_x^3 D_t - 1/5\, D_t^2) f \cdot g = 0$$
$$(D_x^3 + 1/5\, D_t) f \cdot g = 0 \quad (13)$$

$$(D_x^3 + D_t) f \cdot g = 0$$
$$(D_x D_t) f \cdot g = 0 \quad (14)$$

both tests suggest a connection with inverse scattering, but no eigenvalue problem is known at this point. Unfortunately, we do not know how to find them, but we believe that they do, indeed, exist.

Conclusion

When the Painlevé conjecture was formulated, I saw it as a way to tell rapidly that an equation could not be solved by IST and that there was no point wasting one's time and energy trying to find an eigenvalue problem related to it.

Passing the Painlevé test, however, though it is a necessary condition (if the conjecture is indeed true), is so far from being a sufficient one that one would not be ready to bet that an equation passing it was really IST-type, unless it closely resembled some classical IST-type equation. But the concept of close resemblance is very ill defined. Does Burger's equation

$$u_t + u_{xx} + 6uu_x = 0$$

resemble the Korteweg–de Vries equation

$$u_t + u_{xxx} + 6uu_x = 0$$

closely enough? Superficially, one would be tempted to answer in the affirmative. It does pass the test, but it cannot be solved by IST! In fact, it is "too simple" (see Appendix B).

By putting the Painlevé test and Hirota's bilinear formalism together, however, one can feel more secure. Equations obtained through Hirota's formalism can safely be said to closely resemble IST-type equations. If they have N-solitons solutions, the resemblance is even closer. Since this property, highly nongeneric, is linked to passing the Painlevé test in all cases known so far, one feels confident enough to claim that such equations are really IST-type. In particular, Equations 10–14 look very promising, and the author would go as far as to predict that eigenvalue problems for at least some of them could soon be found if some attention were given to them.

Acknowledgement

The author wants to thank Professor Ryogo Hirota for illuminating discussions followed by a continued fruitful correspondence without which this work would have been utterly impossible.

References

1. GARDNER, C. S., J. M. GREENE, M. D. KRUSKAL & R. M. MIURA. 1967. Phys. Rev. Lett. **19**: 1095.
2. LAX, P. D. 1968. Commun. Pure Appl. Math. **21**: 467.
3. ZAKHAROV, V. E. & A. B. SHABAT. 1972. Sov. Phys. JETP **34**: 62.
4. SCOTT, A. C., F. Y. F. CHU & D. W. MCLAUGHLIN, 1973. Proc. IEEE **61**: 1443, and references therein.
5. ABLOWITZ, M. J., D. J. KAUP, A. C. NEWELL & H. SEGUR. 1974. Stud. Appl. Math. **53**: 249.
6. CORNILLE, H. 1976. J. Math. Phys. (NY) **17**: 2143; 1977. J. Math. Phys. (NY) **18**: 1855; 1978. J. Math. Phys. (NY) **19**: 1463.
7. HIROTA, R. 1976. *In* Bäcklund Transformations. Lecture Notes in Mathematics, Vol. 515. R. M. Miura, Ed. Springer-Verlag. Berlin.
8. HIROTA, R. & J. SATSUMA. 1976. Prog. Theor. Phys. Suppl. **59**: 64.
9. HIROTA, R. 1980. *In* Solitons. Topics in Current Physics, Vol. 17. R. Bullough and P. Caudrey, Eds. Springer-Verlag. Berlin.
10. HIROTA, R. 1979. Tech. Rep. A-1. Department of Applied Mathematics, Hiroshima University, Japan.
11. ABLOWITZ, M. J. & H. SEGUR. 1977. Phys. Rev. Lett. **38**: 1103.
12. ABLOWITZ, M. J., H. SEGUR & A. RAMANI. 1978. Lett. Nuovo Cimento **23**: 333; 1980. J. Math. Phys. (NY) **21**: 715 and 1006.
13. ITO, M. 1980. J. Phys. Soc. Jpn. **49**: 771.
14. SAWADA, K. & T. KOTERA, 1974. Prog. Theor. Phys. **51**: 1355.
15. PAINLEVÉ, P. 1900. Bull. Soc. Math. France **28**: 227 and many others.
16. INCE, E. L. 1956. Ordinary Differential Equations. Dover Publications. New York, and references therein.
17. FLASCHKA, H. & A. C. NEWELL. 1980. Commun. Math. Phys. **76**: 67.
18. HIROTA, R. 1980. Private communication.

Appendix A

There are some equations that do have solutions where an arbitrarily large number of "bumps" separate asymptotically when time goes to infinity for both positive and negative values, and for which one can still say that such "bumps" are not solitons. Since the proper definition of a soliton is a nonlinear wave that re-emerges after crossing anything, it is enough to show that they are solutions where such a bump, well separated as $t \to -\infty$, interacts at some finite time and does not re-emerge. It is outside the scope of the bilinear formalism to investigate all possible solutions, but, certainly, if one does find explicit solutions where "bumps" do not re-emerge, even if there are some other solutions where they do re-emerge, one can say that they should not be called solitons and that their existence does not suggest a connection with IST.

This would happen if an equation is automatically satisfied by any solution of some IST-type equation, but is not solvable by IST itself. An example of such an equation is

$$2u_{tt} + u_{txxx} + 3(u_{tx} u_x + u_t u_{xx}) - u_{6x} - 15(u_{xx}u_{xxx} + u_x u_{4x}) - 45u_x^2 u_{xx} = 0, \quad \text{(A-1)}$$

which comes from the bilinear equation[18]

$$(2D_t^2 + D_t D_x^3 - D_x^6) f \cdot f = 0,$$

with $u = 2 \partial/(\partial x) \ln f$. This equation is automatically satisfied by any solution of the equation

$$u_t + u_{xxx} + 3u_x^2 = 0,$$

which can be solved by inverse scattering because the derivative u_x of u satisfies the Korteweg–de Vries (KdV) equation. It is actually this derivative u_x that exhibits solitons, not u, but one cannot write a differential equation for u_x because of the $u_t u_{xx}$ term.

Among the solutions of (**A-1**) are the "traveling steps"

$$u = \frac{2p}{1 + e^{\Omega t + px}} \quad \text{with } \Omega = p^3, p \text{ arbitrary constant,}$$

the derivatives of which are solitary waves of KdV, and solutions the derivative of which are the N-solitons solution of KdV for all N. These solutions describe an arbitrary number of "traveling steps" that re-emerge asymptotically after crossing each other.

There are, however, other solutions to (**A-1**). In particular, there is a second kind of traveling step

$$u = \frac{2p}{1 + e^{\Omega t + pt}} \quad \text{with } \Omega = - p^3/2.$$

But there are no solutions where more than two of the traveling steps of the second kind re-emerge after crossing each other or traveling steps of the first kind. This means not only that the former, but also the latter do not exhibit soliton behavior (even though their derivatives would be solitons of KdV) for (**A-1**) because there are solutions of that equation (which are not such that their derivatives satisfy KdV) that traveling steps of the first kind cannot cross.

Before one is satisfied that an equation passes the test for N-solitons solutions, one should thus demand that the existence of such solutions be proved for every combination of solitary waves (or traveling steps).

Appendix B

The conjecture says that, if a partial differential equation can be solved by inverse scattering, all its reductions to ordinary differential equations are Painlevé-type, but it does not say that the reciprocal is true. Indeed, it is easy to construct nonlinear PDEs such that all their reductions to ODEs are Painlevé-type, but not related to IST. One way to do so is to transform a linear PDE through some local change of the dependent variable that can, at most, introduce movable poles.

As a very simple example, let us take the equation

$$i u_t = u_{xx} \tag{B-1}$$

and transform it through

$$u = 1/v \tag{B-2}$$

into the nonlinear equation

$$i\, v_t = v_{xx} - \frac{2 v_x^2}{v} \tag{B-3}$$

For any ansatz that reduces (**B-3**) into an ODE, there is a corresponding ansatz that reduces (**B-1**) into a linear ODE. Since the solutions of the latter can have no movable singularities at all, the only movable singularities of the nonlinear ODEs are poles that arise from the zeroes of the solution of the linear ODE. For instance, the ansätze

$$u = U(x/(2t)^{1/2}), \qquad v = V(x/(2t)^{1/2}), \tag{B-4}$$

with $U = 1/V$, reduce (**B1**) and (**B3**) into

$$i\,\xi\, U' = U'' \tag{B-5}$$

$$i\,\xi\, V' = V'' - \frac{2\, V'^2}{V}, \tag{B-6}$$

$$\xi = \frac{x}{(2t)^{1/2}}, \qquad ' = \frac{d}{d\xi}.$$

Equation B-6 is, of course, Painlevé-type.

Granted, this is quite trivial and, given (**B-3**) or (**B-6**), one would, after a few manipulations, rapidly be led to use (**B-2**) in order to simplify the calculations. Indeed, it would be the first step in the reduction of (**B-6**) into its canonical form by the method described by Ince,[16] the canonical form being, of course, (**B-5**).

However, one could imagine some more subtle changes of the dependent variable that would not be so easy to "disentangle." Besides, if one insists on working on (**B-3**) and forbids oneself to use u as the variable, one can still find many beautiful results, including the existence of an infinite number of conserved quantities, namely

$$\int_{-\infty}^{+\infty} (1/v)^{(p)}\, 1/v^*\, dx \tag{B-7}$$

where $(1/v)^{(p)}$ stands for the pth derivative of $1/v$. This infinite number of conserved quantities, which is also present in the cases of IST, is not related to IST here, but rather to the existence of an underlying linear equation that has somehow been "disguised." Here the disguise is quite trivial and, in that sense, one could say that (**B-3**) is "too simple" to be solvable by IST, as it can be solved by a local change of dependent variable. A similar case is that of Burger's equation,

$$u_t + u_{xx} + 2u\, u_x = 0,$$

which is related to the linear heat equation

$$v_t + v_{xx} = 0$$

by the local transformation $u = v_x/v$. It, also, is "too simple" to be IST-type, although any reduction of it to an ODE has, of course, no movable singularities.

The inverse scattering transform can be considered as a more elaborate, nonlocal "disguise," the linear evolution of the scattering data being hidden under the nonlinear evolution of the "potential." This is quite apparent for equations that can be solved through a Golfand-Levitan-Marchenko-type (GLM) integral equation.[3,12] Here, the kernel F of the GLM equation, being, essentially, the Fourier transform of the scattering data, obeys a linear time evolution equation, while the solution K of the GLM equation, which is related to the potential, obeys a nonlinear evolution equation.

Just as the conserved quantities of form (**B-7**) are obvious consequences of (**B-1**), all the nice properties, like the infinite number of conservation laws, of IST-type equations, can be traced down and appear to be consequences of the linear equation satisfied by F, which underlies the nonlinear evolution of K.

Conversely, these nice properties suggest a connection with some linear problem, either by a local transformation, by IST, or by some more elaborate transformation, as suggested in Reference 17.

EXPLOSIVELY HEATED GAUSSIAN OBJECTS*

F. J. Mayer and D. J. Tanner

*KMS Fusion, Inc.
Ann Arbor, Michigan 48106*

INTRODUCTION

There are three energy delivery mechanisms currently being examined or proposed for inertial confinement fusion: lasers, electron beams, and ion beams. Preliminary experiments usually involve delivering large amounts of energy in a short time to simple targets: foils, wires, and spheres. This energy is rapidly distributed throughout the material and "explodes" the target. These explosions are usually modeled with complicated heat transport, hydrodynamic codes that are not easily accessible to those actually doing the experiments. As a hopefully usable tool, we have developed a similarity solution to the one-fluid hydrodynamic system and obtained some simple but quantitative estimates of the hydrodynamic behavior of these targets in this exploding mode. We assume that the target has a uniform temperature (isothermal approximation) and that the power absorbed from the beam is proportional to the local matter density. We expect the expansion to remain nearly one-dimensional for foils and wires until the thickness has expanded to the approximate size of the depositing beam spot.

The similarity method for solving the fluid equations is discussed extensively by Sedov.[1] Previously, similarity solutions to the hydrodynamic equations have been used in laser fusion research to calculate basic plasma properties (see, for example, the early work of Dawson,[2] Basov and Krokhin,[3] and Haught and Polk[4]). Our present approach differs in that we assume uniform energy deposition and a one-temperature isothermal fluid. Some features of this model have been worked out for spherical symmetry by various authors[5-9] and for planar symmetry by Nemchinov.[10]

HYDRODYNAMIC MODEL IN PLANAR SYMMETRY

We first develop the solution for an explosively heated foil and then generalize to cylindrical and spherical geometry. Also, we use an energy deposition term appropriate to an ion beam. Generalization to other beams is straightforward. The one-fluid isothermal hydrodynamic equations in planar geometry are given by

$$\frac{\partial n}{\partial t} + \frac{\partial}{\partial x}(nv) = 0 \tag{1}$$

$$m_i n \left(\frac{\partial v}{\partial t} + v \frac{\partial v}{\partial x}\right) = -\frac{\partial p}{\partial x} \tag{2}$$

---

*This research was supported by a contract from the US Department of Energy, no. ED-78-C-08-1598.

$$n\left(\frac{\partial\theta}{\partial t}+v\frac{\partial\theta}{\partial x}\right)+\frac{2}{3}p\frac{\partial v}{\partial x}=\frac{2}{3}P_a(x,t), \tag{3}$$

where n is the number density of plasma ions ($p = n\theta$), n_0 is the central ion density at $t = 0$, $P_a(x, t)$ is the deposited ion beam power per unit volume, θ is the weighted average of the electron and ion temperatures ($\theta = \overline{Z}\theta_e + \theta_i$), and $\overline{Z}$ is the average ionization.

We take the absorbed power per unit volume, $P_a(x,t)$, to be given by

$$P_a(x, t) = \frac{Rf(t)}{A_b n_0}\frac{dE}{dx}n(x, t) = P_0(t)n(x, t),$$

where $Rf(t)$ is the number of beam ions per unit time in the heating pulse, $f(t)$ is the temporal shape of the ion beam (normalized to one), and A_b is the beam area. The energy deposition rate, dE/dx, can be a function of time (we will assume it to be constant), but not space. Although (1–3) can be solved for arbitrary $f(t)$, we simplify the subsequent analysis by choosing a simple pulse shape,

$$\begin{aligned} f(t) &= \sin^2 \pi\tau & 0 \le \tau \le 1 \\ &= 0 & \tau > 1, \end{aligned} \tag{4}$$

where $\tau = t/t_p$ and t_p is the pulse length.

We choose the following notation and look for solutions to (1–3) of the form

$$\begin{aligned} \theta &= \theta(\tau) & p &= n\theta \\ n &= n_0 h\left(\frac{x}{L}\right)\psi(\tau) & v &= \frac{x}{t_p L}\dot{L} \\ & & L &= L_0 y(\tau), \end{aligned} \tag{5}$$

where θ is independent of x, the dot signifies differentiation with respect to τ, and L_0 is the scale length at $\tau = 0$. Substituting these into (1) and integrating with the boundary condition $y(0) = 1$, we find

$$\psi y = 1. \tag{6}$$

This result and the momentum equation, (2), result in the usual separation of variables:

$$(h\xi)^{-1}\frac{dh}{d\xi} = \alpha \text{ and } y\ddot{y} = -\frac{\alpha t_p^2}{m_i L_0^2}\theta, \tag{7}$$

where $\xi = x/L$ and α is a constant.

Without loss of generality, we can take $\alpha = -2$ and integrate the spatial part of (7) to give

$$h = \exp\left(-\left(\frac{x}{L}\right)^2\right). \tag{8}$$

We see that the conditions imposed on the form of the solution result in a Gaussian profile in space, where the central density, $n_0\psi$, and the characteristic thickness, L, vary with time. Kidder has noted previously that a Gaussian density profile is required for an isothermal self-similar solution in spherical geometry.[7] His solution did not include the heating phase of the explosion.

Substituting n and v from (5) into (3) and using (6) and (7) to eliminate ψ, θ, and L, we have the following ordinary differential equation:

$$y\ddot{y} + \frac{5}{3}\dot{y}\dot{y} = A \sin^2 \pi\tau, \qquad (9)$$

where $A = 2P_0 t_p / 3\theta_0$, $P_0 = R(dE/dx)/A_b n_0$, and $\theta_0 = m_i L_0^2 / 2t_p^2$ is a scale temperature; when $\tau > 1$, the right hand side is equal to zero.

The constant A is more meaningful when it is written in terms of the absorbed energy per ion. The total absorbed energy, E_a, is the integral of P_a over space and time:

$$E_a = t_p \int_0^1 d\tau \int P_a(x,\tau) \, d^3x = \frac{\sqrt{\pi}}{2} L_0 t_p R \frac{dE}{dx}. \qquad (10)$$

The number of ions absorbing energy from the beam is the integral of n:

$$N_i = \int n \, d^3x = \sqrt{\pi} \, n_0 A_b L_0 \qquad (11)$$

Using these relations, A is seen to be

$$A = \frac{4 E_a}{3 N_i \theta_0} = \frac{4}{3} \frac{\theta_a}{\theta_0}, \qquad (12)$$

where $\theta_a = E_a/N_i$ is the absorbed energy per ion. Although θ_0 and θ_a have the units of temperature, it is difficult to associate them with physical temperatures. It is possible to interpret θ_a as the temperature rise in the plasma if all the energy is absorbed only by the ions and no hydrodynamic expansion occurs. However, for ionized high Z materials, if very much energy is transferred to the electrons, the resulting temperature will be significantly lower than θ_a, even in the absence of expansion.

Equation 9 can be checked by looking at the kinetic and thermal energy in the flow. Integrating $\frac{1}{2} m_i n v^2$ over space gives the kinetic energy

$$E_k = \frac{1}{2} N_i \theta_0 (\dot{y})^2.$$

Likewise, the thermal energy is the integral of $\frac{3}{2} p$ over space:

$$E_{\text{th}} = \frac{3}{2} N_i \theta_0 \ddot{y} y$$

Integrating (9) from zero to a time greater than t_p ($\tau > 1$), we have

$$E_k + E_{\text{th}} = \frac{3}{4} N_i \theta_0 A = E_a,$$

showing that (9) is, simply, the statement of energy conservation.

GENERALIZATION TO CYLINDRICAL AND SPHERICAL SYMMETRY

It is simple to generalize the above analysis of exploding planar objects to include explosive heating of cylindrical and spherical objects. Equation 6 becomes

$$\psi y^d = 1, \tag{13}$$

where $d = 1$, 2, or 3 for planar, cylindrical, or spherical symmetry, and (9) becomes

$$y\ddot{y} + \left(\frac{3 + 2d}{3}\right)\dot{y}\ddot{y} = A \sin^2 \pi\tau. \tag{14}$$

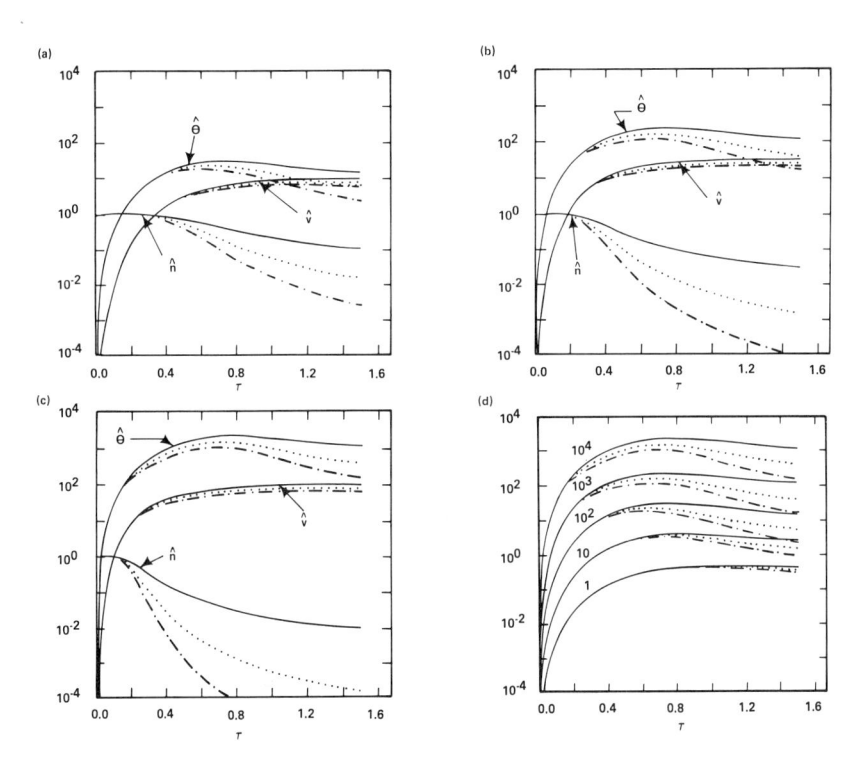

FIGURE 1. Normalized temperature ($\hat{\theta} = \theta/\theta_0$) and velocity ($\hat{v} = v/v_0$) at the scale length and normalized central density ($\hat{n} = n/n_0$) for planar (—), cylindrical ($\cdots$), and spherical ($-\cdot-\cdot-$) symmetry as a function of τ. (This line coding applies to both FIGURES 1 and 2.) (a) $A = 10^2$. (b) $A = 10^3$. (c) $A = 10^4$. (d) Normalized temperature as a function of τ for $A = 1$, 10^1, 10^2, 10^3, and 10^4.

Since $n \simeq \psi = y^{-d}$, the first relation shows the effect of spherical or cylindrical divergence on the density. The value of d has a much smaller effect on y in (14) than on ψ in (13). As seen below, the velocity, determined by y, is weakly dependent on d, while the density, determined by ψ, is strongly dependent on d. Several authors have obtained **(14)** for the spherical case;[5,6,8] Nemchinov has derived the planar case.[10] The expression for the constant A, **(12)**, remains unchanged, although the calculation of the integrals introduces different geometric factors in N_i and E_a. The kinetic energy is modified to

$$E_k = \frac{1}{2} d\, N_i\, \theta_0 (\dot{y})^2, \tag{15}$$

while the thermal energy is unchanged. As before, adding these together produces the first integral of **(14)**, demonstrating energy conservation.

NUMERICAL RESULTS

Now let us proceed to the results. Equation 14 was integrated numerically to produce the graphs of FIGURE 1. In FIGURE 1a–c, we present, as functions of time, the normalized temperature θ/θ_0, the normalized central density $n(0)/n_0$, and the normalized velocity v/v_0 at $x = L$ in each kind of symmetry for three values of $4\theta_a/3\theta_0$. The scale velocity v_0 is L_0/t_p. The effects of cylindrical and spherical divergence on the density are clearly seen. In fact, the densities for cylindrical and spherical symmetry are, approximately, the square and cube of the density for planar symmetry. As expected, the velocity is much less sensitive to geometry. The decrease in temperature in going from planar to spherical geometry occurs because a greater fraction of the energy is kinetic, as seen in **(15)**. The maximum in temperature occurs roughly midway between the maximum in the pulse at $\tau = 0.5$ and the pulse cutoff. For a given energy input, maximum thermal energy density occurs for planar targets. FIGURE 1d presents normalized temperature as a function of time with $4\theta_a/3\theta_0$ as a parameter. Above the value of 10^2, the ratio of the temperature for planar geometry to that for spherical geometry is independent of θ_a/θ_0.

FIGURE 2 shows the dependence of the maximum temperature and the central density at that temperature as a function of $3\theta_a/4\theta_0$. At large deposited energies, it is clear that exploding foils maintain higher densities than cylinders or spheres. The slope of the maximum normalized temperature line is nearly 1. From the definitions $\hat{\theta}_m = \theta_m/\theta_0$ and $A = 4\theta_a/3\theta_0$, we see that $\hat{\theta}_m \simeq \theta_a = E_a/N_i$. The maximum temperature achieved is then proportional to the energy deposited per ion and independent of the pulse length. This is true as long as θ_a/θ_0 is large. However, if higher densities are desired, θ_a/θ_0 should be kept small.

The isothermal explosion of any foil, wire, or sphere is represented by the solution of **(14)** if the density profile is Gaussian. The solution is determined by the ratio of two parameters: θ_a, the absorbed energy per ion, and θ_0, the kinetic energy of an ion that moves through the scale length, L_0, in the scale time, t_p. In TABLE 1, we present formulae for calculating θ_0 and θ_a from experimental parameters in practical units. We also present the relationship between the characteristic length, L_0, and the dimension of a physical object of uniform density. This was obtained by taking a Gaussian object with a central density n_0 equal that of the physical object, integrating the density n from **(5)** and **(8)** over space and requiring the resulting mass to be equal to the mass of the physical object.

As a practical example, consider a polyethylene foil 20 μm thick struck by a heavy ion beam with a diameter of 2 mm, which deposits 0.5 kJ in the foil. The scale

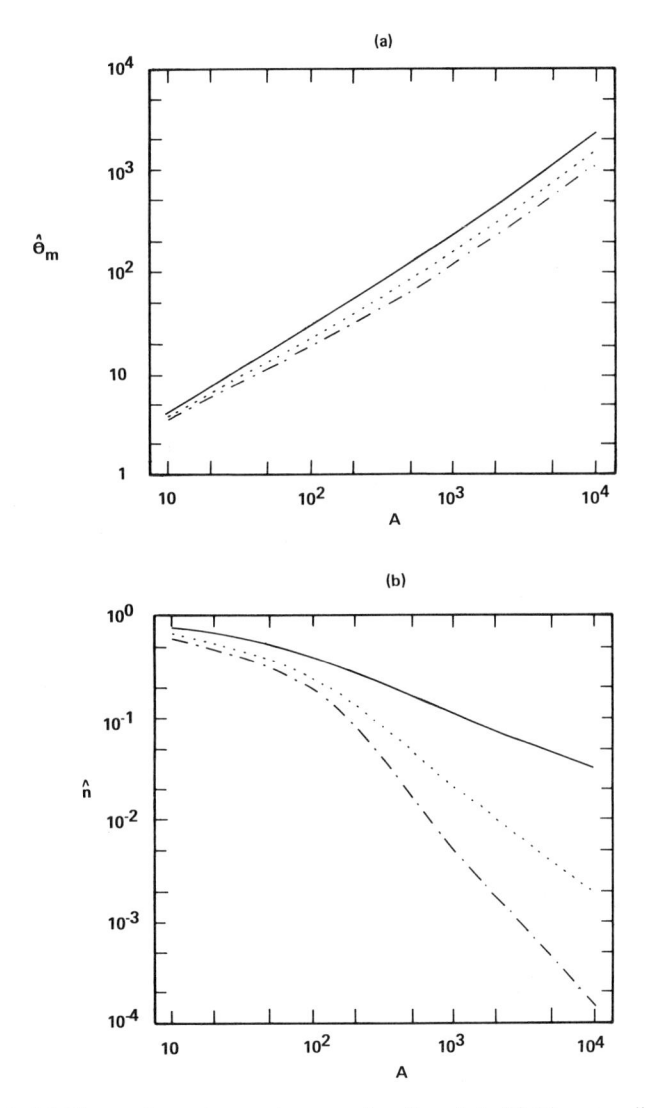

FIGURE 2. (a) The maximum temperature as a function of A. (b) The normalized central density at the time of maximum temperature.

temperature θ_0 is 2.5 eV, θ_a is 400 eV, and A is 200. Referring to FIGURE 2, we have $\theta_{max} = 50\theta_0$ or 125 eV and the central density at this temperature is 20% of the initial density.

TABLE 1
Characteristic Quantities for Real (Square) Objects

	Foil	Wire	Sphere
θ_a	$\dfrac{3.3\, E_a\, A_n}{\rho x_0 r_d^2}$	$\dfrac{3.3 \times 10^3\, E_a\, A_n}{\rho l x_0^2}$	$\dfrac{2.5 \times 10^6\, E_a\, A_n}{\rho x_0^3}$
L_0	$1.13\, x_0/2$	x_0	$0.91\, x_0$
θ_0	$\dfrac{A_n L_0^2}{192\, t_p^2}$		

NOTE: θ_a is in keV, θ_0 in eV, L_0 in μm, t_p in ns, E_a in kJ, and ρ in g/cm³. A_n is the atomic number; x_0 is the thickness of the foil, the radius of the wire, or the radius of the sphere in μm; and r_d is the radius of the spot size on the foil in mm or l is the length of the spot on a wire in mm. θ_0 is independent of geometry.

Comparison with a Hydrodynamic Code

In real experiments, targets do not initially have a Gaussian shape. An important question to ask is, When (if ever) does an exploding foil, wire, or sphere obtain a Gaussian shape and how is the parameter L_0 related to the initial half-thickness of the target? We have examined this question with our one-dimensional heat transport hydrodynamic code TRHYD. The code calculations were made for a lead foil that has a density of 19.3 g/cm³ and a thickness of 40 μm and absorbs 40 kJ in 1 ns from a beam 2 mm in diameter. Energy was deposited proportionally to mass and the electron and ion conductivities were artificially made large enough to keep the plasma isothermal to within about 5%. Using these parameters in the relations of TABLE 1, we find that $L_0 = 22.6$ μm, $\theta_0 = 0.55$ keV, and $\theta_a = 35$ keV; the resulting value of the parameter A is 84.

A value for y is required in order to compare the density profile of the simulation

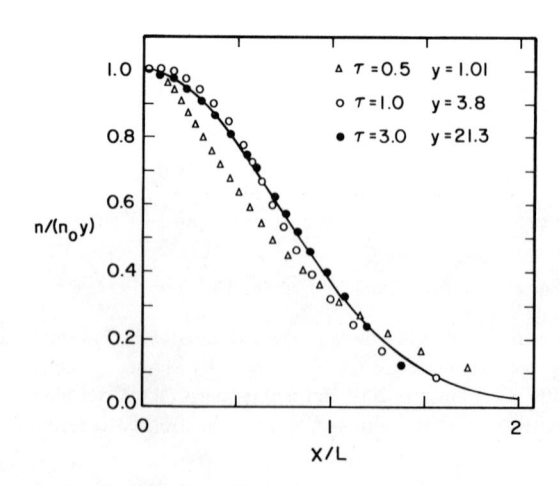

FIGURE 3. Density profile for the simulation of the explosion of a lead foil at three values of τ. Density is normalized by the central density and distance is normalized by $L_0 y$, as discussed in the text. The solid line is the Gaussian shape predicted by the model.

with our model. We obtain y at a given time by equating the central density to $n_0 y$, where n_0 is the initial foil density. In FIGURE 3, we plot the Gaussian shape (8) of our model and simulation profiles at three times. The density is normalized by the central density and the distance is normalized by $L_0 y$. We see that, by the end of the energy pulse, $\tau = 1$, the profile is close to Gaussian and the agreement improves with time (we have run the simulation to $\tau = 10$). Note that, for the three times shown, the central density varies from 19.3 g/cm^3 to 0.9 g/cm^3 and the outer boundary varies from 41 μm to 630 μm. Since the two parameters that characterize a Gaussian—value at zero and half-width—are determined by y, the agreement with the model is further strengthened.

FIGURE 4. Comparison of the Gaussian width y predicted by the model and that determined by the simulation.

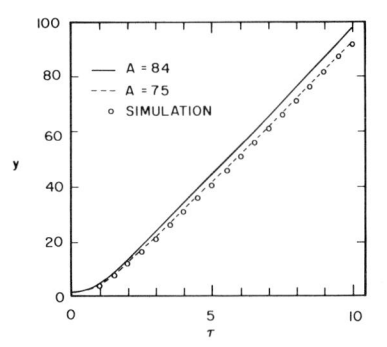

In FIGURE 4, we plot y as a function of τ for the model and for the simulation. A better fit to the model is obtained using a value of 75 for A. We feel the agreement between the simulation and our model shows that the model can be applied to objects that are not initially Gaussian shaped. We further suggest that an exploding object of any initial shape will asymptotically approach a Gaussian shape.

CONCLUSION

We have developed a simple similarity solution that may be very useful for approximating some hydrodynamic quantities in experiments on explosively heated objects, particularly in ion beam experiments. All uniformly heated explosions are characterized by just two parameters, θ_a and θ_0. Given these parameters, it is possible to calculate how far one is from the ideal case in which all the deposited energy is deposited as thermal energy at temperature θ_a in the ions. The similarity solutions show that explosions of variously sized objects with appropriate deposited energies are hydrodynamically identical.

These solutions are most useful in determining the approximate hydrodynamic character of experiments employing rapid energy input to regularly shaped objects. We have shown that these solutions become accurate representations of square profile objects in the latter stages of expansion. These solutions, also, may be useful as routine test problems for one-dimensional hydrodynamic codes.

Summary

The hydrodynamic expansion of Gaussian-shaped objects subject to uniform rapid heating was exactly calculated using the similarity technique. Such explosions were found to be characterized by only two energy parameters. The dependence of hydrodynamic quantities on the heating rate and deposited energy was determined. Finally, we compared the similarity solutions to a numerical solution for a square object explosion and found good agreement is reached by a time t_p and continues thereafter. Apparently, the natural shape for an exploding object is a Gaussian density profile. It will asymptotically approach this profile regardless of its initial shape.

References

1. Sedov, L. I. 1959. Similarity and Dimensional Methods in Mechanics, Chapter 4. Academic Press. New York.
2. Dawson, J. M. 1964. Phys. Fluids **7**: 981.
3. Basov, N. G. & O. N. Krokhin. 1964. Sov. Phys. JETP **19**: 123.
4. Haught, A. F. & D. H. Polk. 1966. Phys. Fluids **9**: 2049.
5. Fader, W. J. 1968. Phys. Fluids **11**: 2200.
6. Haught, A. F. & D. H. Polk. 1970. Phys. Fluids **13**: 2825.
7. Kidder, R. E. 1971. Proc. Int. Sch. Phys. Eurico Fermi **49**: 306.
8. Dzung, L. S. 1972. Z. Angew. Math. Phys. **23**: 301.
9. Lengyel, L. & M. Salvat. 1975. Z. Naturforsch. Teil A **30**: 1577.
10. Nemchinov, I. V. 1961. Zh. Prikl. Mekh. Tekh. Fiz. **1**: 17.

SURFACE STUDIES OF FUSION REACTOR WALL MATERIALS AT AFI

T. Fried, B. Emmoth, and M. Braun

Research Institute of Physics (AFI)
104 05 Stockholm, Sweden

INTRODUCTION

In recent years, greater efforts have been made in fusion research. A great number of challenging problems face scientists all over the world. At the present time, one of the fields where a further understanding of basic mechanisms is required most urgently is surface physics and metallurgy. The inner walls of the vacuum vessel of the reactor, or, rather, the metallic parts directly facing the plasma in the reactor chamber, will be exposed to a violently eroding flux of particles. The mean kinetic energy of these particles when they are close to the walls is expected to be equal to, or somewhat less than, the thermal energy of the particles, i.e., around 10 keV. There is, at present, no agreement as to which materials are best suited to compose the different parts that will be directly exposed to the hot plasma.

There are a number of requirements that these parts have to fulfill. They should erode at a low rate, so as not to contaminate the plasma and thus cool it. The contaminations that will inevitably occur should be of low atomic mass, in order to reduce their undesirable cooling effect. The hydrogen diffusion properties should be such that it is possible to control the amounts of deuterium and tritium in the plasma, as well as in all parts surrounding it, at all times. The amount of tritium in the reactor walls has to be kept at a minimum. Additionally, materials that do not develop induced radioactivity by the fast neurons from the fusion process would be preferred. Lastly, obvious requirements for all materials are that they should be reasonably accessible, workable, and cheap.

The surface physics group at AFI is engaged in a series of projects in some areas of fusion materials research. These projects will be described separately, following a brief description of the experimental facilities.

EXPERIMENTAL EQUIPMENT

The surface physics laboratory is centered around a group of accelerators. Ion beam experiments are carried out at two accelerators: one small isotope separator, which is capable of deflecting all elements at 10–100 keV, and one small accelerator for low energies, which handles beams of elements up to krypton at energies between 1 and 20 keV. For analysis, we use a 2 MeV Van de Graaff accelerator, with which we employ various techniques, such as Rutherford back-scattering (RBS), induced x-ray analysis, and nuclear reactions analysis. The Van de Graaff beam path is intersected by the low energy accelerator beam path. Thus, both experiment and analysis can be done in the same chamber without breaking vacuum. This chamber is equipped with a target holder on a manipulator that allows motion in five degrees of freedom and a temperature control from LN_2 temperature to 1000 K. The vacuum system is compatible with UHV requirements down to the 10^{-10} torr range. Additionally, there are two separate experiment chambers along the Van de Graaff beam path; one is used for standard RBS measurements and channeling experiments and one, which is kept at UHV, is specially designed for hydrogen permeation experiments. All chambers are separated by differential pumping stages. The isotope separator is connected to a separate system and used for less complex ion beam irradiation experiments. Standard nuclear electronics are used throughout, including solid state detectors and a 4096-channel multi-channel analyzer, which, together with a terminal, is connected to a computer. Samples that need to be prepared before the experiment are polished in silicon carbide and diamond spray and washed and rinsed ultrasonically in a degreasing agent and in alcohol. In some cases, they are treated with electropolishing, as well.

Sputtering Experiments

Developing fusion reactor materials requires a detailed knowledge of the sputtering process, i.e., the process by which particles are eroded from a surface due to incident energetic ions or neutral atoms. This process is believed to become the dominant factor in erosion of the materials that are exposed to the plasma and, thus, will probably make the largest contribution to the contamination of the plasma. One project is devoted to the study of angular distributions of sputtered particles from fusion reactor materials candidates, such as stainless steels, Inconels, refractory metals, thin films of various materials on stainless steel, etc.

A sample is mounted on the target holder of the manipulator, which holds it in the center of the main target chamber where the two beams intersect. An aluminum or carbon strip bent in a semicircle is mounted around it and in front of it. This semicircle can be rotated 360° in the horizontal plane around its center, which coincides with the center of the target chamber, thus allowing both beams to reach both the sample and the semicircle everywhere. The semicircle, which has a radius of 35 mm and a width of 8 mm, collects the particles ejected from the target surface.

Following irradiation of the sample with a helium or hydrogen beam through a small hole in the collector, the collector is rotated so that it will face the analyzing beam of 1.8 MeV alpha particles from the Van de Graaff accelerator. A solid state surface barrier detector is mounted so that it will detect the scattered alpha particles at a scattering angle of 160°. RBS analysis[1] will then give information on the absolute amount of sputtered atoms that have been collected at each angle of the collector. The angular resolution is 1.5°. The distribution thus obtained can also be integrated to give the total sputtering yield.

The same procedure described above is used to investigate dry methods of etching (sputter-etching) relevant to the semiconductor industry.

The current mathematical model for sputtering from isotropic materials suggests a cosine distribution for the ejected particles,[2] an idea that originates in the Knudsen law.[3] We have found significant deviations from this distribution with various beams, energies, and targets.[4] One class of deviating distributions arises in crystalline targets

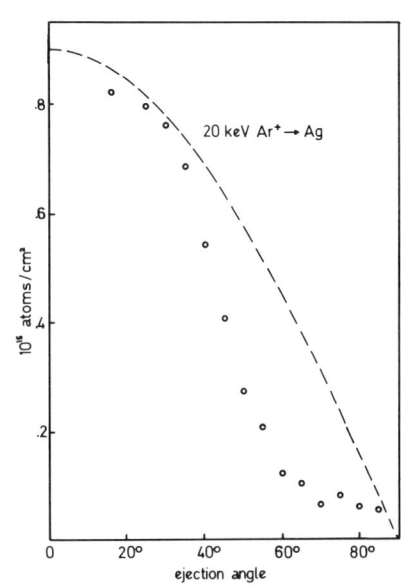

FIGURE 1. Angular distribution from silver sputtered by 20 keV argon. From Emmoth et al.,[4] by permission of the North Holland Publishing Company.

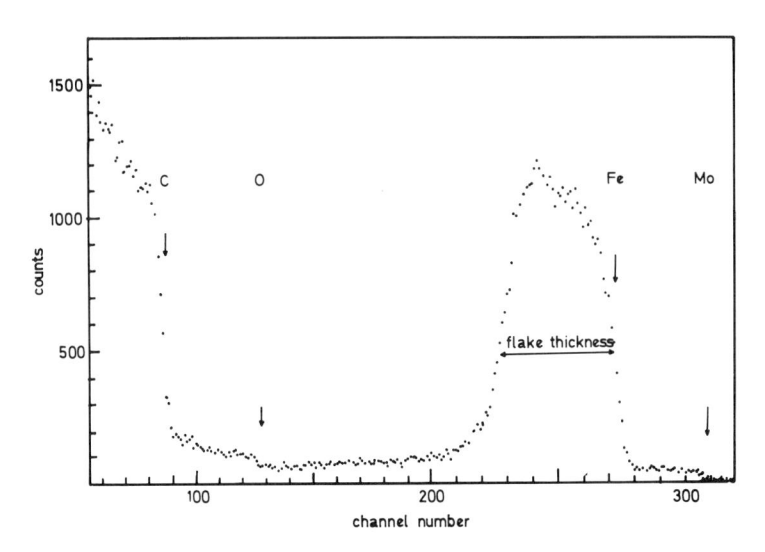

FIGURE 2. A typical back-scattering spectrum obtained from 304 stainless steel flakes isolated on carbon tape after 50 keV irradiation. The front edges of different elements in the flakes and carbon tape are indicated. The flake thickness (in atoms per cm^2) is obtained from the energy difference derived from the spectrum of the stainless steel film. From Braun et al.,[8] by permission of the North Holland Publishing Company.

and shows peaks superposed on a cosine distribution. One explanation of this phenomenon is that variations in the surface potential give rise to preferred directions.[5] Another explanation postulates "focusons," which are particles from the primary collision cascade that are focused due to head-on collision sequences in a close-packed crystallographic direction.[6] Other deviating distributions suggest that processes more complicated than those considered in the simple Knudsen theory are responsible. FIGURE 1 shows the angular distribution of silver atoms sputtered by 20 keV argon. It can be nicely fitted to a square cosine, which is a rare and unusually large deviation from the cosine. Similar results have been obtained for the distribution of ions measured by secondary ion mass spectrometry.[7] Additional measurements are necessary for a more complete understanding. The use of single crystals combined with quantitative measurements may give unambiguous information.

BLISTERING AND FLAKING STUDIES

Extended helium irradiation causes a build up of helium atoms in the materials to the extent that bubbles form. The bubbles can rupture, causing blisters to appear or causing exfoliation of large areas of the metal surface. Knowledge of these processes is decisively important in the planning of fusion reactor walls.

The highest concentration of helium in the metal after some irradiation will be found at a distance below the metal surface corresponding to the mean projected range for the implanted helium. As helium has a low sputtering rate, low mobility, and low solubility in metals, the helium will be trapped and can build up to high enough concentrations to create bubbles. Ultimately, the gas pressure and the induced stress exerted on the metal in the surroundings of the bubble can cause the surface to open. Direct measurements of the flake thickness in a scanning electron microscope (SEM) yield values larger than expected, i.e., larger than the mean projected range. At AFI, we have isolated the flakes on carbon tapes and measured the thicknesses with RBS (FIGURE 2). In this application, the method gives correct thicknesses in terms of areal density, i.e., numbers of atomic layers or atoms per cm^2. It is, however, insensitive to variations in the bulk density of the material. Comparing with SEM, we have found a considerable swelling (see FIGURE 3 and REFERENCE 8). If the irradiation is continued, more layers will peel off, but recent results indicate that this will only progress to a certain point. This fact seems to be connected to the observation that only the first layer swells appreciably.[9] After successive blistering, deep pits appear all over the surface. The helium that is implanted can then move sideways and leak out into the pit, so that, after some time, there will be no more blistering. From then on the surface will be eroded only by sputtering. If this interpretation is correct, it will have a profound impact on fusion reactor wall design.

HYDROGEN SURFACE RECOMBINATION EXPERIMENTS

A stainless steel membrane separates a chamber pressurized with deuterium from a vacuum chamber. The deuterium molecules in the pressure chamber will dissociate into atoms at the membrane surface, diffuse through the membrane, recombine into molecules at the other surface, and immediately leave it. A phenomenological recombination rate constant can be defined by relating the near surface deuterium concentration on the downstream side to the release rate of deuterium molecules from the surface.[10] The release rate is determined experimentally by measuring the deuterium partial pressure on the downstream side of the membrane. The surface concentration of deuterium is measured by the nuclear reaction D(^{3}He,p)^{4}He. A 990 keV ^{3}He beam from the Van de Graaff accelerator is incident on the membrane and

FIGURE 3. The measured skin thickness of flakes as a function of the helium implantation energy, determined by SEM studies (crosses) and RBS analyses (circles). The solid curve shows the calculated mean projected range. From Braun et al.,[8] by permission of the North Holland Publishing Company.

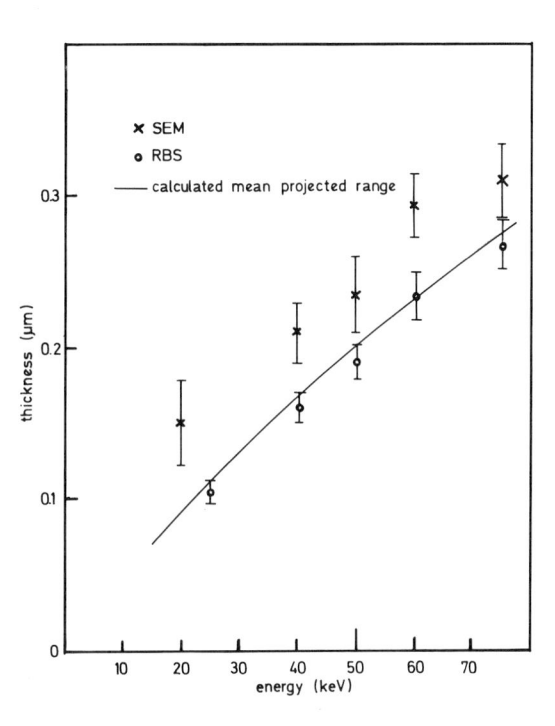

the protons (around 13 meV) from the nuclear reaction are detected by a special solid state surface barrier detector.[11] In addition, the diffusion constant can be determined by the measured permeation flux density, the upstream pressure, and the solubility of deuterium in the metal.[11] FIGURE 4 shows a plot of the recombination rate constant at different temperatures of the membrane. Work is in progress to investigate how the recombination rate constant varies with radiation damage and different surface conditions.

THIN FILM STUDIES

An obvious method of obtaining desired reactor wall properties with rare or exotic materials is to deposit them on a workable construction material, e.g., stainless steel. However, the deposited films must not only demonstrate sufficient adherence to the

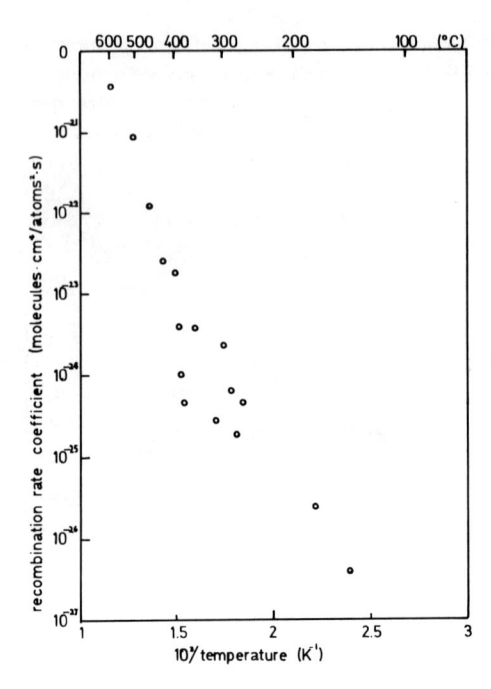

FIGURE 4. Arrhenius plot of the surface recombination rate coefficient. From Braun et al.,[11] by permission of the North Holland Publishing Company.

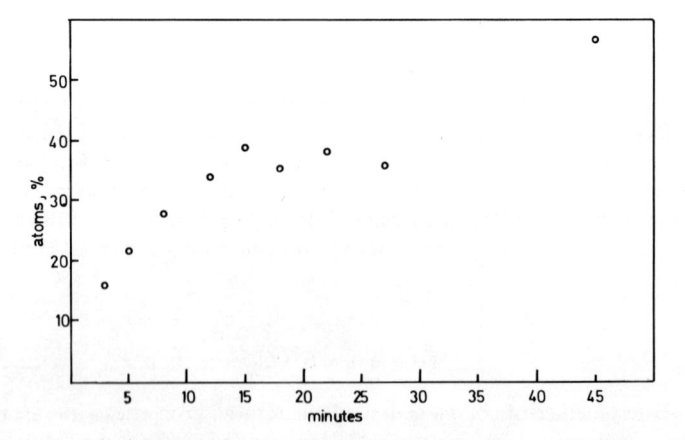

FIGURE 5. The relative stainless steel concentration at the surface of samples of stainless steel covered with 1000 Å palladium and annealed at 607°C. From Fried et al.,[12] by permission of the North Holland Publishing Company.

underlying material, but the compound structure must also be able to withstand elevated temperatures for prolonged periods of time without changing composition or other properties.

The thin film project serves to investigate interdiffusion of thin films with stainless steels and possible ways to control the interdiffusion. Polished pieces of stainless steel are mounted in the central target chamber. They can be covered with a thin film by using the low energy accelerator as a sputter-deposition system and treated with ion implantation, again, with the low energy accelerator. Following the desired treatment, the sample is simultaneously annealed and RBS-analyzed and the depth distributions of the constituent atoms are determined.

Previous investigations have shown that stainless steel covered with palladium exhibits interdiffusion at modest annealing times, so that the surface no longer consists exclusively of palladium after only 10 min at 530°C, with an initial palladium thickness of 1000 Å. FIGURE 5 shows the composition evolution of 1000 Å palladium on stainless steel as a function of time at 607°C. Even pure stainless steel will exhibit a change in the surface concentrations of the constituents of the alloy at elevated temperatures. A possible explanation of the chromium enrichment observed is that a residual partial oxygen pressure will oxidize the chromium, the oxygen thus creating a sink for chromium at the surface.[13] Further experiments in UHV will reveal more details about these processes.

REFERENCES

1. MAYER, J. W. & E. RIMINI, Eds. 1977. Ion Beam Handbook. Academic Press. New York.
2. SIGMUND, P. 1969. Phys. Rev. **184**: 383.
3. KNUDSEN, M. 1909. Ann. Phys. (N.Y.) **28**: 75.
4. EMMOTH, B., T. FRIED & M. BRAUN. 1978. J. Nucl. Mater. **76**: 129.
5. LEHMANN, C. & P. SIGMUND. 1966. Phys. Status Solidi **16**: 507.
6. SILSBEE, R. H. 1957. J. Appl. Phys. **28**: 1246.
7. SCHOOTBRUGGE, G. A., von der A. G. J. de WIT & J. M. FLUIT. 1976. Nucl. Instr. Meth. **132**: 321.
8. BRAUN, M., B. EMMOTH & J. L. WHITTON. 1980. J. Nucl. Mater. **94 & 95**: 728.
9. WHITTON, J. L., C. H. MING, U. LITTMARK & B. EMMOTH. 1981. Nucl. Instr. Meth. **182 & 183**: 291.
10. WAELBROEK, F., I. ALI-KHAN, K. J. DIETZ & P. WEINHOLD. 1979. J. Nucl. Mater. **85 & 86**: 345.
11. BRAUN, M., B. EMMOTH, F. G. WAELBROEK & P. WEINHOLD. 1980. J. Nucl. Mater. **94 & 95**: 861.
12. FRIED, T., B. EMMOTH & M. BRAUN. 1979. J. Nucl. Mater. **85 & 86**: 1155.
13. LEYGRAF, C., G. HULTQUIST & S. EKELUND. 1975. Surf. Sci. **46**: 157.

A QUANTUM MODEL OF DOUBT*

Yuri F. Orlov

Prison Camp No. 37-2
The Urals, USSR

INTRODUCTION—LEVELS OF CONSCIOUSNESS—INCOMPATIBLE LOGICS—DOUBT

Doubt in the correctness of a proposition is a component of the process of thinking. Since this process is, to a considerable extent, unconscious, it is natural to assume that the phenomenon of doubt already exists on a subconscious level. In a "subconscious" state, we are aware of the presence of doubt, but it is not yet formulated as a proposition. In such a state, we are able in some form to fix the presence of doubt. Then we can realize, with some effort, that this fixation has taken place. Thus, each ascending stage of self-consciousness requires ever greater effort. In accordance with this scheme, I formulate the following—in the present paper only qualitative—assumption: there are different *levels of consciousness*. Each level of this hierarchical system is "consciousness," in the narrow sense of the word, for the lower levels, and "subconsciousness" for the upper levels. The process of self-consciousness involves the stimulation of ever higher levels of consciousness. If, at a given level, some logical system is used, then the operations performed at the next level up will purport to justify the postulates of this logical system. The transition from a given level to the next higher level is what V. F. Turchin calls "a metasystem transition."[1]

When a conflict occurs at a given level of consciousness, an arbitration at a higher level becomes necessary. Specifically, we speak of a conflict if: (1) a decision must be taken, (2) there are different viewpoints about how to cope with the situation, (3) different points of view lead to different and incompatible commands, and (4) the incompatibility of the points of view cannot be resolved at this level, rendering decision making impossible.

There exist situations where the reason a decision cannot be taken is not that there are too many points of view, but, on the contrary, that there are too few. In such cases we will say that the logical system has insufficient resolving power. In the simplest case, the ultimate reason may be that insufficient information has been received from the sensor organs. Or it may be that the question is such that it is intrinsically impossible to select one "true" criterion (e.g., truth, beauty, or morality) from an irreducible set of criteria, that it is a matter of the personal convictions of the individual.

If the individual is still obliged to make some decision in the case of insufficient resolving power, the resulting state will be called, by definition, a *doubt state*. The actual decision in this situation will be taken by a spontaneous (non-logical) act of will—by a free choice.

It is also possible in this situation that there exist several mutually incompatible approaches, which make up incompatible logical subsystems of the full logical system

*Translated by Valentin F. Turchin, The City College of the City University of New York, New York, NY 10031.

of the individual. Each of these subsystems may be of insufficient resolving power with respect to a given decision problem. For instance, considerations of duty, conscience, and truth may compete because of irreducible concepts that, essentially, cannot be strictly enough defined. It is at the higher metalevels of consciousness that a choice is made of the "correct" approach. In particular, a logic may be chosen that excludes doubt, i.e., that has the requisite resolving power.

One possible physical model of decision making in doubt states is considered below.

ELEMENTARY DOUBT CELL—
NONCOMMUTABLE OPERATIONS OF INCOMPATIBLE LOGICS

The discussion that follows is based on the principles of wave logic.[2,3]

An objective description of the logical system of an individual is based on a set of propositions. We use the word "proposition" to denote a quantity that takes discrete values and is used in the system of logical inference. An elementary proposition can, by definition, take one of two values (YES and NO) and is perceived as "something whole." We assume that there exists an elementary doubt cell corresponding to every elementary proposition. This logical system is a quantum system that has energy E_0 in the ground state, and energy E on the first doubly degenerate level. It can have higher levels, but they are not significant in this model. Proposition σ_n (where n is the sequential number of the proposition and its cell, which we shall drop in the following) may appear in the logical system of an individual with different original postulates, i.e., σ_n may be used in different logical subsystems p. Consequently, logical tests YES and NO appearing in different places in a logical discourse may have different meanings and should be marked with a subscript p indicating the number of the logical subsystem. If a YES or a NO decision is taken with respect to a proposition σ, then the nature of the final commands will depend on which point of view p was used to make the decision. One of the basic assumptions of the present theory is that labels p exist (although they may be "wrong" from the viewpoint of an outside observer).

On every YES(p) test a signal reaches the cell; a special polarizer transforms it into a quantum mechanical perturbation $V^{(p)}$. The polarization p, the meaning of which will be discussed in detail later, is in a one-to-one correspondence with the viewpoint p in the framework of which the YES test for the proposition σ has been made. Analogously, the NO(p) test is transformed into the perturbation $V_1^{(p)}$. Perturbations V and V_1 have matrix elements of transitions $E_0 \rightleftharpoons E$. On the output of the cell, the transitions $E \rightarrow E_0$ are transformed by a special analyzer into YES(p)/NO(p) signals that are no longer tests but YES/NO decisions or evaluations about the proposition σ.

Let us choose a pair $\psi_1(q)$, $\psi_2(q)$ of orthogonal and normalized eigenfunctions of the doubly degenerate level E, and let us denote by ψ_0 the wavefunction of the ground state E_0. All of the functions include their time dependence. If the level E is excited, the general state of the cell is the superposition

$$\psi = a_0\psi_0 + R[a_1\psi_1(q) + a_2\psi_2(q)], \qquad (1)$$

$$|a_0|^2 + R^2 = 1, \qquad |a_1|^2 + |a_2|^2 = 1. \qquad (2)$$

Let us switch to the energy representation by introducing

$$\Phi = \begin{pmatrix} a_0 \\ R\phi \end{pmatrix} \quad \text{and} \quad \phi = \begin{pmatrix} a_1 \\ a_2 \end{pmatrix}. \tag{3}$$

As we shall see later, the spinor ϕ is identical to the logical state function of proposition σ that was introduced in Reference 2. Let us rewrite ϕ as a superposition of two orthogonal spinors:

$$\phi = a_1\phi_1(q) + a_2\phi_2(q),$$

$$\phi_1(q) = \begin{pmatrix} 1 \\ 0 \end{pmatrix}, \quad \phi_2(q) = \begin{pmatrix} p \\ 1 \end{pmatrix}. \tag{4}$$

We shall say that ϕ is written in the q-representation. By definition, $\phi_{1,2}$ have the following meaning: if a cell is in the state $\phi_1(q)$ and is "asked" from the point of view of the orientation $p = q$ then it will give the answer YES(q) with probability 1, and the answer NO(q) with probability 0. In the state $\phi_2(q)$ YES and NO exchange places.

In the state ϕ, the probability of the answer YES(q) is $|a_1|^2$, while the probability of the answer NO(q) is $|a_2|^2$. The viewpoint $p = q$ is specified with the help of an analyzer. Two pairs of orthnormalized functions, $\phi_1(q')$, $\phi_2(q')$ and $\phi_1(q'')$, $\phi_2(q'')$, are related through a unitary transformation

$$\phi_i(q) = S(q, p)\, \phi_i(p), \quad i = 1, 2$$
$$(S^{-1}(q, p) = S(q, p), \quad S^+ = S^{-1}, \quad \|S\| = 1), \tag{5}$$

$$|S_{11}| = |S_{22}|, \quad S_{21} = -S_{12}, \tag{6}$$

where $\|S\|$ is the determinant of S and S_{ik} are matrix elements.

By the definition of the index p and the corresponding perturbations, the matrix element of the transition $\psi_0 \rightleftharpoons \psi_1(p)$ is equal to zero for $V_1^{(p)}$, while that of $\psi_0 \rightleftharpoons \psi_2(p)$ is equal to zero for $V^{(p)}$, i.e., the YES(p) test leads to transitions $\psi_0 \rightleftharpoons \psi_1(p)$, and the NO($p$) test leads to transitions $\psi_0 \rightleftharpoons \psi_2(p)$.

Suppose that a cell in a state Φ receives the YES(p) signal. Let us express Φ in the representation ϕ_0, ϕ_1, ϕ_2. Then the evolution of Φ in time will obey the Schrödinger wave equation:

$$i\hbar \frac{\partial \Phi}{\partial t} = U^{(p)}\Phi, \tag{7}$$

$$U^{(p)} = \begin{pmatrix} 0 & V_{01}^{(p)}(t) & 0 \\ V_{10}^{(p)}(t) & 0 & 0 \\ 0 & 0 & 0 \end{pmatrix}, \quad V_{ik}^{(p)}(t) = \int \psi_i^*(t)\, V^{(p)}(t)\psi_k(t)\, dx. \tag{8}$$

The integral here is over the configuration space of the cell. If the NO(p) signal is received, then, in the same representation,

$$ih\frac{\partial\Phi}{\partial t}=U_1^{(p)}\Phi,\qquad U_1(p)=\begin{pmatrix}0 & 0 & V_1^{(p)}(t)_{02}\\ 0 & 0 & 0\\ V_1^{(p)}(t)_{20} & 0 & 0\end{pmatrix}. \tag{9}$$

Now let us suppose that the signal to the cell comes from a subsystem p' such that $p' \neq p$. If Φ is expressed in the representation p' (let us denote it Φ'), then the equation for Φ' will have the form (7–8) with the substitutions $p \to p'$ and $\Phi \to \Phi'$. We can now express Φ' in terms of Φ in the p-representation by means of a unitary transformation. We can thus obtain the following equations for Φ in the situation in which the signal is coming from subsystem p':

$$ih\frac{\partial\Phi}{\partial t}=U'(p')\,\Phi,\qquad k=1,2, \tag{10}$$

$$U_k'^{(p')}=\begin{pmatrix}1 & 0\\ 0 & S(p',p)\end{pmatrix}U_k^{(p')}\begin{pmatrix}1 & 0\\ 0 & S^+(p',p)\end{pmatrix}, \tag{11}$$

where $S(p',p)$ is a two-dimensional unitary matrix over the doubly degenerate space of the excited state and $U_k^{(p')}$ has the form (7–8), except for the substitution $p \to p'$. The subscript k indicates whether it is the case YES(p),

$$U_k(p)=U(p)$$

or NO(p),

$$U_k(p)=U_1(p)$$

The meaning of the perturbations $U_k'^{(p')}$ will become clearer if we decompose them into the two polarization directions of the logic p (which provide the basis for the representation of Φ):

$$U'(p')=\begin{pmatrix}0 & S_{11}^*V_{01}^{(p')} & 0\\ S_{11}V_{10}^{(p')} & 0 & 0\\ 0 & 0 & 0\end{pmatrix}+\begin{pmatrix}0 & 0 & S_{21}^*V_{01}^{(p')}\\ 0 & 0 & 0\\ S_{21}V_{10}^{(p')} & 0 & 0\end{pmatrix}, \tag{12}$$

$$U_1'^{(p')}=\begin{pmatrix}0 & S_{12}^*(V_1^{(p')})_{02} & 0\\ S_{12}(V_1^{(p')})_{20} & 0 & 0\\ 0 & 0 & 0\end{pmatrix}+\begin{pmatrix}0 & 0 & S_{22}^*(V_1^{(p')})_{02}\\ 0 & 0 & 0\\ S_{22}(V_1^{(p')})_{20} & 0 & 0\end{pmatrix} \tag{13}$$

(the matrix elements $V_{ik}^{(p')}$ are taken by using the functions $\psi_i^{(p')}$).

It follows from (12) that, e.g., with the YES(p) signal, the amplitudes in (1) obey the equations:

$$ih\frac{\partial Ra_1}{\partial t}=S_{11}V_{10}^{(p')}a_0,\qquad ih\frac{\partial Ra_2}{\partial t}=S_{21}V_{10}^{(p')}a_0. \tag{14}$$

This means that the YES(p') signal is a superposition of two signals of the logic

p: YES(p), with an amplitude proportional to S, and NO(p), with an amplitude proportional to S_{21}. This is exactly what should be expected. The following function corresponds to the YES(p) state in the p representation

$$\psi_1(p) = \begin{pmatrix} 1 \\ 0 \end{pmatrix}.$$

Therefore, the following column corresponds to the YES state in the logic p', as expressed in the p representation:

$$\begin{pmatrix} a_1 \\ a_2 \end{pmatrix} = \begin{pmatrix} S_{11} & S_{12} \\ S_{21} & S_{22} \end{pmatrix} \begin{pmatrix} 1 \\ 0 \end{pmatrix} = \begin{pmatrix} S_{11} \\ S_{21} \end{pmatrix}, \qquad |S_{11}|^2 + |S_{21}|^2 = 1, \tag{15}$$

i.e., S_{11} is the amplitude of the probability that the cell will "say" YES(p) and S_{21} is the amplitude of the probability of the answer NO(p)—provided that the cell is originally in the YES(p') state. The meaning of the perturbation $U'^{(p')}_1$ is explained in an analogous manner. Since $U^{(p)}$ and $U_1^{(p)}$ do not commute, the perturbations resulting from different logics also prove to be noncommutative:

$$U'^{(p')} U'^{(p'')} - U''^{(p'')} U''^{(p')} \neq 0, \qquad p' \neq p''. \tag{16}$$

A doubt cell develops in time as described above until it is forced to give an answer to the question about the truthfulness of the corresponding proposition. The answer (decision making in a doubt state) results in a transition of the cell from an excited state into the ground state.

EXPERIENCE AND COGNIZANCE OF DOUBT— DECISION MAKING IN A STATE OF DOUBT— RANGE OF APPLICABILITY OF THE MODEL

The state of a doubt cell cannot be measured without changing that state. Since a measurement performed by the individual himself is equated in this model with making a decision, we must look for other mechanisms to explain the experience (the feeling of having) doubt. Our hypothesis is that the experience of doubt is not of a quantum mechanical nature, and stems from those signals which the logical system sends when it discovers both the impossibility and the necessity of making a decision in a state with insufficient resolving power.

These signals are also sent to the higher levels of consciousness. It is natural to assume that a logical analysis related to the situation performed at higher levels of consciousness would be related to the cognizance of doubt.

The process of decision making begins with choosing a logical subsystem in order to specify the viewpoint from which a given proposition will be evaluated as YES or NO. In other words, one first chooses a polarization direction p, which is supposed to reflect a certain meaning attached to the proposition.

When the polarization is chosen, it is set in the analyzers, i.e., in the outputs of the cell, sending signals to the neurons, which transfer them along. In the case of polarization p, the YES signal corresponds to the transition $\psi_1(p) \rightarrow \psi_0$, while the NO signal corresponds to the transition $\psi_2(p) \to \psi_0$. With a fixed polarization p, these YES and NO signals have a definite meaning only in logic p.

We use here one of the most important properties of quantum mechanical systems: the result of a measurement depends on how (in what representation) the question is formulated. In our case, the answer depends on the direction of the polarization of the instrument, i.e., on how the instrument violates the symmetry of the cell. This symmetry-breaking brings about the reduction or degradation of the excited state. Thus, the symmetry of the instrument determines those channels in which the acts of decay of the excited state are registered. Therefore, $|a_1|^2$ is the a priori probability of receiving the YES signal, $|a_2|^2$ is the probability of the NO signal, where a_1 and a_2 are relative amplitudes in the expansion of the excited state of the cell by means of the basis states $\psi_1^{(p)}$ and $\psi_2^{(p)}$, and p is the logic used to formulate the question.

The evaluation of this or that answer to the problem means a transition of the cell into the ground state, i.e., a termination of the state of doubt.† Two observable consequences follow:

1. It is found (by the individual) that a decision is no longer necessary.
2. A new postulate emerges, which is the affirmation or negation of the p' proposition within the framework of one of the logical subsystems p of the individual.

Thus, making a decision with respect to a given proposition leads not only to a change in the state of one doubt cell, but also to a restructuring of the individual's logical system at a definite level of consciousness. During the whole time that the individual retains the proposition in his mind as unquestionable, no signals requesting a YES/NO test are sent to the doubt cell of that proposition.

A decision made as described above will be referred to as a free will decision, or a free choice. We will attempt to clarify some of the details.

What perturbations bring about the observed transition of the cell into the ground state? In the simplest version, they are the same perturbations that excite the cell if the input of signals does not stop till the moment a YES or NO decision is made. As another example, one may imagine a more precise measurement, one by which the analyzers could provide the boundary conditions for the cell's transition into the ground state in addition to their own perturbations: $V^{(p)}$ for the first analyzer and $V_1^{(p)}$ for the second one. Perturbations $V^{(p)}$ and $V_1^{(p)}$ must be such that $V_{01}^{(p)} = (V_1^{(p)})$, so that the relative probabilities of transitions $\psi_1^{(p)} \to \psi_0$ and $\psi_2^{(p)} \to \psi_0$ induced by the analyzers are proportional to $|a_1|^2$ and $|a_2|^2$. If, however, input of signals continues, then one cannot separate the transitions induced by the analyzers from those induced by the input signals; therefore, the measurement remains unprecise. But is the requirement of precision necessary for the consciousness of the individual?

---

†Since, in the general case, $a_0 \neq 0$, there is always a possibility that the cell is in the ground, not the excited, state. However, posing the question (i.e., stimulating output neurons) does not necessarily terminate the sending of input signals to the cell. Consequently, after a span of time that is inversely proportional to R^2, a transition from the excited into the ground state will occur. On the other hand, the source of the input signals to the cell may be stopped the moment the question is posed and, after a prolonged absence of an answer, it may be found that a decision is no longer necessary.

What are the decision making mechanisms involved at higher levels of consciousness that lead to the choice of a logic p? First of all, a higher level uses more profound principles, which may provide the necessary logical calculation. Secondly, if the resolving power still does not provide for such a calculation, a decision can be made through the arbitration of a still higher level, in a manner similar to that given above. Thirdly, if transitions to ever higher levels of consciousness do not help (e.g., if higher levels just do not exist), then some system of usual (classic) random choice may be used. Another mechanism may consist of comparing the measures of the degree of conviction $Y^{(p)}$ of the individual in the correctness of a given proposition from the viewpoint of different frames of reference (logics) p.

This experiment is a test by the individual of the extent of his conviction from different viewpoints. For this purpose, a one by one examination of logics is performed at a higher level of consciousness, by repeatedly posing the same questions and obtaining the answers determined by each of the different logics. In this process, consciousness tries to recover its doubt state just before each question is asked, i.e., to reinitialize the state Φ with sufficient accuracy (if accuracy is needed). Specifically, the purpose of this experiment is to obtain the values

$$Y^{(p)} = |a_1|^2 - |a_2|^2, \qquad \psi = a_0 \psi_0 + R(a_1 \psi_1^{(p)} + a_2 \psi_2^{(p)}), \tag{17}$$

for each of the competing logics p (if, of course, there is more than one logic). The logic p that is selected is the one with a magnitude of the degree of conviction, $|Y^{(p)}|$, closest to the classical value of unity. It will then be natural to make the YES decision (formulated in the language of the logic p) if $Y^{(p)} > 0$ and the NO decision if $Y^{(p)} < 0$.

According to our assumption, the YES or NO decision made in the frame of reference of one logic is different from the same decision made from the viewpoint of a different and incompatible logic. Let us now examine how these logics differ linguistically, i.e., in their verbal expressions.

We should note, first of all, that, generally speaking, it is impossible to give strict verbal expressions (i.e., expressions having strict meanings in formal logic) of the same proposition considered from the viewpoints of two different (incompatible) logics. This results from the fact that a proposition $\sigma^{(p)}$ expressed in logic p and the same proposition $\sigma^{(p')}$ expressed in logic p' do not commute:[2]

$$\sigma^{(p')} = S^{(p'p)} \sigma^{(p)} S^{(pp')}, \tag{18}$$

$$\sigma^{(p')} \sigma^{(p)} - \sigma^{(p)} \sigma^{(p')} \neq 0, \qquad p' \neq p. \tag{19}$$

Since the YES (NO) decision can be dichotomized as an affirmative, not tentative, proposition "YES, σ" ("NO, not σ"), the verbal expressions of decisions in different logics cannot, generally, be strictly formal.

Before considering a specific example, let us clarify the range of applicability of the proposed theory of doubt.

The model that we posit is applicable to a consciousness with few levels of organization. In particular, it applies to a consciousness with a single level. In the case of such a consciousness, arbitration of the doubt cells becomes necessary when there is a conflict between the necessity to make a decision and the lack of resolving power of the sensor organs.

In the case of a highly developed consciousness, the main role played by the doubt states must be in the criteria used to assess facts, not in the facts themselves. In other words, the main field of application of this model for a human being is in the sphere of his inner convictions: moral criteria, aesthetic criteria, etc.

Suppose, for example, that the individual has to make a moral judgment about the fact of the punishment of another individual. Suppose that two different moral approaches to phenomena coexist in his consciousness, which is reflected in the existence of two different logical subsystems: the morality of duty p_1 and a version of popular or religious morality p_2. The formal proposition to be decided may look, for example, like this:

$$\sigma \equiv \text{"it is good (right)."} \tag{20}$$

Suppose that neither p_1 nor p_2 contains any uniquely interpreted formulations that would provide a definitive judgment about σ—this is the way it usually happens. Then the individual has to make a free will decision. His subsequent behavior depends on what decision he has made, and what logic, p_1 or p_2, was used. Let the logic (the viewpoint) p_1 be used, and let the decision be YES; we shall refer to this as YES$_1$. From the viewpoint of logic p_2, this decision will be a superposition of YES$_2$ and NO$_2$, which, however, cannot be expressed precisely in terms of formal propositional logic. Informally, it may be described as something like:

$$\sigma^1 \equiv \text{"YES, but ... maybe also NO"}. \tag{21}$$

Unlike σ, operator σ^1 is not diagonal:[2]

$$\sigma^1 = \begin{pmatrix} S_{11} & S_{12} \\ S_{21} & S_{22} \end{pmatrix} \sigma \begin{pmatrix} S_{11}^* & S_{21}^* \\ S_{12}^* & S_{22}^* \end{pmatrix}, \qquad \sigma = \begin{pmatrix} 1 & 0 \\ 0 & -1 \end{pmatrix}. \tag{22}$$

Apparently, the extent to which two logics differ is mainly determined by the angle θ between the axes Z' and Z in the truth value space. If $\sigma^1 = $ NO, then $\theta = 180°$ (and the corresponding basis spinors for YES in the different logics will be orthogonal). For (21), $\theta \leq 90°$; but, of course, there is no clear relation between σ^1 and θ because of the intrinsic ambiguity in σ^1 itself. Obviously, p_1 and p_2 can be interchanged in the above reasoning.

Let us finally note that σ in (22) is an unequivocal (classic) proposition of the logic p_2; $+1$ corresponds to "right" and -1 corresponds to "wrong" in logic p_2, in whose language (21) is also expressed.

Conclusion

In this article, only a single doubt cell has been considered. In real world decisions, we deal with the interaction of a great number of elementary cells, all of which play a role in composite propositions. Also, we have not considered effects caused by the interference of cell perturbations in the intervals between decision making acts. Wave effects in complex systems must lead to phenomena analogous to diffraction. I hope to consider these problems in the future.

Many discussions I had on the problems connected with the hypothesis of wave logic were of great help to me, and I would like to thank most cordially Yu. A. and Ya. Yu. Golfand, V. F. Turchin, V. Finn, my sons A. Yu. and D. Yu. Orlov, E. K. Tarasov, N. N. Meiman, G. Rosenstein, and my wife Irina for her faith in me, which, without doubt, exceeds my actual capabilities, but which helps tremendously.

REFERENCES

1. TURCHIN, V. F. 1977. The Phenomenon of Science. Columbia University Press. New York.
2. ORLOV, YU. F. 1978. Group theory approach to logic. The wave logic. Philosophia Naturalis **17**: 120.
3. Translator's note: See also ORLOV, YU. F. 1978. Wave Calculus Based Upon Wave Logic. Int. J. Theor. Phys. **8**: 585–98.

AGREEMENT THROUGH FAIR PLAY

Aleksandr Ya. Lerner

Moscow, USSR

INTRODUCTION

The author and his coworkers have suggested constructing control of the production activity of hierarchically structured active systems on the Fair Play Principle of Control.[1-3] They demonstrated substantial advantages of such controls. "Active systems" are systems in which some of the elements are stimulated by the participants' own subjectively assessed interests.

The Fair Play Principle of Control consists of ensuring that every decision taken by the controlling body will be advantageous for all participants, both executives and executors. In such a case, neither of these two groups would be interested in concealing their goals and preferences. Under such conditions, the executives would reveal their control algorithms, while the executors would reveal their possibilities. This would ensure that this system is more efficient than similar systems in which it would be more advantageous to conceal certain algorithms of control and certain possibilities.[4]

It will be shown here that the Fair Play Principle of Control can be useful for finding the optimal version of an agreement by negotiating parties, a task that can be considered one of the tasks of the theory of active systems control.

A SPECIFIC STATEMENT OF THE TASK

Negotiations involve two or more independent sides, with different goals and preferences, interested in reaching agreement, or at least negotiating, on a definite range of issues whose solution depends on the negotiating sides. The purpose of the talks is to achieve an agreement containing a certain number of elements or points, thereby formulating decisions accepted by all the negotiating parties.

For each party the purpose of the negotiations is to reach an agreement corresponding most to its interests and preferences, or to continue the negotiations.

Agreement can be reached if each party tends to a definite degree to take into consideration the interests of the other parties. The possibility of tolerating the interests of the other parties stems from the fact that the interests of all parties are not polar, that the weight attached to a given decision on a question discussed is not the same for all participants. Because of this, there are opportunities for compromise, which lies, as a rule, at the foundation of any agreement. Moreover, the very fact that agreement is reached, as well as the fact that certain clauses are defined, can prove of interest, i.e., be mutually advantageous, to all sides; this, in addition, stimulates the tendency of each side to accede to concessions for the sake of reaching agreement as a whole.

The usual practice of conducting negotiations (at all levels, from those between suppliers and buyers to those between nations) is complicated by the spontaneous nature of the negotiating process, the desire of parties to achieve unilateral advantages by bluffing, concealing their interests and preferences, and the absence of coordinated procedures for conducting negotiations. As a result, talks often drag out, while either agreement is not reached when there is an objective possibility of agreement or the agreement reached is not the best for all parties.

It can, therefore, be of interest to attempt to use certain formal procedures to facilitate and accelerate the search for agreement, and, if agreement is possible, to find its optimal version.

A Formal Statement

Let the negotiating parties $X, Y, Z, \ldots$ introduce n proposals as elements of an agreement $a_1, a_2, \ldots a_n$. Then any version of the agreement $A_p (p = 1, \ldots 2^n)$ represents a definite composition of these elements. If we attribute the measure 1 to elements introduced in this set, and 0 to those not introduced, the set A_p of versions of the agreement can be represented by a set of vertices of an n-dimensional cube. Each of the 2^n vertices of the cube will correspond to a definite version of the agreement.

However, not every version is worth considering as a claim in the agreement sought. To be so considered, the version must also possess the properties of completeness, with no contradictions, and must be permissible, i.e., must contain essential proposals, but none that would be mutually exclusive, contradictory, or incompatible; it must not contain impermissible proposals categorically unacceptable to any party. The version of an agreement that is complete, noncontradictory, and permissible will be termed permissible. Obviously, there are far fewer permissible versions than 2^n. We will discuss only permissible versions below.

It will evidently be advisable to regard each of the negotiating parties as unanimous and able to evaluate the utility of every version of the agreement and to assume that their estimates are transitive.

Let U_{ij} represent the value estimate of a version of the agreement A_j for participating party i; i.e., $U_{ij} = U_i(A_j)$. Obviously, all versions of A_j in which the estimate of value is no less than a certain threshold α_i can be regarded as acceptable. I.e.,

$$A_j = \tilde{A}_j \text{ if } U_{ij} \geq \alpha_i.$$

Thus, let the subsets $\tilde{A}_X, \tilde{A}_Y, \ldots$ be acceptable versions of an agreement for the corresponding parties $X, Y, \ldots$. Then the existence of an intersection for all these subsets testifies to the possibility of reaching agreement, and, on the contrary, the absence of an intersection would indicate the impossibility of reaching agreement, at least until the parties alter their estimates of the versions and/or revise their thresholds of acceptance.

The tactics of negotiation are often such that the sides first put up such high acceptability thresholds α_i that the sets $\tilde{A}_X, \tilde{A}_Y, \ldots$ do not intersect. Consequently, each side tries to have the other negotiating sides revise their thresholds, while lowering its own threshold till all the sets do intersect in some version (Figure 1).

If the intersection of the sets $\tilde{A}_X, \tilde{A}_Y, \ldots$ contains more than a single vertex, the task arises of choosing one optimal version from those acceptable, with respect to a definite criterion. The choice of the criterion is especially difficult, since it must reflect the value of the agreement for all the participants in the negotiations, whose interests clearly do not coincide.

THE ADVISABILITY OF FAIR PLAY

If each of the elements of the agreement can be regarded as independent of all the others, the negotiators can evaluate the utility not only of versions of the agreement,

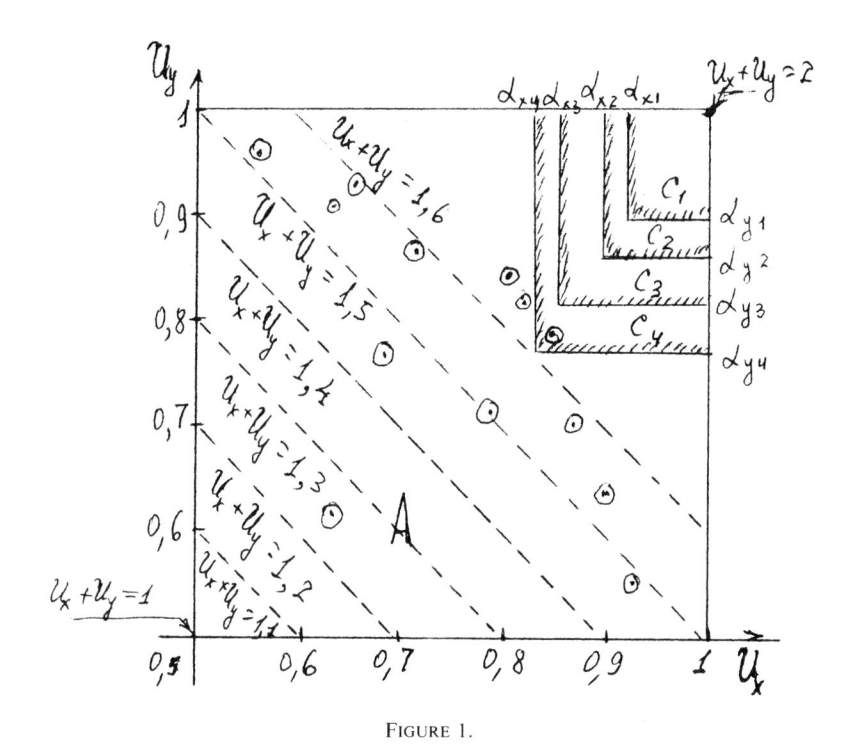

FIGURE 1.

but also of each separate element a_i with its j-estimate U_{ij}. The value of each version can be expressed as a definite function of the utility of its elements. For these (relative) estimates of the elements submitted by the sides to be compared, they must be normalized. It is convenient for our purpose to normalize them by restricting the module sums of the estimates of all elements for every side, i.e., by establishing

$$\sum_{i=1}^{n} |U_{ij}| \leq 1 \tag{1}$$

for all $j = 1, \ldots N$.

Since the submission of estimates is each negotiating side's way of getting elements it finds advantageous accepted in the agreement and elements it finds disadvantageous rejected, the sides cannot be interested in insufficient use of the estimate resources. Therefore, each side must always pose estimates with a view to their sum equaling the permissible limit, i.e., unity.

In order to facilitate the iterative procedure of submitting estimates and to avoid vagueness due to possible manipulation with negligibly small changes in assessment values, it is advisable to allow a choice of estimates not from the continuum {0, 1}, but from a discrete series of permissible values defined, for example, by the number of quanta, Δ. Thus, every estimate can be expressed in the form $U_{ij} = \Delta r_{ij}$, where r is an integer.

As a rule, no utility estimates for one or another element of the agreement can be determined by objective methods, but must be worked out by experts or by persons directly interested in the results of the talks. The authenticity of the estimates and their correct submission to the other side are very important, since this would facilitate the working out of formal procedures for determining the possibility of reaching agreement and for seeking a mutually advantageous optimal version.

Let the choice of an optimal agreement A^* be made by maximizing the function of the utility of the agreement U, which belongs to class $U \in \mathbf{I}$, and let this be known to all the negotiating parties. We will assign to this class any function monotonic for each of its arguments U, i.e., one that satisfies the following condition,

$$\frac{\partial U_j}{\partial u_{ij}} \geq 0 \quad \text{or} \quad \frac{\partial U_j}{\partial u_{ij}} \leq 0, \quad (j = 1, \ldots N), (i = 1, \ldots n), \tag{2}$$

throughout its domain of definition. Obviously, class $\mathbf{I}$ includes all the additive and multiplicative functions.

Here, u_{ij} is the value estimate for party i of the particular element a_j.

It is simple to demonstrate that none of the N negotiating parties can profit (increase their value function) $U_j \in \mathbf{I}$ by deliberately falsifying its estimates u_{ij} $(j = 1, \ldots N, i = 1, \ldots n)$ of the value of each of the n elements of the agreement, if the optimal agreement is achieved by maximizing the function $U \in \mathbf{I}$ of the value of the agreement to all participants. Indeed, suppose one of the participants, X for instance, deliberately falsifies its estimate of at least a couple* of elements from the set $\{U_x(a_i)\}$, so that estimate $u^f_{xj} = U_{xj} + \Delta$ stands for element a_j and estimate $u^f_{xk} = u_{xk} - \Delta$ stands for element a_x.

If this falsification of estimates has not brought about a difference between the contents of the optimal agreement A^{*f} and the contents when the estimates are not falsified A^*, then it will turn out that participant X gained nothing from the falsification.

If, as a result of overestimating the value of element a_j, it has been incorporated in agreement A^{*f} instead of another element, say a_r, for any utility function $U \in \mathbf{I}$, then it must be that

---

*It is obviously disadvantageous to falsify an assessment of a single element, since lowering an assessment cannot be advantageous, while raising it cannot be permitted without simultaneously reducing the value of another element, since, according to (1), the sum of the assessments is limited.

$$u_{xj} < u_{xr} < u^f_{xj},$$

in view of the montony of utility function U_x. But participant X would lose out from this value U_x, since $u_{xj} < u_{xr}$, and the value u_x^{*f} of agreement A^{*f} for participant X would be less than agreement A^*. If, as a result of the lowering of the estimate of element a_k, it were to be eliminated from the optimal agreement A^{*f}, and exchanged for any other element, for example a_s, then, for any value function $U_x \in \mathbf{I}$, it could only result from

$$u^f_{xk} < u_{xs} < u_{xk},$$

in view of the monotony of utility function U_x. But then participant X would also lose out in the value of its utility function $U_x \in \mathbf{I}$, since $u_{xs} < u_{xk}$, and the value u_x^{*f} of agreement A^{*f} would be less than the value of agreement A^*.

It thus turns out that any distortion of estimates is disadvantageous, and there is no sense in the participants concealing their true evaluations. The situation arising at the talks makes it advisable for the sides to *play fair,* i.e., to be prepared to reveal their assessments, preferences, and the nature of their utility functions to their partners in the talks.

A SIMPLE CASE

Let us examine a simple situation where negotiations are underway between two sides, X and Y, the estimates of all elements of the agreement are independent,† and the utility functions satisfying condition 2 are, for participant X,

$$U_x = \sum_{i=1}^{n} k_i u_{xi} \rightarrow \max,$$

and, for participant Y,

$$U_y = \sum_{i=1}^{n} k_i u_{yi} \rightarrow \max,$$

where

$$k_i = \begin{cases} 0 & \text{for } a_i \\ 1 & \text{for } a_i^* \end{cases}$$

and a_i^* signifies the elements of the agreement contained in the version considered.

This is presented graphically by the location of points representing the estimates of each element of the agreement on the Estimation Plane (FIGURE 2). The plane divides naturally into six domains:

1. Mutually advantageous elements
2. Elements substantially advantageous for participant X and somewhat disadvantageous for participant Y, $(U_x + U_y) > 0$

†Separate elements may, however, be incompatible.

3. Elements somewhat advantageous for participant X and substantially disadvantageous for participant Y, $(U_x + U_y) < 0$
4. Mutually disadvantageous elements
5. Elements substantially disadvantageous for participant X and somewhat advantageous for participant Y, $(U_x + U_y) < 0$
6. Elements somewhat disadvantageous for participant X and substantially advantageous for participant Y, $(U_x + U_y) > 0$.

It is clear that elements for which the estimation point lies in domain 4 have no chance of being included in the agreement; those for which the estimation point lies in domain 1 will certainly be accepted. The points in domains 2 and 6 represent elements with a good chance of being accepted in the agreement, while those in domains 3 and 5 have a poor chance.

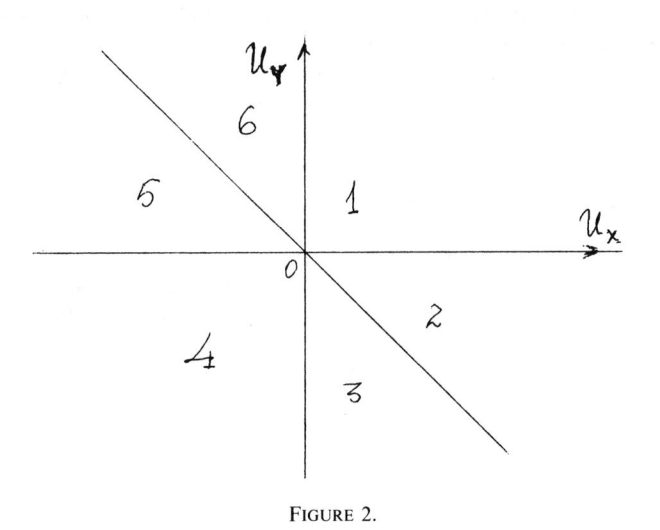

FIGURE 2.

In this case, it would be reasonable to choose a version with the greatest chance of being approved by both sides, using the criterion of sum total value:

$$U_{xy} = U_x + U_y \to \max,$$

guided by the obvious condition that there be no incompatible elements in it.

Obviously, the greatest total utility value will be for the version of the agreement containing all those elements whose total estimate will be $(U_x + U_y) > 0$. Other versions can be chosen by discarding elements of minimum sum total value, till the total utility version reaches the threshold α.

Each of the versions of the agreement thus obtained can be evaluated, in full, by the participants, taking into account the utility of all the elements incorporated, the choice of a mutually acceptable version being made by gradually reducing thresholds of acceptability till both the following conditions are satisfied:

$$U_x = \sum_{i=1}^{n} k_i u_{xi} \geq \alpha_X$$

and

$$U_y = \sum_{i=1}^{n} k_i u_{yi} \geq \alpha_Y.$$

Thus, the procedure for reaching agreement is as follows:

1. The sides submit their proposals concerning the elements of the agreement, which form the set of elements.

2. The sides work out estimates for each element in the set and exchange information about their estimates.

3. This data is modified to take into account the other side's evaluation. As a result, an iterative procedure takes place, which leads to the ultimate estimate.

4. Those elements for which the sum total of estimates is positive and those alternative elements for which the sum of estimates is maximum are chosen. The elements thus chosen comprise the version with the maximum total utility. To this are added the versions whose total utility is higher than the limit α.

5. The utility of each version of the agreement achieved is evaluated by the sides.

6. They check whether a version is mutually acceptable by comparing the utility values with each side's threshold of acceptability. If there is no such version, the sides lower their acceptability thresholds. Either they establish that no agreement can be reached or they alter their thresholds towards a mutually acceptable version.

7. If it turns out that agreement is not reached with the given set of elements, further attempts are undertaken to reach agreement. For this, the array of elements with a low total sum of estimates is reformulated with a view to raising the sum. It goes without saying that, in the process of seeking agreement, proposals can also be submitted for increasing the mass of elements (new ideas).

INTERDEPENDENCE AMONG THE ELEMENTS OF THE AGREEMENT

Very often, the elements of an agreement are not strictly independent, as a result of which the estimation of utility is not additive. This substantially complicates a solution to the task, yet certain particular cases of interdependence between the elements can be considered and simple heuristic methods can help reduce such tasks to the simple case just examined.

The most substantial interdependence between k elements (in the simplest case $k = 2$) is the relationship of compatibility: positive (allowing only joint inclusion of any given combination of elements in the agreement) or negative (including this combination in the agreement is not allowed). For independent elements it is natural to regard compatibility as zero.

k-Dimensional matrices of compatiblity M_X and M_Y can be devised by sides X and Y, stemming from their ideas on the essence of the propositions contained in the elements a_i.

Since, in the general case, the matrices M_X, and M_Y do not coincide, the question of considering the compatibility relations between some of the elements can be solved on the basis of the following obvious rules. (1) If a certain combination of elements is present in one of the matrices as a positive compatibility and in the other as a zero one, it should be regarded as positive. (2) In a case of negative compatibility in one of the matrices and zero in the other, it should be regarded as negative. (3) If the compatibility relation in a given combination is positive for one side but negative for the other, it means that the elements of this combination cannot be included in the agreement.

But interdependence between elements of an agreement can consist of other things than compatibility relations. Often, a combination of k elements ($k = 2$ in the simplest case) is evaluated by the sides above or below the total utility of each element taken separately. In such a case, additional positive or negative usefulness for each element combination can be represented by the k-dimensional matrices R_X and R_Y submitted by each of the negotiating sides on the basis of its consideration of the elements of each of the k-element combinations.

To avoid haphazard moves when the sides submit additional values and to ensure that the estimates of different sides can be compared, the estimates must be restricted. One possible and reasonable restriction could be to include the additional estimates in the total, restricted by condition 1, which determines the estimate resources for each side.

Considerations of compatibility and the application of the rules for determining compatibility make it possible to choose admissible versions of the agreement when participants X and Y differ. The additional utility of combinations of the elements for the total utility of each of the versions of the agreement can be taken into account by including the additional combinations of usefulness appearing in the matrices R_X and R_Y in the total usefulness of the elements in each version.

REFERENCES

1. BURKOV, V. N. & A. YA. LERNER. 1970. Avtom. Telemekh. **8**.
2. BURKOV, V. N. & A. YA. LERNER. 1971. Differential Games and Related Topics. North-Holland Publishing Co. Amsterdam, London.
3. LERNER, A. YA. 1972. Proc. 5th IFAC World Congress. Paris.
4. LERNER, A. YA. 1975 In Proc. Moscow Seminar July, 1974. Brain Research Publications. Fayetteville, N.C.

NONSMOOTH ANALYSIS AND THE THEORY OF FANS

Aleksandr D. Ioffe

Moscow, USSR

INTRODUCTION

The idea of extending the differential calculus beyond the boundaries of smoothly differentiable maps is, of course, very old. It was the central idea of such theories as the differentiation theory connected with the Lebesque integral and the theory of distributions.

Both these theories deal, essentially, with the nonlocal aspect of the calculus, which is centered on the Newton-Leibniz formula. The notion of a value of the derivative at a given point does not make sense in either of them.

Nonsmooth analysis was developed in the early seventies to carry out an extension of the local aspect of the classical calculus connected with the idea of (linear) approximations about a given point. (Certain attempts were undertaken earlier—one recalls the Dini numbers, for instance—but a systematic attack began some eight or nine years ago.) It is not surprising that the main impulse came from optimization theory, which has natural mechanisms that generate nonsmoothness.

Several approaches were offered,[1-3] all aimed at applications to extremal problems. I do not have the space to discuss them in detail. What must be mentioned is that they all use sets of linear operators for local approximations of maps (this idea was borrowed from convex analysis).

The analysis that had been built thereafter proved to be very efficient for real-valued functions and finite-dimensional maps. This is especially true concerning Clarke's theory (see References 4 and 5 for a survey of the results), though, in certain cases, the methods of Halkin and Warga allow one to establish sharper results.[6] However, for infinite-dimensional mappings, all three methods essentially fail to work.

I think that, to a large extent, the real cause of this and certain other difficulties is that the techniques of sets of linear operators are not completely adequate to the problem. To observe the variety of local behavior of nonsmooth mappings, a more flexible instrument is needed. (This does not mean that sets of operators are completely unnecessary. On the contrary, a derivative container, for instance, can provide a better approximation; the point is that, often, derivative containers cannot be defined.) In attempts to find such an instrument, two more approaches were offered, both connected with set-valued mappings. One of them, recently developed by Aubin,[7] is based on the utilization of Bouligand contingent cones, the other, offered by the author, involves set-valued mappings of a special type called fans.[8-10] (Interrelations between these theories will be briefly discussed in the Conclusion.)

This article surveys some results on fans and their applications obtained in the last two years. Details and more results can be found in References 10–13. Some of these papers could not be published earlier, mainly because of "communication difficulties" (see the footnote in Reference 14). Moreover, neither these nor my other papers could have been completed and prepared for publication without generous encouragement and assistance from my friends and colleagues, especially, Jean-Pierre Aubin, Ivar Ekeland, Terry Rockefeller, and Richard Vinter. I wish to express my deepest gratitude to all of them.

FANS[8–10]

Let X, Y be linear spaces. A set-valued mapping $\mathcal{A}(x)$ from X into Y is called a fan if it has the following properties:
1. $\mathcal{A}(x)$ is nonempty and convex for any $x \in X$
2. $\lambda \mathcal{A}(x) = \mathcal{A}(\lambda x)$, $\lambda > 0$ (homogeneity)
3. $\mathcal{A}(x + u) \subset \mathcal{A}(x) + \mathcal{A}(U)$ (subadditivity).

If Y is endowed with a locally convex topology, then we require, in addition to 1, that $\mathcal{A}(x)$ be closed for any $x \in X$ and replace 3 by $\mathcal{A} \in x + u) \subset \overline{\mathcal{A}(x) + \mathcal{A}(u)}$ (the bar denotes closure).

The fan is odd if
4. $\mathcal{A}(-x) = -\mathcal{A}(x)$ (oddness).

If X, Y are normed spaces, then $\mathcal{A}$ is called bounded if
5. there is some $k > 0$ such that $y \in \mathcal{A}(x) \Rightarrow \|y\| \leq k\|x\|$. (Of course, boundedness can be naturally defined for arbitrary locally convex spaces.)

Here are several examples of fans:
1. If $A: X \to Y$ is a linear operator, then $\mathcal{A}(x) = \{Ax\}$ is an odd fan, bounded if A is bounded.
2. If $\mathcal{N}$ is a convex set of linear operators $X \to Y$, then

$$\mathcal{A}(x) = \{y \mid y = Ax \quad \text{for certain} \quad A \in \mathcal{N}\}.$$

is an odd fan, bounded if $\mathcal{N}$ is a bounded closed set. We say that $\mathcal{A}$ is generated by $\mathcal{N}$.

3. If Y is a vector lattice and $P: X \to Y$, $Q: X \to Y$ are sublinear (i.e., convex and positively homogeneous) and superlinear operators such that $P(x) \geq Q(x)$ for all x, then

$$\mathcal{A}(x) = [Q(x), P(x)] = \{y \mid Q(x) \leq y \leq P(x)\}$$

is a fan (not necessarily odd); in particular,

$$(x) = [-P(-x), P(x)]$$

is an odd fan.

In the last example, none of the fans may be generated by a set of linear operators. In general, even in finite-dimensional spaces, there exist odd bounded fans not generated by linear operators.

Nonetheless, fans seem to be the most natural set-valued extensions of linear operators. TABLE 1 is a comparative list of the main notions and geometrical characteristics of fans and linear operators.

Here, as usual, the asterisk denotes quality and $\langle . , . \rangle$ is the canonical pairing. The crucial notion is (as is usual for convex-valued mappings) the notion of support function. Among the properties of support functions (of bounded fans) we mention the following: s is sublinear in either variable (but not in both!), it is (norm) continuous in both variables (jointly), is weak* l.s.c. in y^*, and satisfies $s(y^*, x) \leq \|\mathcal{A}\| \|y^*\| \|x\|$. Any function having these properties is the support function of a bounded fan. Oddness is equivalent to $s(y^*, -x) = s(-y^*, x)$.

It is important to observe that each of these notions can be expressed in terms of support functions. This circumstance is not surprising, but, nevertheless, the formula for the Banach constant seems to be new even for operators (the proof based on the Ky Fan saddle point theorem is actually very easy) and, as we shall see, is very useful.

TABLE 1

Fans	Linear Operators
Support function $s(y^*, x) = \sup_{y \in \mathcal{A}(x)} \langle y^*, y \rangle$	Bilinear form $\langle y^*, Ax \rangle$
Adjoint fan $\mathcal{A}^*: Y^* \to X^*$ $\mathcal{A}^*(y^*) = \{x^* \mid s(y^*, x) \geq \langle x^*, x \rangle, \forall x\}$	Adjoint operator $A^*: Y^* \to X^*$ $\langle A^*y^*, x \rangle = \langle y^*, Ax \rangle$
Norm $\|\mathcal{A}\| = \sup\{\|y\| \mid y \in \mathcal{A}(x), \|x\| \leq 1\}$ $= \sup\{s(y^*, x) \mid \|y^*\| \leq 1, \|x\| \leq 1\}$	Norm $\|A\| = \{\sup \|Ax\| \mid \|x\| \leq 1\}$
Banach constant $C(\mathcal{A}) = \inf\{\|x^*\| \mid x^* \in \mathcal{A}^*(y^*), \|y^*\| \leq 1\}$ $= -\sup_{\|y^*\|=1} \inf_{\|x\| \leq 1} s(y^*, x)$	Banach constant (modulus of surjectivity) $C(A) = \inf\{\|A^*y^*\| \mid \|y^*\| \leq 1\}$
Diameter $\operatorname{diam} \mathcal{A} = \sup_{\|x\| \leq 1} \sup_{y_i \in \mathcal{A}(x)} \|y_1 - y_2\|$ $= \sup \sup (s(y^*, x) - s(-y^*, x))$	0
Pointwise convergence $\mathcal{A}_n \to \mathcal{A} \iff h(\mathcal{A}_n(x), \mathcal{A}(x)) \to 0, \forall x$ (h = Hausdorff distance)	Pointwise convergence $A_n \to A \iff \|A_n x - Ax\| \to 0, \forall x$
Hilbert spaces (odd fans)	
Self-adjoint fan $\mathcal{A} = \mathcal{A}^* \iff s(y, x) = s(x, y)$	Self-adjoint operator $A = A^* \iff \langle Ay, x \rangle = \langle y, Ax \rangle$
Positive semidefiniteness $s(x, x) \geq 0, \forall x$	Positive semidefiniteness $\langle x, Ax \rangle \geq 0, \forall x$
Positive definiteness $s(-x, x) \leq -k \|x\|^2, \forall x$	Positive definiteness $\langle x, Ax \rangle \geq k \|x\|^2, \forall x$

TABLE 2 presents a comparison of certain main formulae and theorems. Note that the summation and composition of fans are naturally defined by

$$s_{\mathcal{A}_1 + \mathcal{A}_2}(y^*, x) = s_{\mathcal{A}_1}(y^*, x) + s_{\mathcal{A}_2}(y^*, x)$$

$$s_{\mathcal{B} \circ \mathcal{A}}(z^*, x) = \sup\{s_{\mathcal{B}}(z^*, y) \mid y \in \mathcal{A}(x)\}$$
$$= \sup\{s_{\mathcal{A}}(y^*, x) \mid y^* \in \mathcal{B}^*(z^*)\}.$$

In TABLE 2, U_Y and B_X are the open and closed unit balls respectively, in corresponding spaces and $a^+ = \max\{a, 0\}$.

The proof of the Banach-Steinhaus theorem for fans is very similar to that for linear operators; in fact, it is even somewhat simpler because the convexity properties the theorem uses are more explicit in the case of fans. Note that, as for linear operators, the Banach-Steinhaus theorem for fans can be established for arbitrary barrel spaces.

Contrarily, the proof of the Banach open mapping theorem for fans is based on ideas quite different from those used in the proof usually applied in standard operator situations. The key element of the proof is the application of Ekeland's variational principle.[15,16] The role of this principle in the theory of fans and nonsmooth analysis can be compared with the role of the contraction mapping principle in the smooth theory, which is, in fact, an easy corollary from Ekeland's principle.

TABLE 2

Fans	Linear Operators
$\\|\mathcal{A}_1 + \mathcal{A}_2\\| \leq \\|\mathcal{A}_1\\| + \\|\mathcal{A}_2\\|$	$\\|A_1 + A_2\\| \leq \\|A_1\\| + \\|A_2\\|$
$\\|\mathcal{B} \circ \mathcal{A}\\| \leq \\|\mathcal{B}\\| \\|\mathcal{A}\\|$	$\\|BA\\| \leq \\|B\\| \\|A\\|$
$C(\mathcal{A}_1 + \mathcal{A}_2) \geq C(\mathcal{A}_1) - \\|\mathcal{A}_2\\|$	$C(A_1 + A_2) \geq C(A_1) - A_2$
diam $\mathcal{A}$ = diam $\mathcal{A}^*$	$0 = 0$
$(\mathcal{B} \circ \mathcal{A})^* = \mathcal{A}^* \circ \mathcal{B}^*$	$(BA)^* = A^*B^*$

Banach-Steinhaus theorem
  Given a sequence $\{\mathcal{A}_n\}$ of bounded fans such that, for any $x \in X$, $\mathcal{A}_n(x)$ Hausdorff converge (as sets). Then $\\|\mathcal{A}_n\\| \leq k < \infty$ for all n, $\mathcal{A}(x) = \lim \mathcal{A}_n(x)$ is a bounded fan, and $\\|\mathcal{A}\\| \leq \lim \inf \\|\mathcal{A}_n\\|$

Banach-Steinhaus theorem
  Given a sequence $\{A_n\}$ of bounded linear operators such that, for any $x \in X$, $A_n(x)$ norm converge. Then $\\|A_n\\| \leq k < \infty$ for all n, $Ax = \lim A_n x$ is a bounded linear operator and $\\|A\\| \leq \lim \inf \\|A_n\\|$.

Banach open mapping theorem
  Given a bounded fan $\mathcal{A}$. Then 1. $(C(\mathcal{A}) - \text{diam }\mathcal{A})^+ U_Y \subset \mathcal{A}(B_X)$ 2. If $\mathcal{A}$ is norm-compact-valued and Y has an equivalent norm with strictly convex dual, then $C(\mathcal{A})U_Y \subset \mathcal{A}(B_X)$. 3. If $\mathcal{A}$ is weak-compact-valued and Y has an equivalent norm with locally uniform convex dual, then $C(\mathcal{A})U_Y \subset \mathcal{A}(B_X)$.

Banach open mapping theorem
  Given a bounded linear operator A. Then $C(A)U_Y \subset A(B_X)$.

Two more remarks should be made in connection with the Banach open mapping theorem. Usually, it is stated as follows: if $A(X) = Y$, then $0 \in \text{int } A(B_X)$. In such a form, it cannot be extended to fans. For linear operators, the inequality $C(A) > 0$ is an immediate corollary of $C(X) = Y$; however, for linear operators, $C(A)$ is the maximal radius of a ball about the origin that is contained in the closure of $A(B_X)$. This is not true for fans.

Note also that, for fans, three independent theorems are presented under the common title. The first of them is a direct extension of the operator theorem (indeed, diam $A = 0$ and $C(A) \geq 0$). The other two are spacial theorems for fans containing a considerably stronger statement. It is not clear if, or how much, the conditions of the last two theorems can be weakened.

EXTENSIONS OF LINEAR OPERATORS[12]

We have deliberately separated the Hahn-Banach theorem from the other two main Banach principles of linear analysis. This is where the first profit from fans comes.

The Hahn-Banach theorem is a theorem on the majorized extension of linear functionals. During the last three decades, since the works of Kantorovich[17] and Nachbin,[18] there has been a constant interest in the extension problem for linear operators, which is immeasurably more complex.

It turns out that, in many cases, the operator extension problem can be naturally formulated in terms of fans as follows: Given linear spaces X, Y, a fan $\mathcal{A}: X \to Y$, a subspace $L \subset X$, and a linear operator $B: L \to X$ such that $Bx \in \mathcal{A}(x)$, $\forall x \in L$. Does there exist a linear operator A from all of X into Y that coincides with B on L and satisfies $Ax \in \mathcal{A}(x)$ for all x?

Indeed, if we deal with a norm-preserving extension, then we can take $\mathcal{A}(x) = \|x\| B_Y$; if we deal with a majorized extension with a given sublinear operator P, then we can take $\mathcal{A}(x) = [-P(-x), P(x)]$, etc.

If the answer to the question is affirmative for any L and B, then we shall say that the fan has the Hahn-Banach extension property. The following surprising theorem characterizes fans having such a property.

We shall say that a set $C \subset Y$ is bounded if, for any $y \in Y$, $y \neq 0$, there is $t_0 > 0$ such that $C \cap (C + ty) = \emptyset$ whenever $t \geq T_0$.

THEOREM 1. Let $\mathcal{C}$ be a collection of convex bounded subsets of Y closed under algebraic summation and scalar multiplication. Then the following two statements are equivalent:

1. Any $\mathcal{C}$-valued odd fan into Y has the Hahn-Banach extension property.
2. $\mathcal{C}$ is a collection of bounded order intervals relative to some certain conditionally complete vector order in Y.

In particular, this theorem implies the characterization theorem of Nachbin[18] (for norm-preserving extensions) and the theorems of Bonnice-Silverman[19] and To[20] (for majorized extensions).

Using this fact (or, more precisely, this method of proof), Kutateladze established another surprising result for majorized extensions of operators in ordered moduli.[21]

Let M be a lattice-ordered ring with a positive unit and Y an ordered unitary M-modulus. We shall say that Y has the M-extension property if, for any M-modulus X, any M-sublinear operator $P: X \to Y$, any M-submodulus $L \subset X$, and any M-linear operator (homeomorphism of moduli) $B: L \to Y$ majorized by P, there is an M-linear operator A from all of X into Y that is also majorized by P and coincides with B on L.

THEOREM 2. Assume that Y has the M-extension property. Then Y can be obtained from a conditionally complete vector ordered space by ignoring the scalar multiplication.

This and another result of Kutateladze show that, as far as the majorized extension properties are concerned, only vector spaces (i.e., moduli over reals) may appear and should be taken into account.

Convex Analysis[11]

The theory of fans has a powerful effect on convex analysis. On the one hand, it allows one to obtain elementary proofs of all elements of the duality theory of convex mappings into conditionally complete vector lattices. On the other hand, fans appear to be very suitable as candidates for subdifferentials of convex mappings into arbitrary ordered spaces (which cannot be defined by way of linear operators) and for building a satisfactory theory that, again, turns out to be almost trivially simple.

For brevity we shall consider only those mappings defined everywhere. (The extension to the general case requires the additional concept of orientation, introduced in References 8 and 9.)

So let Y be a conditionally complete vector lattice and X a linear space.

THEOREM 3. *Let $\mathcal{A}$ be a fan from X into Y whose values are bounded order intervals. Then there is an odd fan $\mathcal{A}'$ from X into Y that also takes values in the collection of bounded order intervals such that $\mathcal{A}'(x) \subset \mathcal{A}(x)$ for all x. In particular, there is a linear operator $A: X \to Y$ such that $Ax \in \mathcal{A}(x)$ for all x.*

As an immediate corollary of this theorem, we obtain the so-called "sandwich theorem": If $P: X \to Y$ is sublinear, $Q: X \to Y$ is superlinear, and $P(x) \geq Q(x)$ for all x, then there is a linear operator $A: X \to Y$ separating P and Q (i.e., such that $P(x) \geq Ax \geq Q(x)$).

This theorem, in turn, allows one to establish, without difficulties, all the necessary duality formulae for subdifferentials. Let $F: X \to Y$ be a convex mapping from X into Y. Changing the accepted terminology slightly, we shall say that a linear operator $A: X \to Y$ is a *linear subgradient* of F at $z \in X$ if

$$F(x) - F(z) \geq A(x - z), \qquad \forall x \in X.$$

The collection of all linear subgradients at z will be called the *linear subdifferential* of F at z and denoted $\partial F(z)$. We have the following formulae:

$$\partial(F + G)(z) = \partial F(z) + \partial G(z);$$

if A is a linear operator,

$$\partial(F \circ A)(z) = (\partial F(Az)) \circ A;$$

if G is a nondecreasing convex mapping,

$$\partial(G \circ F)(z) = \bigcup_{A \in \partial G(F(z))} \partial(A \circ F)(z);$$

here, the range space of F need not be a lattice,

$$\partial(G \vee F)(z) = \bigcup_{m \in \mathcal{M}} (m \partial G(z) + (I - m)\partial F(z)),$$

where $\mathcal{M}$ is the collection of multiplicators, i.e., positive operators $Y \to Y$ such that $0 \leq my \leq y$ for any y, and I is the identity operator.

All these formulae were established by Kutateladze[22] with the help of the "canonical sublinear operator" he introduced. Here, as in the scalar case, no special device is needed.

Now what does one do if Y is an ordered space but not a conditionally complete vector lattice? Then, as is well known, no linear subgradient can exist, so the concept of the linear subdifferential makes no sense.

Again, let $F: X \to Y$ be a convex mapping. Then, if Y^* is a dual space, we can consider, for any positive y^*, the function

$$F_{y^*}(x) = \langle y^*, F(x) \rangle.$$

This is a real-valued function and, hence, we can define the (linear) subdifferential $\partial F_{y^*}(x)$. It turns out that the mapping $y^* \to \partial F_{y^*}(z)$ (if we take it to be equal to Y^* for $y^* \not\geq 0$) is a fan, having the additional property

$$\partial F_{y^* + v^*}(z) = \partial F_{y^*}(z) + \partial F_{v^*}(z), \quad y^*, v^* \geq 0,$$

which we shall call the *subdifferential* of F at z. There is no problem with extending all the formulae of the standard convex analysis of real-valued functions thanks to the "scalarization" procedure we used to define the subdifferentials.

A similar scalarization procedure is the key device in the nonsmooth analysis to be considered in the next section.

NONSMOOTH ANALYSIS[10,13]

Now we again approach the starting point of our research. Let X, Y be Banach spaces and $F: X \to Y$ Lipschitz spaces in a neighborhood of z X.

A bounded fan $\mathcal{A}$ from X into Y is a *prederivative* of F at z if

$$F(z + h) \subset F(z) + \mathcal{A}(h) + r(h) \|h\| B_Y,$$

where $r(h) \to 0$ if $\|H\| \to 0$. It is a *strict prederivative* of F at z if

$$F(x + h) \subset F(x) + \mathcal{A}(h) + r(x, h) \|h\| B_Y,$$

where $r(x, h) \to 0$ if $\|x - z\| \to 0$, $\|h\| \to 0$. (A mapping need not be Lipschitz to have a prederivative, but it must be Lipschitz if it has a strict prederivative.)

There is a natural one-parametric family of strict prederivatives. For any $\epsilon > 0$, consider the function

$$F^0_\epsilon(z; y^*, h) = \sup \{t^{-1} \langle y^*, F(x + th) - F(x) \rangle \mid t > 0,$$

$$\|x - z\| < \epsilon, \|x + th - z\| < \epsilon \}.$$

It turns out that this function is the support function of a bounded fan (in particular, it is convex in h), which we shall call the ϵ-*prederivative* of F at z and denote $F'_\epsilon(z)$.

The functions F^0_ϵ decrease as $\epsilon \downarrow 0$ and converge to what we shall denote by $F^0(z; y^*, h)$ and call the *upper directional derivative* of F at z. It may not be the support function of any bounded fan (it may fail to be weak* l.s.c. in y^*) but, if it is, then the corresponding fan is called the *upper derivative* of F at z and denoted $F'(z)$.

On the other hand, the fan

$$F'^*(z)(y^*) = \{x^* \mid F^0(z; y^*, h) \geq \langle x^*, h \rangle, \quad \forall h \in X$$

(adjoint to $F'(z)$ if it exists) is always well defined. It is called the *generalized gradient* of F at z (or perhaps it would be better to call it the *upper coderivative*, following Aubin[7]).

Here are several facts concerning these newly introduced notions.

1. A strict derivative (see Bourbaki[23]) is also a strict prederivative; a Fréchet derivative is also a prederivative.

2. If there is a strict prederivative with weakly compact values, then the upper derivative exists. This is true, in particular, if Y is a reflexive space.

3. If $\mathcal{A}$ is a strict prederivative, then $F'(z)(h) \subset \mathcal{A}(h)$ for all h. In other words, $F'(z)$ is the lower bound of the collection of all strict prederivatives.

4. If both X and Y are finite-dimensional, then the upper derivative is also a strict prederivative.

5. $F'^*(z)(y^*)$ is the Clarke generalized gradient of the real-valued function $x \to \langle y^*, F(x) \rangle$ at z. In particular, for a real-valued function f, we have $\partial f(z) = f'^*(z)(1)$.

6. The sum of (strict) prederivatives is a (strict) prederivative of the sum of the mappings; the composition of (strict) prederivatives is a (strict) prederivative of the composition of the mappings.

7. If $F: X \to Y$, $G: Y \to W$, and F has a strict prederivative at z with norm compact values (or if G has a strict prederivative at $F(z)$ whose adjoint has norm compact values), then $(G \circ F)'(z)(h) \subset (G'(F(z)) \circ F'(z))(h)$ for all h. This is true, in particular, if $\dim Y < \infty$.

8. If Y is a reflexive space, then any ϵ-prederivative and the upper derivative are generated by convex bounded sets of linear operators $X \to Y$.

9. The upper coderivative is generated by a convex bounded set of linear operators $Y^* \to X^*$ (none of which may be an adjoint to an operator from X into Y).

We conclude this section by stating extensions of main theorems of differential calculus (more general results are given in References 10 and 13).

INTERIOR POINT THEOREM 1. Let $\mathcal{A}$ be a strict prederivative of F at z. Then $F(z) \in \text{int } F(z + \delta B_X)$ for any $\delta > 0$ if any of the following three conditions is satisfied:

1. $C(\mathcal{A}) > \text{diam } \mathcal{A}$
2. Y has an equivalent norm with locally uniformly convex dual, $\mathcal{A}$ is weak-compact-valued, and $C(\mathcal{A}) > 0$
3. Y has an equivalent norm with strictly convex dual, $\mathcal{A}$ is norm-compact-valued, and $C(F'(z)) > 0$.

INVERSE MAPPING THEOREM. If, in addition to the assumptions of the preceding theorem, $C(\mathcal{A}^*) > 0$, then the restriction of F on a neighborhood of z is a homeomorphism and the inverse mapping is Lipschitz about $F(z)$.

Assume now that $X = U \times W$, $z = (u_0, w_0)$, and $\mathcal{A}$ is a strict prederivative of F at z. Then we can consider "partial" strict prederivatives $\mathcal{A}_u$ and $\mathcal{A}_w$ defined by $\mathcal{A}_u(u) = \mathcal{A}(u, 0)$ and $\mathcal{A}_w(w) = \mathcal{A}(0, w)$. Partial upper derivatives are defined similarly.

IMPLICIT FUNCTION THEOREM. Let $X = U \times W$, let $\mathcal{A}$ be a strict prederivative of F at $z = (u_0, w_0)$, let $C(\mathcal{A}_w^*) > 0$, and let one of the conditions of Interior Point Theorem 1 be satisfied for $\mathcal{A}_w$, $(F'(z))_w$, and Y. Then there are $\epsilon > 0$ and a Lipschitz mapping f from $u_0 + \epsilon B_U$ into W such that, for u, w satisfying $\|u - u_0\| < \epsilon$, $\|w - w_0\| < \epsilon$, relations

$$w = f(u) \quad \text{and} \quad F(u, w) = F(u_0, w_0)$$

are equivalent.

MEAN VALUE THEOREM. Let $F: X \to Y$ be Lipschitz in a neighborhood of the line segment $[x, x + h]$. Then, for any $y^* \in Y^*$, there are $\xi \in [0, 1]$ and $x^* \in F'^*(x + \xi h)(y^*)$ such that

$$\langle y^*, F(x + h) - F(x) \rangle = \langle x^*, h \rangle.$$

If, in addition, F has upper derivatives at any point of the line segment, then $F(x + h) - F(x)$ belongs to the closure of the convex hull of the union of $F'(x + \xi h)(h)$ ($\xi \in [0, 1]$).

If $\dim Y < \infty$, the assumptions of the theorems can be weakened.

INTERIOR POINT THEOREM 2. Let $\dim Y < \infty$ and let F be continuous in a neighborhood of $z \in X$. Let $\mathcal{A}$ be a prederivative of F at z (not necessarily strict) such that $C(\mathcal{A}) > 0$. Then $F(z) \in \text{int } F(z + \in B_X)$ for any $\in > 0$.

The latter result is closed for a theorem of Halkin.[24] Interrelations between prederivatives and approximations used by Halkin (and, by a remarkable coincidence, also called fans) were thoroughly discussed in Reference 13. Briefly, Halkin's fans are sets of linear operators that may be nonconvex—this is, to a certain extent, an advantage of Halkin's approach. On the other hand, if Halkin's fan is convex, it generates a prederivative, but the collection of all prederivatives is much richer. (To be more precise, in this comparison we considered, not prederivatives, but somewhat different objects called weak prederivatives.)

APPLICATION TO EXTREMAL PROBLEMS

To conclude, we shall quote, in a simplified form, a Lagrange multiplier rule proved in Reference 13. We shall consider the problem of minimizing $f_0(x)$, subject to

$$F(x) = 0, \quad f_i(x) \leq 0, \quad i = 1, \ldots n,$$

where F is a mapping from X into Y (X, Y being Banach spaces) and f_i, $i = 0, \ldots n$ are real-valued functions. Letting $z \, X$, we set

$$I = \{0\} \cup \{i \mid 1 \leq i \leq n, f_i(z) = 0\}.$$

The Lagrangian of the problem is

$$L(\lambda_0, \ldots \lambda_n, y^*, x) = \sum_{i=1}^{n} \lambda_i f_i(x) + \langle y^*, F(x) \rangle.$$

We shall denote by $\partial_x L(\lambda_0, \ldots \lambda_n, y^*, z)$ the Clarke generalized derivative of L (as a function of x) at z.

To state the result, one more notion is needed. Let $\mathcal{A}$ be a bounded fan from X into Y. We shall say that it has the finite condimension property if there is a projection $\pi: Y \to Y$ onto a closed subspace of finite codimension such that $C(\pi \circ \mathcal{A}) > 0$.

LAGRANGE MULTIPLIER RULE. Let F and f_i, $i = 0, \ldots n$ be Lipschitz in a neighborhood of Z. Assume that Y has an equivalent norm with strictly convex dual and that there is a norm-compact-valued strict prederivative of F at z having the finite codimension property.

If z is a local solution to this problem, then there are Lagrange multipliers, $\lambda_0, \ldots \lambda_n$, y^*, not all equal to zero, such that

$$0 \in \partial_x L(\lambda_0, \ldots \lambda_n, y^*, z) \qquad \lambda_i f_i(z) = 0, \quad i = 1, \ldots n, \qquad \lambda_i \geq 0, \quad i = 0, \ldots n.$$

This theorem seems to be the first general Lagrange multiplier rule for infinite-dimensional nonsmooth problems in mathematical programming. It extends earlier finite-dimensional results of Clarke[25] and Hiriart-Urruty[26] (up to nonfunctional constraints, which are, however, incorporated in Reference 13, where a more general result is proved).

Again, we want to point out how naturally the analysis of fans enters the proof of the theorem, which is but a mere reformulation of the fact that the Banach constant of the upper derivative of the mapping $G(x) = (f_0(x), \ldots f_n(x), F(x))$ at z equals zero. (In view of the interior point theorem, the latter is a necessary condition for z to be a local minimum.) Indeed, the condition of the Lagrange Multipler Rule means that

$$G^0(z; (\lambda_0, \ldots \lambda_n, y^*), h) \geq 0, \qquad \forall h,$$

so that

$$0 = C(G'(z)) = \max_{(\lambda_0, \ldots \lambda_n, y^*)} \inf_{\|h\| \leq 1} G^0.$$

The finite codimension property guarantees that the upper bound in the definition of the Banach constant is attained.

CONCLUSION

We hope that the results and discussions presented here are sufficient for the reader to conclude that fans are useful objects that naturally arise in different situations, have a wide range of applications, and are convenient and flexible to work with.

At the same time, one must understand, as far as applications to nonsmooth analysis are concerned, that the quality of approximation fans (or the generalized gradients of Clarke) provide is not always very high: being convex-valued, a strict prederivative may be too large to approximate a mapping closely. This is the price we pay for the very nice analytical properties of fans.

In the smooth calculus, all types of derivatives are naturally ordered so that, if, say, a mapping is strictly differentiable, the strict derivative is also the Fréchet derivative and the directional derivative and so on. In the nonsmooth case, we observe a different situation: a mapping may have different approximating objects not coinciding with each other and, ideally, the choice should be defined by convenience and purpose. As an analytical apparatus, fans are, perhaps, the best, but, if we do not need a sophisticated analysis to obtain the necessary result, other approximations may be preferable.

In particular, we can refer to Reference 7 for a demonstration of the usefulness and workability of contingent derivatives. They have been designed to substitute for derivatives (not strict) and they are better than prederivatives, at least in two respects: they provide for a better approximation and they are intrinsically defined by the mapping (while no such definition exists for prederivatives). On the other hand, their structure is far from being as regular as the structure of fans and it is rather unlikely that they can be applied to obtain, say, a kind of multiplier rule for constraint optimization problems.

Thus, the multiplicity of derivatives in nonsmooth analysis is more a positive factor than an indication of imperfection of our present knowledge, even if one admits (and we do) that there is still much to be understood and clarified.

References

1. CLARKE, F. H. 1975. Generalized gradients and their applications. Trans. Am. Math. Soc. **205**: 245–62.
2. HALKIN, H. 1976. Mathematical programming without differentiability. *In* Calculus of Variations and Control Theory. Academic Press. New York.
3. WARGA, J. 1976. Derivative containers, inverse functions and controllability. *In* Calculus of Variations and Opitmal Control. Academic Press. New York.
4. AUBIN, J. P. 1978. Gradients généralisées de Clarke. Ann. Sci. Math. Que. **2**: 197–254.
5. CLARKE, F. H. 1978. Nonsmooth analysis and optimization. Int. Congr. Mathematicians. Helsinki.
6. WARGA, J. 1981. Fat homeomorphisms and unbounded derivative containers. In press.
7. AUBIN, J. P. 1981. Contingent derivatives of set-valued maps and existence of solutions to nonlinear equations and differential inclusions. In press.
8. IOFFE, A. D. 1978. A new approach to nonsmooth analysis. Unpublished.
9. IOFFE, A. D. 1979. Différentielles généralisées d'applications localement Lipschitziennes d'un espace de Banach dans un autre. C. R. Acad. Sci. **289**: 637–39.
10. IOFFE, A. D. 1981. Nonsmooth analysis: differential calculus of nondifferentiable mappings. In press.
11. IOFFE, A. D. 1980. On foundations of convex analysis. Ann N. Y. Acad. Sci. **337**: 103–17.
12. IOFFE, A. D. 1981. A new proof of the equivalence of the Hahn-Banach extension and the least upper bound properties. In press.
13. IOFFE, A. D. 1981. Necessary conditions in nonsmooth optimization. In press.
14. IOFFE, A. D. 1979. Regular points of Lipschitz functions. Trans. Am. Math. Soc. **251**: 61–69.
15. EKELAND, I. 1974. On the variational principle. J. Math. Anal. Appl. **47**: 324–53.
16. EKELAND, I. 1979. Nonconvex minimization problems. Bull. Am. Math. Soc. **1**: 443–74.
17. KANTOROVICH, L. V. 1938. Sur la continuité et sur le prolongement des opérations linéares. C. R. Acad. Sci. **206**: 833–35.
18. NACHBIN, L. 1950. A theorem of the Hahn-Banach type for linear tranformations. Trans. Am. Math. Soc. **68**: 28–46.
19. BONNICE, W. & R. J. SILVERMAN. 1967. The Hahn-Banach extension and the least upper bound properties are equivalent. Proc. Am. Math. Soc. **18**: 843–50.
20. TO, T. O. 1971. The equivalence of the least upper bound property and the Hahn-Banach extension property, in ordered vector spaces. Proc. Am. Math. Soc. **30**: 287–96.

21. KUTATELADZE, S. S. 1981. Convex analysis in ordered moduli. In press.
22. KUTATELADZE, S. S. 1979. Convex operators. Usp. Mat. Nauk **34**(1): 167–96. (In Russian.)
23. BOURBAKI, N. Variétés différentielles et analytiques. Hermann. Paris.
24. HALKIN, H. 1978. Necessary conditions for optimal control problems with differentiable and nondifferentiable data. *In* Lecture Notes in Mathematics, Vol. 68. Springer-Verlag. Berlin.
25. CLARKE, F. H. 1976. A new approach to Lagrange multipliers. Math. Oper. Res. **1**: 165–74.
26. HIRIART-URRUTY, J. B. 1979. Refinements of necessary conditions in nondifferentiable programming. Appl. Math. Opt. **5**: 63–82.
27. WARGA, J. 1978. Controllability and a multiplier rule for nondifferentiable optimization problems. SIAM J. Control Optimization **16**: 803–12.

ON MULTIPLE REGRESSION FOR THE CASE WITH ERROR IN BOTH DEPENDENT AND INDEPENDENT VARIABLES

Viktor Brailovsky

Moscow, USSR

INTRODUCTION

As is well known, the least mean square (LMS) method is often used in processing experimental data. With the help of this method, the parameters of a linear model may be estimated and an approximation function may be found, provided that the values of independent variables are known precisely (without errors). If this is not the case, the LMS method leads to inconsistent estimates of the parameters of the model and other approaches must be considered.[1] The problems of fitting a straight line ($y = \alpha + \beta x$) and finding consistent parameter estimators when both variables are subject to error was considered in references 2–4. Parameter estimates may vary widely among these methods and some requirements cannot be easily satisfied. There are also difficulties in extending these methods to the case of multiple regression.[5] But their great advantage is that no assumption concerning the distribution of the independent variables is needed. Their other advantage is that all the parameters of the model, as well as the variances of the errors in both dependent and independent variables, may be estimated from experimental data and no additional assumption concerning their values is needed.

The other approach is based on the method of maximum likelihood. To apply this approach, one should know the probability distribution on the space of independent variables (see Chapter 9 of Reference 1) as well as some additional information concerning the parameters of the model. Such information may make it possible for one to find the values of the variances of the errors in both dependent and independent variables to a known precision (i.e., the precise ratio of any pair of these variances) or the values of the variances of the errors in independent variables only (no information concerning the variance in dependent variables is needed). A discussion of which requirement is preferable is given in Chapter 9 of Reference 1. The conclusion is that the nature of many practical problems is such that the latter requirement is the more adequate one.

The great disadvantage of the maximum likelihood approach in the case considered here is the requirement that the probability distribution on the space of the independent variables be known. This is an impossible requirement for the vast majority of practical problems. On the other hand, this approach can provide consistent estimates of good quality with the help of rather simple calculations.

In this article, an approach rather close to the LMS method is considered. Consistent estimates of unknown parameters of the model are obtained. This article considers, especially, the properties of approximation functions and estimators of approximation quality obtained in such a way. This study is performed under the assumption that a probability distribution on the space of the independent variables exists, but may be unknown, and that the variances of the errors in the independent variables are known. We will consider, below, the case where different observations have different errors and recommend how to process such data.

THE MODEL OF FUNCTION APPROXIMATION

Let R ($x \in R$) be a finite-dimensional space of arguments and let $f(x)$ be a function to be approximated. Let there be an observer who can observe the values of

$$f(x,\omega_0) = f(x) + \delta_0(\omega_i), \qquad (1)$$

instead of those of $f(x)$. $\delta_0(\omega_0)$ stands for error random variable with a mean of 0 and an unknown variance σ_0^2.

Let $\phi_1(x)$, $\phi_2(x)$, ... $\phi_n(x)$ be a set of n linear independent functions, such that

$$f(x) = \sum_{i=1}^{n} a_i \phi_i(x). \qquad (2)$$

In the model considered, one assumes that the observer can observe the values of

$$\phi_i(x_i, \omega_i) = \phi_i(x) + \delta_i(\omega_i), \qquad (3)$$

instead of those of $\phi_i(x)$. $\delta_i(\omega_i)$ stand for normally distributed error random variables with means of 0 and known variances σ_i^2: $N(0, \sigma_i^2)$, $i = 1, 2, \ldots n$.

The requirement that random variables, $\delta_i(\omega_i)$, be normally distributed is not very important. In our calculations below (i.e., **(19)**, **(21)**, and **(25)**), we shall use the facts that the third central moment of the random variables is zero and that the fourth central moment is $3\sigma_i^4$.[8] If their distribution has a different form, these formulae, as well as formulae derived therefrom, will have another form, but the approach remains the same.

Let $\mu(x)$ be a probability distribution on R such that the first and second moments of functions $\{\phi_i(x)\}$ exist with respect to the distribution $\mu(x)$.

Now let us move on to the problem of approximating the function $f(x)$, i.e., finding the values of unknown coefficients $\{a_i\}$. To begin, let us imagine for a moment that we know the following quantities:

$$P_{ij} = \int_R \phi_i(x) \phi_j(x) \, d\mu(x), \quad i,j = 1, 2, \ldots n \qquad (4)$$

and

$$C_j = \int_R f(x) \phi_j(x) \, d\mu(x), \quad j = 1, 2, \ldots n. \qquad (5)$$

After multiplying both sides of **(2)** by $\phi_j(x)$ and integrating over the space R with the probability distribution $\mu(x)$ from **(2)**, **(4)**, and **(5)**, it follows that

$$\sum_{i=1}^{n} a_i P_{ij} = C_j, \quad j = 1, 2, \ldots n. \qquad (6)$$

The matrix of this system is the Grammian matrix for the system of linear independent functions $\{\phi_i\}$ and, as a result,

$$d = \det \| p_{ij} \| > 0.$$

Using Cramer's rule, one obtains the solution of this system, which is

$$a_i = \frac{d_i}{d}, \quad i = 1, 2, \ldots n, \tag{7}$$

where d_i is the determinant obtained from the determinant d after replacing its ith column with the column of constant terms of the system of (6), $C_1, C_2, \ldots C_n$. From (2) and (7), it follows

$$f(x) = \sum_{i=1}^{n} \frac{d_i}{d} \phi_i(x). \tag{8}$$

One cannot, however, use this method to obtain $f(x)$, for lack of the necessary information. But one can estimate the values of p_{ij} and C_i by a sample set and, as a result, obtain estimates of the determinants d and d_i and the function $f(x)$.

One assumes that there is a sample set, as follows.

Let $x^1, x^2, \ldots x^N$ be N independently drawn samples from the space R, according to the probability law $\mu(x)$. One does not know all the values of x^i, but, for each x^i, one obtains $(n + 1)$ values $f(x^i, \omega_0^i), \phi_1(x^i, \omega_1^i), \phi_2(x^i, \omega_2^i), \ldots \phi_n(x^i, \omega_n^i)$ (see (1) and (3)), such that $x^i, \delta_0(\omega_0^i), \delta_1(\omega_1^i), \delta_2(\omega_2^i), \ldots \delta_n(\omega_n^i)$ are independent random variables. We shall call this sample set Q.

If $\sigma_1^2 = \sigma_2^2 = \cdots \sigma_n^2 = 0$, one can use estimators of the values of p_{ij} and C_j (see (4) and (5)), such as the following:

$$\hat{C}_j = \frac{1}{N} \sum_{i=1}^{N} f(x^i, \omega_0^i) \phi_j(x^i, \omega_j^i), \tag{9}$$

$$\hat{p}_{ij} = \frac{1}{N} \sum_{r=1}^{N} \phi_i(x^r, \omega_i^r) \phi_j(x^r, \omega_j^r), \quad i \neq j, \tag{10}$$

$$\hat{p}_{ii} = \frac{1}{N} \sum_{r=1}^{N} \phi_i^r(x^r, \omega_i^r). \tag{11}$$

It is easy to see that these conditions (9–11) are unbiased consistent estimators of the values of (4) and (5). Using these estimators instead of the corresponding values in (6–8), one obtains the LMS approximation of $f(x)$, which has well-known properties.[1,6,7]

If $\sigma_i^2 \neq 0$ $(i = 1, 2, \ldots n)$, it is easy to prove, with the help of direct calculations, that the estimators of (9) and (10) remain unbiased. But this is not so for the estimator of p_{ii} (11). Taking (3) into account, one obtains

$$E\hat{p}_{ii} = Eo_i^r \frac{1}{N} \sum_{r=1}^{N} E_x \phi_i^2(x^2, o_i)$$

$$= Eo_i^r [\phi_i(x^r) + 2\phi_i(x^r) E_x \delta_i(o_i^r) + E_x \delta_i^2(o_i^r)]$$

$$= p_{ii} + \sigma_i^2,$$

where E is an averaging operation over all possible sample sets Q and $E_{x'}$ ($E_{\omega_i^r}$) is an averaging operation for which $x'(\omega_i^r)$ is fixed.

Let us now consider the other estimator of the value of p_{ii},

$$\hat{p}_{ij} = \frac{1}{N} \sum_{r=1}^{N} \phi_i^2(x^r, \omega_i^r) - \sigma_i^2. \tag{12}$$

This estimator is unbiased.*

Thus we shall use the following method of approximation.

Let us use (9), (10), and (12) instead of the corresponding values of (4) and (5) in (6) and (7). We shall consider the following approximation function

$$\hat{f}(Q, x) = \sum_{i=1}^{n} \frac{\hat{d}_i}{\hat{d}} \phi_i(x) = \sum_{i=1}^{n} \hat{a}_i \phi_i(x), \tag{13}$$

where $\hat{d} = \det \| \hat{p}_{ij} \|$ and $\hat{d}_i$ is the determinant obtained from the determinant $\hat{d}$ after replacing its ith column with the column of estimates of constant terms $\hat{c}_1, \hat{c}_2, \ldots \hat{c}_n$. In the next section, some properties of the approximation quality of (13) will be investigated.

THE PROPERTIES OF COEFFICIENT ESTIMATES $\{\hat{a}_i\}$ AND THE CALCULATION OF THE APPROXIMATION QUALITY OF EQUATION 13

Introduction

In the last section, unbiased parameter estimators ((9), (10), and (22)) were introduced. It is easy to demonstrate that the variances and covariances of the corresponding estimates are on the order of $1/N$. These calculations will be performed in this section (see (17–27)). It follows that the estimators are consistent ones. As a result, the estimators $\{\hat{a}_i\}$ (see (13)) are consistent as well. In Appendix 1, we prove that the estimates of coefficients $\{\hat{a}_i\}$ are asymptotically normal, with a mean value of $\{a_i\}$. From the proof it is clear how the covariance matrix of this normal distribution can be obtained. We shall use it for further calculations.

Asymptotic Expression of Approximation Quality

We shall estimate the approximation quality of (13) as follows:

$$D = E \int_R [\hat{f}(Q, x) - f(x)]^2 \, d\mu(x), \tag{14}$$

where E is the average over all possible sample sets Q with sample size N. We shall be interested in estimating the asymptotic value of D, i.e., the value obtained from the

---

*Estimators of such a form were considered in Chapter 9 of Reference 1, where the method of maximum likelihood of solving the problem, provided that $\mu(x)$ is a normal probability distribution, was discussed.

asymptotic distribution for $\{\hat{a}_i\}$.† Equation 14 may be written in the following manner (see **(2)**, **(4)**, and **(13)**):

$$D = E \int_R [\hat{f}(Q, x) - f(x)]^2 \, d\mu(x)$$

$$= E \int_R \left[\sum_i (a_i - \hat{a}_i) \phi_i(x) \right]^2 d\mu(x) \quad (15)$$

$$= \sum_{i=1}^n p_{ii} \operatorname{var} \hat{a}_i + 2 \sum_{\substack{i,j=1 \\ i>j}}^n p_{ij} \operatorname{cov}(\hat{a}_i, \hat{a}_j).$$

From the results of Appendix 1 ((**A5**) and (**A6**)), it follows that variances and covariances from (**15**) may be expressed through variances and covariances of estimates $\hat{C}_i$ and $\hat{p}_{ij}$. As a result, after some algebra, the asymptotic expression for the approximation quality takes the form

$$D = \sum_{i=1}^n \left[\int_R \left(\frac{\partial \hat{f}}{\partial c_i} \right)^2 d\mu(x) \right] \operatorname{var} \hat{c}_i + \sum_{\substack{i,j=1 \\ i \geq j}}^n \left[\int_R \left(\frac{\partial \hat{f}}{\partial p_{ij}} \right)^2 d\mu(x) \right] \operatorname{var} \hat{p}_{ij}$$

$$+ 2 \sum_{\substack{i,j=1 \\ i>j}}^n \left[\int_R \frac{\partial \hat{f}}{\partial c_i} \frac{\partial \hat{f}}{\partial c_j} d\mu(x) \right] \operatorname{cov}(\hat{c}_i, \hat{c}_j)$$

$$+ 2 \sum_{\substack{i,j,k,r=1 \\ i \geq j, \, k \geq r, \, [(i>k) \cup (i=k)(j>r)]}}^n \left[\int_R \frac{\partial \hat{f}}{\partial p_{ij}} \frac{\partial \hat{f}}{\partial p_{kr}} d\mu(x) \right] \operatorname{cov}(\hat{p}_{ij}, \hat{p}_{kr}) \quad (16)$$

$$+ 2 \sum_{\substack{i,k,r=1 \\ k \geq r}}^n \left[\int_R \frac{\partial \hat{f}}{\partial c_i} \frac{\partial \hat{f}}{\partial p_{kr}} d\mu(x) \right] \operatorname{cov}(\hat{c}_i, \hat{p}_{kr}) + \bar{O}\left(\frac{1}{N} \right).$$

All indices, i, j, k, and r, vary from 1 to n, special restrictions being indicated under each summation symbol. The restrictions on the fourth term on the right-hand side of (**16**) are as follows: $i \geq j$, $k \geq r$, and either $i > k$ or $i = k$ and $j > r$. All partial derivatives in (**16**) are to be calculated at the point $\hat{c}_i = c_i$, $\hat{p}_{ij} = p_{ij}$ ($i, j = 1, 2, \ldots n$).

Calculation of Equation 16

Let us now calculate equation 16. To begin, let us calculate the expressions for variances and covariances. From (**1–5**), (**9**), (**10**), and (**12**), it follows that

---

†We are interested here in estimates of D obtained from asymptotic distributions for $\{\hat{a}_i\}$ because of the fact that, in the model considered, D may not exist for any finite N. The cause of this phenomenon is the singularity of $\hat{f}(Q, x)$, but the probability that the samples are drawn from the neighborhood of a singularity converges exponentially to zero as $N \to \infty$ (see Appendix 1). As is usual under such circumstances, we are interested in estimates obtained from asymptotic distributions.

$$\text{var } \hat{c}_j = E\left\{\left[\frac{1}{N}\sum_{i=1}^{N} f(x^i, \omega_0^i)\phi_j(x^i, o_j^i)\right] - c_j\right\}^2$$

$$= \frac{1}{N}\int_R \int_{\Omega_0} \int_{\Omega_j} [f(x, \omega_0)\phi_j(x, \omega_j) - c_j]^2 \, d\mu(x) \, d\mu(\omega_0) \, d\mu(\omega_j) \quad (17)$$

$$= \frac{1}{N}\int_R \int_{\Omega_0} \int_{\Omega_j} \left\{\left[\sum_i a_i \psi_i(x) + \delta_D(\omega_0)\right][\phi_j(x) + \delta_j(\omega_j)] - c_j\right\}^2 d\mu(x) d\mu(\omega_0) d\mu(\omega_j),$$

where Ω_0, Ω_j, $\mu(\omega_0)$, and $\mu(\omega_j)$ are the spaces and the corresponding probability distributions associated with random variables ω_0 and ω_j, respectively. Taking (6) into account, one obtains, after some algebra,

$$\text{var } \hat{c}_j = \frac{1}{N}\left\{\left[\sum_{i=1}^{n} a_i^2 \text{ var }(\phi_i\phi_j) + 2\sum_{\substack{k,m=1 \\ k>m}}^{n} a_k a_m \text{ cov }(\phi_k\phi_j, \phi_m\phi_j) + p_{jj}\sigma_0^2\right]\right.$$

$$\left. + \sigma_j^2 \sum_{k,m=1}^{n} a_k a_m p_{km} + \sigma_0^2 \sigma_j^2 \right\}.$$

In the same way, one obtains

$$\text{var } \hat{p}_{rs} = \frac{1}{N}\{[\text{var }(\phi_r, \phi_s)] + \sigma_r^2 p_{ss} + \sigma_s^2 p_{rr} + \sigma_r^2 \sigma_s^2\}, \quad r \neq s, \quad (18)$$

$$\text{var } \hat{p}_{rr} = \frac{1}{N}\{[\text{var }(\phi_r^2)] + 4\sigma_r^2 p_{rr} + 2\sigma_r^4\}. \quad (19)$$

In calculating this expression, we assumed that the random variable $\delta_r(\omega_r)$ is normally distributed and, therefore, that its third central moment equals zero and its fourth central moment equals $3\sigma_r^4$.

$$\text{cov }(\hat{c}_i, \hat{c}_j) = \frac{1}{N}\left[a_i a_j \text{ var }(\phi_i\phi_j) + \sum_{\substack{m,l=1 \\ m\neq j \text{ or } l\neq i}}^{n} a_m a_l \text{ cov }(\phi_m\phi_i, \phi_l\phi_j) + \sigma_0^2 p_{ij}\right]. \quad (20)$$

The indices m and l under the summation symbol on the right-hand side of (20) vary from 1 to n and either m is not equal to j or l is not equal to i.

$$\text{cov }(\hat{p}_{rr}, \hat{p}_{rs}) = \frac{1}{N}\{[\text{cov }(\phi_r^2, \phi_r\phi_s)] + 2p_{rs}\sigma_r^2\}, \quad s \neq r, \quad (21)$$

$$\text{cov }(\hat{p}_{rr}, \hat{p}_{ij}) = \frac{1}{N}[\text{cov }(\phi_r^2, \phi_i\phi_j)], \quad r \neq i, r \neq j, \quad (22)$$

$$\text{cov}\,(\hat{p}_{ij}, \hat{p}_{ik}) = \frac{1}{N}\{[\text{cov}\,(\phi_i\phi_j, \phi_i\phi_k) + \sigma_i^2 p_{jk}\}, \qquad i \neq j, i \neq k, j \neq k, \quad (23)$$

$$\text{cov}\,(\hat{p}_{ij}, \hat{p}_{lk}) = \frac{1}{N}[\text{cov}\,(\phi_i\phi_j, \phi_k\phi_l)], \qquad i \neq j \neq k \neq l, \quad (24)$$

$$\text{cov}\,(\hat{c}_i, \hat{p}_{ii}) = \frac{1}{N}\left\{\left[\sum_{m=1}^{n} a_m \text{cov}\,(\phi_m\phi_i, \phi_i^2)\right] + 2\sigma_i^2 \sum_{m=1}^{n} a_m p_{im}\right\}, \quad (25)$$

$$\text{cov}\,(\hat{c}_i, \hat{p}_{ir}) = \frac{1}{N}\left\{\left[\sum_{m=1}^{n} a_m \text{cov}\,(\phi_m\phi_i, \phi_i\phi_r)\right] + \sigma_i^2 \sum_{m=1}^{n} a_m p_{rm}\right\}, \qquad i \neq r. \quad (26)$$

$$\text{cov}\,(\hat{c}_i, \hat{p}_{kl}) = \frac{1}{N}\left[\sum_{m=1}^{n} a_m \text{cov}\,(\phi_m\phi_i, \phi_k\phi_l)\right], \qquad i \neq k, i \neq l. \quad (27)$$

In calculating (21) and (25), we used the fact that the third central moments of the random variables, $\delta_r(\omega_r)$ ($r = 1, 2, \ldots n$), equal zero.

Let us return to our calculation of the value of D (16). The approximation function (13) depends on the values of $\hat{c}_i$ and $\hat{p}_{ij}$ in the same way an LMS approximation function does.[7] That is why both functions have the same partial derivatives with respect to these arguments and the same integral expressions in square brackets on the right-hand side of (16). These expressions are calculated in Appendices 3 and 4 of Reference 7. Results of these calculations are presented in Appendix 2 of this paper.

Comparing the expressions for variances and covariances (17–27) with those for the LMS approximation,[7] one can see that they coincide for (20), (22), (24), and (27). We have indicated variance and covariance by square brackets. For (17–19), (21), (23), (25), and (26), the expressions in square brackets coincide with corresponding expressions for the case of LMS approximation,[7] but all the equations have some additional terms. Therefore, multiplying the expressions in square brackets in (17–27) by their respective coefficients (see the expressions in square brackets in (16) and the corresponding formulae in Appendix 2) and summing up the results, one obtains the expression for the asymptotic estimation of the LMS approximation quality of a sample of size N. Such an expression was obtained in Reference 7 (see, especially, equations (3–14) of Reference 7) and may be written as follows:

$$D_{\text{LMS}} = \frac{n\sigma_0^2}{N} + \bar{\bar{O}}\,(1/N). \quad (28)$$

As is clear from the previous discussion and (16), in this case,

$$D = D_{\text{LMS}} + \Delta D + \bar{\bar{O}}\left(\frac{1}{N}\right). \quad (29)$$

To calculate ΔD, one must multiply the additional terms (the expressions outside the square brackets) of (17–19), (22), (23), (25), and (26) by their respective coefficients (see (16) and Appendix 2) and sum up the results. After some algebra, one obtains

$$D = D_{\text{LMS}} + \Delta D + \bar{\bar{O}}\left(\frac{1}{N}\right)$$

$$= \frac{n\sigma_0^2}{N} + \frac{n\left(\sum_{j=1}^{n} \sigma_j^2 a_j^2\right)}{N} + \frac{1}{Nd}\left[\sigma_0^2\left(\sum_{j=1}^{n} \sigma_j^2 A_{jj}\right)\right] \qquad (30)$$

$$+ \sum_{r,s=1}^{n} A_{rs} a_r a_s \sigma_r^2 \sigma_s^2$$

$$+ \left(\sum_{j=1}^{n} \sigma_j^2 A_{jj}\right)\left(\sum_{r=1}^{n} \sigma_r^2 a_r^2\right)\right] + \bar{\bar{O}}\left(\frac{1}{N}\right),$$

A_{rs} is the cofactor of the element p_{rs} of the matrix $\|p_{ij}\|$. Taking into account the fact that both the matrix $\|p_{ij}\|$ and its inverse $\|A_{ij}/d\|$ are positively defined, one sees that all the components of the term ΔD are non-negative.

It is interesting to note that if one assumes that the values of experimental errors (i.e., the values of $\sigma_0^2, \sigma_1^2, \ldots \sigma_n^2$) are small enough that the terms of order σ^4 in the expression for D (30) can be neglected, then this expression takes the following form:

$$D \simeq \frac{n\sigma_0^2}{N} + \frac{n\left(\sum_{j=1}^{n} \sigma_j^2 a_j^2\right)}{N}. \qquad (31)$$

This expression does not depend on either the form or the parameters of the probability distribution $\mu(x)$ in the space R and, therefore, it gives a distribution-free asymptotic estimate of the approximation quality of (23).

A Remark

For some purposes, it is helpful to have an estimate of unknown variances, σ_0^2. Let us consider the following expression:

$$\hat{\sigma}_0^2 = \frac{1}{N}\sum_{r=1}^{N}[f(x^r, \omega_0^r) - \overline{f(x, \omega_0)}]^2$$

$$- \sum_{r=1}^{N} \hat{a}_i \left\{\frac{1}{N}\sum_{r=1}^{N}[f(x^r, \omega_0^r) - \overline{f(x, \omega_0)}] \right. \qquad (32)$$

$$\left. - [\phi_i(x^r, \omega_i^r) - \overline{\phi_i(x, \omega_i)}]\right\},$$

where

$$\overline{f(x, \omega_0)} = \frac{1}{N}\sum_{r=1}^{N} f(x^r, \omega_0^r) \quad \text{and} \quad \overline{\phi_i(x, \omega_i)} = \frac{1}{N}\sum_{r \neq 1}^{N} \phi_i(x^r, \omega_i^r).$$

It is easy to show that the first component on the right-hand side of (32) is an unbiased consistent estimator of the expression

$$(\sigma_0^2 + \sum_i a_i^2 \text{Var}(\phi_i^2) + 2 \sum_{i>j} a_i a_j \text{cov}(\phi_i, \phi_j).$$

The expression inside the braces in (32) is an unbiased consistent estimator of the formula

$$\left[a_i \text{var}(\phi_i^2) + \sum_{\substack{j \\ (j=i)}} a_j \text{cov}(\phi_i, \phi_j) \right].$$

Taking into account the fact that $\hat{a}_i$, is a consistent estimator of a_i one realizes that (32) gives a consistent estimator of the unknown parameter σ_1^2.

ON APPROXIMATIONS BASED UPON THE USE OF A MIXTURE OF DATA WITH DIFFERENT ERRORS IN DIFFERENT OBSERVATIONS

Introduction

While processing experimental data, it may happen that one obtains various values of the same variable $f(x, \omega_0), \phi_1(x, \omega_1), \ldots \phi_N(x, \omega_n)$ with different errors in different observations. To study how to process such data, let us consider the following addition to the model described above. Let there be a space of possible values of variances $\sigma_0^2, \sigma_1^2, \ldots \sigma_n^2$, with a probability distribution $\mu(\sigma_0^2, \sigma_1^2, \ldots \sigma_n^2)$, which may be unknown. For each sample x^r drawn from R, independently draw a vector $\sigma_{0r}^2, \sigma_{1r}^2, \ldots \sigma_{nr}^2$, according to the distribution $\mu(\sigma_0^2, \sigma_1^2, \ldots \sigma_n^2)$. Using this vector of variances, one draws $(n + 1)$ values of independent random variables $\delta_0(\sigma_{0r}^2, \omega_0^r), \delta_1(\sigma_{1r}^2, \omega_1^r), \ldots \delta_n(\sigma_{Nr}^2, \omega_n^r)$;‡ the observer then obtains vectors of experimental data (observations) $f(x^r, \omega_0^r), \phi_1(x^r, \omega_1^r), \ldots \phi_n(x^r, \omega_n^r)$, which are connected with the aforementioned values of independent random variables, $\delta_i(\sigma_{ir}^2, \omega_i^r)$ ($i = 0, 1, 2, \ldots n$), with the help of (1) and (3). We assume that, for any vector of observations, the values of $\sigma_{1r}^2, \sigma_{2r}^2, \ldots \sigma_{nr}^2$ are known.

A First Method of Processing Heterogeneous Data

It is easy to show that, for the model considered, the unbiased consistent estimators of the values of c_j, p_{ij} ($i = j$), and p_{ij} (14) and (5) have the following forms:

$$\hat{c}_j = \frac{1}{N} \sum_{r=1}^N f(x^r, \omega_0^r) \phi_j(x^r, \omega_j^r), \tag{33}$$

$$\hat{p}_{ij} = \frac{1}{N} \sum_{r=1}^N \phi_i(x^r, \omega_i^r) \phi_j(x^r, \omega_j^r)_j \quad i \neq j \tag{34}$$

‡Of course, these values are unknown to the observer.

$$\hat{p}_{ii} = \frac{1}{N} \sum_{r=1}^{N} [\phi_i^2(x^r, w_i^r) - \sigma_{ir}^2]. \tag{35}$$

Let us consider the approximation function in the form of (13), where the determinants d and d_i are based on the estimates above (33–35). After calculations similar to the ones performed in the previous section, we obtain an expression for asymptotic approximation quality (see 14–26):

$$D_1 = \frac{n}{N}\overline{(\sigma_0^2)} + \frac{n}{N}\sum_{j=1}^{n} \overline{a_j^2(\sigma_j^2)} + \frac{1}{Nd}\left[\sum_{j=1}^{n} A_{jj}\overline{(\sigma_0^2 \sigma_j^2)} + \sum_{r,s=1}^{n} A_{rs}a_r a_s \overline{(\sigma_r^2 \sigma_s^2)}\right.$$
$$\left. + \sum_{j,r=1}^{n} A_{jj}a_r^2\overline{(\sigma_j^2 \sigma_r^2)}\right] + \overline{\overline{O}}\left(\frac{1}{N}\right). \tag{36}$$

Let us note that, in calculating (36), we also averaged over the space of the possible values of $\sigma_0^2, \sigma_1^2, \ldots, \sigma_n^2$, with the probability distribution $\mu(\sigma_0^2, \sigma_1^2, \ldots \sigma_n^2)$; $\overline{(\sigma_j^2)}$ and $\overline{(\sigma_j^2 \cdot \sigma_s^2)}$ ($j, s \pm 0, 1, 2, \ldots n$) are the expected values of the corresponding quantities with respect to this distribution. Of course, (36) is the direct generalization of (30). Just as was done for the latter formula, let us assume that the values $\sigma_0^2, \sigma_1^2, \ldots \sigma_n^2$ are small enough that the terms of order σ^4 in (36) can be neglected. Then (36) takes the following form:

$$D_1 \simeq \frac{n}{N}\left[\overline{(\sigma_0^2)} + \sum_{j=1}^{n} \overline{a_j^2(\sigma_j^2)}\right] = \frac{n}{N}\overline{\left(\sum^2\right)}. \tag{37}$$

Using this assumption and (37), it is easy to show that the method of processing heterogeneous data considered (i.e., data with different errors in different observations) leads to undesirable results—for instance, when there are two homogeneous sample sets and the first has variances σ_i^2 less than those of the second. It is easy to demonstrate that this relation leads to the conclusion that the use of the first sample set only (the approximation quality of which was calculated by (31)) gives a better result than the use of both sample sets (the approximation quality of which were calculated by (37)).

This means that this method of processing data is not the optimal one.

A Second Method of Processing Heterogeneous Data

Let us consider another approach to heterogeneous data processing. Let us imagine for a moment that, for any observation, one knows σ_{0r}^2 ($r = 1, 2, \ldots N$) and the values of coefficients $\{a_i\}$. (Below, of course, one will use estimates instead.) Let us consider the following estimates, where each observation has weight

$$\omega_r = \left(\sigma_{0r}^2 + \sum_{i=1}^{n} a_i^2 \sigma_{ir}^2\right)^{-1}$$

$$\hat{c}_j^* = \frac{1}{N}\sum_{r=1}^{N}\left(\sigma_{0r}^2 + \sum_{i=1}^{n} a_i^2 \sigma_{ir}^2\right)^{-1} f(x^r, w_0^r)\phi_j(x^r, w_j^r), \tag{38}$$

$$\hat{p}_{ij}^* = \frac{1}{N} \sum_{r=1}^{N} \left(\sigma_{0r}^2 + \sum_{l=1}^{n} a_l^2 \sigma_{lr}^2 \right)^{-1} \phi_i(x^r, w_i^r) \phi_j(x^r, w_j^r), \qquad i \neq j, \qquad (39)$$

$$\hat{p}_{ii}^* = \frac{1}{N} \sum_{r=1}^{N} \left(\sigma_{0r}^2 + \sum_{l=1}^{n} a_l^2 \sigma_{lr}^2 \right)^{-1} [\phi_i^2(x^r, w_i^r) - \sigma_{ir}^2]. \qquad (40)$$

Let us now consider the approximation function of the form of (13), with the determinants $\hat{d}$, $\hat{d}_i$ based on (38–40). With our previous assumption that the values of $\sigma_0^2, \sigma_1^2, \ldots \sigma_n^2$ are small enough, it is possible to show that the formula for the asymptotic approximation quality takes the form

$$D^* \cong \frac{n}{N} \overline{\left[\left(\sigma_0^2 + \sum_{i=1}^{n} a_i^2 \sigma_i^2 \right)^{-1} \right]^{-1}} = \frac{n}{N} \left(\overline{\left(\sum^2 \right)^{-1}} \right)^{-1}, \qquad (41)$$

where

$$\overline{\left(\sum^2 \right)^{-1}} = \overline{\left(\sigma_0^2 + \sum_{i=1}^{n} a_i^2 \sigma_i^2 \right)^{-1}}$$

stands for the expected value of

$$\left(\sigma_0^2 + \sum_{i=1}^{n} a_i^2 \sigma_i^2 \right)^{-1}.$$

Comparing (41) with (37), we can show that D^* is always less than or equal to D_1 (they are equal for the case of homogeneous sample sets) and using any kind of heterogeneous data is better than using only some part of the data. So no undesirable effect similar to one described above is possible for this method of processing data.

But now the problem is determining the values of σ_0^2 and the coefficients $\{a_i\}$. The value of σ_0^2 may be known a priori and, if so, one may use some estimate of it. Namely, the heterogeneous sample set often consists of a number of homogeneous parts; estimates of σ_0^2 may be performed for each of the parts separately with (32). The values of coefficients $\{a_i\}$ can be estimated by calculations performed either with the use of some homogeneous part of the sample set or with the use of the whole sample set by the first method mentioned in this section.

Of course, one may use an iterative procedure, the values of coefficients $\{a_i\}$ obtained in the previous step being used in the next step for a more precise determination of the weight coefficients in (38–40).

Some numerical experiments performed by the author showed that the procedure described here is not very sensitive to the precision of the determination of $\{a_i\}$ and σ_0^2; using the aforementioned estimates of them instead of their true values in (38–40) may lead to an approximation function, the quality of which is close to that of (41).

In conclusion, it is important to emphasize once more that all the results reported in this article are asymptotic and valid in large sample sets only.

REFERENCES

1. JOHNSTON, J. Econometric methods, 2nd Edit. McGraw-Hill, Kogakusha Ltd. Tokyo.
2. WALD, A. 1940. The fitting of straight lines if both variables are subject to error. Ann. Math. Stat. **11**: 284–300.

3. BARTLETT, M. S. 1949. Fitting a straight line when both variables are subject to error. Biometrics **5:** 207–12.
4. DURBIN, J. 1954. Errors in variables. Rev. Int. Stat. Inst. **22:** 23–32.
5. HOOPER, J. W. & H. THEIL. 1958. The extension of Wald's method of fitting straight lines to multiple regression. Rev. Int. Sta. Inst. **26:** 37–47.
6. VAPNIK V. N. 1979. Restoration of Dependencies by Empirical Data. Nauka, Moscow. (In Russian).
7. BRAILOVSKY, V. 1982. Comparative analysis of some procedures of function approximation, based on use of sample data. In press.
8. CRAMER, H. 1946. Mathematical Methods of Statistics. Princeton University Press, Princeton, N.J.

APPENDIX 1

Let us consider the space of estimates $\hat{c}_i$ ($i = 1, 2, \ldots n$) and $\hat{p}_{ij}$ (one assumes $i \geq j$ for $\hat{p}_{ij} = \hat{p}_{ji}$ [see (**10**) and (**12**)], as well as $p_{ij} = p_{ji}$ [see (**4**)]). The total dimension of the space equals $2n + n(n - 1)/2$. Let us consider a vector of estimates $\hat{a}$ (see (**13**); $\hat{a}_i = \hat{d}_i/\hat{d}^2, i = 1, 2, \ldots n$) as a function on the space. Let c_i, p_{ij} be the point corresponding to the true values of the respective parameters of (**5**) and (**4**). Further, we shall consider the value $\hat{a}$ in the form of an expansion near the center point c_i, p_{ij}.

To simplify further expressions let us denote

$$y_i = c_i, \quad i = 1, 2, \ldots n; \qquad y_{n+1} = p_{11}, y_{n+2} = p_{21}, \ldots y_{2n} = p_{n1},$$
$$y_{2n+1} = p_{22}, y_{2n+2} = p_{32}, \ldots y_{3n-1} = p_{n2}, \qquad y_{2n+n(n-1)/2} = p_{nn}.$$
(A1)

The corresponding estimates will be denoted $\hat{y}_i$.

Let us consider the expansion for the value of $\hat{a}$ in the neighborhood of the center point y. From (**7**), (**13**), and (**A1**), it is clear that $\hat{a}(y) = a$.

$$\hat{a}(\hat{y}) = a + \sum_{k=1}^{2n+n(n-1)/2} \frac{\partial \hat{a}}{\partial y_k} (\hat{y}_k - y_k) + \frac{1}{2} \sum_{k,l=1}^{2n+n(n-1)/2} \frac{\partial^2 \hat{a}}{\partial \hat{y}_k \partial \hat{y}_l} (\hat{y}_k - y_k)(\hat{y}_l - y_l). \quad (A2)$$

The partial derivatives of the first order are calculated at the center point. The point at which the partial derivative of the second order (in the remainder) is calculated depends upon the sample set.

Let us first consider the second component on the right side of (**A2**).

Equations **9, 10,** and **12,** with the use of (**A1**), take the form

$$\hat{y}_k = \frac{1}{N} \sum_{j=1}^{N} \hat{y}_k^j, \quad (A3)$$

$$\hat{y}_k - y_k = \frac{1}{N} \sum_{j=1}^{N} (\hat{y}_k^j - y_k). \quad (A4)$$

From (**A3**) and (**A4**), it follows that, for the second component of (**A2**),

$$\sum_{1} = \sum_{k} \frac{\partial \hat{a}}{\partial y_k} (\hat{y}_k - y_k) = \frac{1}{N} \sum_{j=1}^{N} \sum_{k=1}^{2n+n(n-1)/2} \frac{\partial \hat{a}_k}{\partial y_k} (\hat{y}_k^j - y_k). \quad (A5)$$

The values of the partial derivatives of the first order $\partial \hat{a}/\partial \hat{y}_k$ are calculated in

Appendix 5 of Reference 7. It is clear that, if $d > 0$ (which is true for the center point), the values of these partial derivatives are limited and the right-hand side of (**A5**) is the sum of N independent, equally distributed random variables with means of 0 and finite moments of the second order. So this expression is asymptotically normal and its covariance matrix can be calculated.

Let us consider now the last component of the expression on the right-hand side of (**A2**), i.e., the remainder.

It is easy to see that all singularities of partial derivatives of the second order are in the region of the space of estimates considered, where $\hat{d} = 0$.

On the other hand, it follows from (**9**), (**10**), (**12**), (**A1**), (**A3**), and (**A4**) that the random variable $(\hat{y} - \hat{y})$ with $2n + n(n - 1)/2$ components is distributed asymptotically normally with a mean of 0 and a covariance matrix, each component of which is on the order of $1/N$. The components of the matrix have been calculated above.

Let us consider an ellipsoid, the center of which is at the center point c_i, p_{ij} and the parameters of which correspond to components of the covariance matrix, i.e., an ellipsoid of equal probability for this asymptotically normal distribution such that the region $\hat{d} = 0$ is completely outside the ellipsoid. This is possible because, at the center point, $d > 0$. Let us denote the region inside the ellipsoid (including the boundary surface) by M. From Reference 8, it is easy to see that, under the conditions considered, the probability of obtaining estimates lying outside region M equals $e^{-a^2 N}$, where a^2 is a positive constant. So, taking (**A4**) into account, let us transform (**A2**) into the following form:

$$(\hat{a}(\hat{y}) - a)\sqrt{N} = \frac{1}{\sqrt{N}} \sum_{j=1}^{N} \sum_{k} \frac{\partial \hat{a}}{\partial y_k} (\hat{y}_k^j - y_k) + \frac{1}{N^{3/2}} \sum_{k,l} \frac{\partial^2 \hat{a}}{\partial \hat{y}_k \partial \hat{y}_l}$$

$$- \left[\sum_{i=1}^{N} (\hat{y}_i^i - y_k) \right] \left[\sum_{j=1}^{N} (\hat{y}_l^j - y_l) \right]. \quad (A6)$$

Let us prove that the probability of the second component on the right-hand side of (**A6**) converges to zero. Let us choose $E > 0$ and find N_0 such that, for any $N > N_0$, the probability of obtaining estimates $\hat{c}_i$, $\hat{p}_{ij}$ lying outside M is $e^{-a^2 N} < E$. Let us consider estimates inside region M. In this case, partial derivatives of the second order must be calculated at a point inside M. Inside M, all partial derivatives of the second order are continuous and limited. Let us estimate the average value of the second component (which is denoted $\sqrt{N} \cdot R$) inside M:

$$E(\sqrt{N} \cdot R) \leq \frac{1}{N^{3/2}} \sum_{k,l} \left. \frac{\partial^2 \hat{a}}{\partial \hat{y}_k \partial \hat{y}_l} \right|_{\max} \cdot E \left\{ \left[\sum_{i=1}^{N} (\hat{y}_k^i - y_k) \right] \left[\sum_{j=1}^{N} (\hat{y}_l^j - y_l) \right] \right\}, \quad (A7)$$

where

$$\left. \frac{\partial^2 \hat{a}}{\partial \hat{y}_k \partial \hat{y}_l} \right|_{\max}$$

is the maximum of the respective partial derivative inside M.

From the Schwartz inequality, it follows that

$$\left| E\left[\sum_{i=1}^{N}(\hat{y}_k^i - y_k)\right]\left[\sum_{j=1}^{N}(\hat{y}_l^j - y_l)\right] \right| \le \sqrt{E\left[\sum_{i=1}^{N}(\hat{y}_k^i - y_k)^2\right]} \times \sqrt{E\left[\sum_{j=1}^{N}(\hat{y}_l^j - y_l)^2\right]} = N\sigma_k^* \sigma_l^* \quad (A8)$$

Here

$$\sigma_k^* = \sqrt{E(\hat{y}_k^i - y_k)^2} \quad \text{for any } i = 1, 2, \ldots N \quad (A9)$$

Taking into (A8) and (A9) account, one obtains the following estimate for (A7):

$$|E(\sqrt{N} \cdot R)| \le \frac{C_1}{N^{1/2}}, \quad (A10)$$

where C_1 is a positive constant.

With the help of the same kind of technique, one obtains, for the variance of the second component inside M, the following estimate:

$$\text{Var}(\sqrt{N} \cdot R) \le \frac{C_2}{N}. \quad (A11)$$

Therefore

$$P(|\sqrt{N} \cdot R| \ge E) \le P(|\sqrt{N} \cdot R| \ge E, \hat{y} \in M) + P(\hat{y} \overline{\in} M)$$

$$\le P\{|\sqrt{N} \cdot R - E(\sqrt{N} \cdot R) + E(\sqrt{N} \cdot R)| \ge E, \hat{y} \in M\} + E. \quad (A12)$$

Taking (A1) into account and choosing $C_1/\sqrt{N} < E/2$, i.e., $N > 4C_1^2/E^2$, one obtains

$$P(|\sqrt{N} \cdot R| \ge E) \le P\left(|\sqrt{N} \cdot R - E(\sqrt{N} \cdot R)| \ge \frac{E}{2}\right) + \frac{3E}{2}. \quad (A12)$$

From Chebyshev's inequality and (A10) and (A11), one obtains the following estimate for the expression of interest:

$$P\left\{|\sqrt{N} \cdot R - E(\sqrt{N} \cdot R)| \ge \frac{E}{2}\right\} \le \frac{4C_2^2}{N^2 E^2}. \quad (A13)$$

It follows that the term $\sqrt{N} \cdot R$ converges in probability to zero.

We have considered here a sketch of the proof rather than the detailed proof itself; however, all the details may be added through the usual techniques.

APPENDIX 2

The formulae for partial derivatives $\partial \hat{f}/\partial c_i$ ($i = 1, 2, \ldots n$), $\partial \hat{f}/\partial p_{jj}$ ($j = 1, 2, \ldots n$), and $\partial \hat{f}/\partial p_{ij}$ ($i, j = 1, 2, \ldots n, i \ne j$) are (see (16), which was rewritten from

Reference 7)

$$\frac{\partial \hat{f}}{\partial c_i} = \frac{1}{d} \sum_{j=1}^{n} A_{ij} \phi_j(x),$$

$$\frac{\partial \hat{f}}{\partial p_{jj}} = -\frac{a_j}{d} \sum_{i=1}^{n} A_{ji} \phi_i(x),$$

$$\frac{\partial \hat{f}}{\partial p_{ij}} = -\sum_{k=1}^{n} \frac{1}{d} (a_j A_{ik} + a_i A_{jk}) \phi_k(x), \qquad i \neq j,$$

where A_{ij} is the cofactor of element p_{ij} in the matrix $\|p_{ij}\|$ (see (**4**)).

The expressions for the coefficients of the variance and covariance terms in (**16**) are

$$k_i = \int_R \left(\frac{\partial \hat{f}}{\partial c_i}\right)^2 d\mu(x) = \frac{A_{ii}}{d},$$

$$k_{rr} = \int_R \left(\frac{\partial \hat{f}}{\partial p_{rr}}\right)^2 d\mu(x) = \frac{a_r^2 A_{rr}}{d},$$

$$k_{jk} = \int_R \left(\frac{\partial \hat{f}}{\partial p_{jk}}\right)^2 d\mu(x) = \frac{1}{d}(a_j^2 A_{kk} + a_k^2 A_{jj} + 2 a_j a_k A_{jk}), \qquad j \neq k,$$

$$c_{ik} = 2 \int_R \frac{\partial \hat{f}}{\partial c_i} \frac{\partial \hat{f}}{\partial c_k} d\mu(x) = 2 \frac{A_{ik}}{d}, \qquad i \neq k,$$

$$c_{rr,ss} = 2 \int_R \frac{\partial \hat{f}}{\partial p_{rr}} \frac{\partial \hat{f}}{\partial p_{ss}} d\mu(x) = 2 \frac{a_r a_s}{d} A_{rs}, \qquad r \neq s,$$

$$c_{i,rr} = 2 \int_R \frac{\partial \hat{f}}{\partial c_i} \frac{\partial \hat{f}}{\partial p_{rr}} d\mu(x) = -2 \frac{a_r A_{ir}}{d},$$

$$c_{i,jk} = 2 \int_R \frac{\partial \hat{f}}{\partial c_i} \frac{\partial \hat{f}}{\partial p_{jk}} d\mu(x) = -2 \frac{2(a_j A_{ik} + a_k A_{ij})}{d}, \qquad j \neq k,$$

$$c_{rr,jk} = 2 \int_R \frac{\partial \hat{f}}{\partial p_{rr}} \frac{\partial \hat{f}}{\partial p_{jk}} d\mu(x) = \frac{2 a_r}{d} (a_j A_{rk} + a_k A_{rj}), \qquad j \neq k,$$

$$c_{rs,jk} = 2 \int_R \frac{\partial \hat{f}}{\partial p_{rs}} \frac{\partial \hat{f}}{\partial p_{jk}} d\mu(x) = \frac{2}{d} (a_r a_j A_{sk} + a_r a_k A_{sj} + a_j a_s A_{rk} + a_s a_k A_{rj}),$$

$$r \neq s, j \neq k.$$

ON MULTIVARIATE LINEAR REGRESSION WITH MISSING DATA AMONG THE INDEPENDENT VARIABLES

Viktor Brailovsky

Moscow, USSR

INTRODUCTION

This paper is concerned with the problem of processing data with missing values among the independent variables in order to obtain a linear regression function. This problem has been broadly discussed in the literature on multivariate statistical analysis[1-6] and several different methods of solving it have been considered. The first method considered is to discard all incomplete observations and use only the complete ones in the usual least mean square (LMS) procedure. The second method is to use all available observations to estimate the unknown parameters and then use the estimates to obtain the regression function. The third method is to use some other data instead of the missing data, such as the mean values of the respective variables. Another possibility is to use values that are results of approximations of a given independent variable by other independent variables. To do this, one must first obtain a regression function that approximates the given parameter with the help of other parameters.* It is also possible to use some "neutral" values instead of the missing values, but this method proves to be equivalent to discarding all incomplete observations.[3] One last method is to use maximum likelihood estimations of the missing observations. But this may be done only if the probability distribution on the space of independent variables is known, and that is a very rare situation in practical applications; however, some results have been obtained for the multivariate normal distribution case.[2,6,7] We shall not consider this approach at all.

Despite the fact that all these methods of obtaining the regression function have been considered in the literature and used in practice, there are almost no theoretical analyses or comparative investigations of the properties of these procedures, although the results of such studies would be very important for applications as well as theory.

This article is concerned with some asymptotic properties of the procedures mentioned above, under the condition that all observations of independent variables be obtained from a space with a fixed (but unknown) probability distribution and that the quality of the approximation obtained may be measured by the expectation of a quadratic criterion function.†

---

*In many practical problems, the independent variables are highly correlated, so it is possible to obtain a precise formula with which to approximate one of them with the help of others.

†In this article, we are concerned with the use of regression functions for the purpose of approximation, so we shall not discuss the properties of regression coefficients in detail. The connection between these properties and the quality of the approximation is considered in Reference 7.

LMS PROCEDURE FOR DATA WITH COMPLETE OBSERVATIONS

Let us consider briefly the statement of the problem and the resulting formulae for the LMS procedure of function approximation.

Let R ($x \in R$) be a space of arguments with the probability distribution $\mu(x)$. Let $f(x)$ be an unknown function on R such that

$$f(x) = \sum_{i=1}^{n} a_i \phi_i(x). \tag{1}$$

Here $\{\phi_i(x)\}$ stands for a given set of linear independent functions, the coefficients $\{a_i\}$ being unknown. The values of x and $\phi_i(x)$ may be obtained exactly (without any errors), but, instead of $f(x)$, an observer may obtain

$$f(x,\omega) = f(x) + \delta(\omega), \tag{2}$$

where $\delta(\omega)$ is an error random variable with a mean of 0 and a variance of σ^2; for different samples, errors are independent.

To estimate the coefficients $\{a_i\}$, we have a sample set in the form of an information matrix (IM) Q. Each sample x^i is drawn independently from the space R according to a probability distribution $\mu(x)$. The sample x^i corresponds to a row of the IM, which consists of $(n + 1)$ values: $\phi_1(x^i), \phi_2(x^i) \ldots \phi_n(x^i); f(x^i, \omega^i)$.

To approximate $f(x)$ under such conditions, we may use the LMS procedure, which leads to an approximation function, $\hat{f}(Q, x)$ with the following form:

$$\hat{p}(Q, x) = \sum_{i=1}^{n} \frac{\hat{d}_i}{\hat{d}} \phi_i(x) = \sum_{i=1}^{n} \hat{a}_i \phi_i(x) \tag{3}$$

$$\hat{d} = \det \|\hat{p}_{ij}\| \tag{4}$$

$$\hat{p}_{ij} = \frac{1}{N} \sum_{r=1}^{N} \phi_i(x^r)\phi_j(x^r), \quad i, j = 1, 2, \ldots r, \tag{5}$$

where $\hat{d}_i$ is the determinant obtained from the determinant $\hat{d}$ after the substitution of its ith column by a column of estimates $\hat{c}_i$,

$$\hat{c}_i = \frac{1}{N} \sum_{r=1}^{N} f(x^r, \omega^r) \phi_i(x^r), \tag{6}$$

where N is the sample size.

The properties of the LMS procedure are well known;[8] the case when all observations of independent variables are drawn according to a probability distribution is considered in Reference 7.

To measure the deviation of an approximation function $\hat{f}(Q, x)$ from $f(x)$, one uses the following criterion function:

$$D = E \int_R [f(x) - \hat{f}(Q, x)]^2 d\mu(x) \tag{7}$$

E stands for averaging over all possible sample sets with size N.

It is easy to prove that the estimated $\hat{p}_{ij}$ and $\hat{c}_i$ ((5) and (6)) are unbiased ones, i.e., that

$$E\hat{p}_{ij} = E\frac{1}{N}\sum_{r=1}^{N}\phi_i(x^r)\phi_j(x^r) = \int_R \phi_i(x)\phi_j(x)\,d\mu(x) = P_{ij}$$

$$E\hat{c}_i = E\frac{1}{N}\sum_{r=1}^{N}f(x^r,\omega^r)\phi_i(x^r) = \int_R \int_\Omega f(x,\omega)\phi_i(x)\,d\mu(x,\omega) \tag{8}$$

$$= \int_R f(x)\phi_i(x)\,d\mu(x) + \int_R \phi_i(x)\left[\int_\Omega \delta(\omega)\,d\mu(\omega)\right]d\mu(x)$$

$$= \int_R f(x)\phi_i(x)\,d\mu(x) = C_i, \tag{9}$$

where Ω is the space associated with the random variable ω. In Reference 9, it is proved that var $\hat{p}_{ij}$, var $\hat{c}_i$, and the covariances of these estimates are on the order of $1/N$, which means that estimates $\hat{p}_{ij}$ and $\hat{c}_i$ converge in probability to their respective quantities p_{ij} and c_i, as $N \to \infty$.[8] It follows that $\hat{d}$ converges in probability to $d = \det \|p_{ij}\|$‡ and that $\hat{d}_i$ converges in probability to d_i. d_i is the determinant obtained from the determinant d after the substitution of its ith column by a column of values C_i (9). It follows that $\hat{a}_i \to a_i$ and $\hat{f}(Q,x) \to f(x)$ for any $x \in R$.§

It can also be proved that the set of estimates $\{\hat{a}_i\}$ is asymptotically normal and that the covariance matrix of this multivariate normal density may be estimated. With the help of this matrix, the asymptotic expression for the criterion function (7) may be obtained as follows:

$$E\int_R [f(x) - \hat{f}(Q,x)]^2\,d\mu(x) = E\int_R \left[\sum_i (a_i - \hat{a}_i)\phi_i(x)\right]^2 d\mu(x)$$

$$= \sum_i p_{ii}\,\text{var}\,\hat{a}_i + 2\sum_{\substack{i,j \\ i>j}} p_{ij}\,\text{cov}(\hat{a}_i,\hat{a}_j). \tag{10}$$

In Reference 9, all these calculations are performed (in a slightly different way) and, as a result, the following asymptotic expression for the criterion function is obtained:

$$D_{\text{LMS}} = \frac{n\sigma^2}{N} + O\left(\frac{1}{N}\right). \tag{11}$$

so $D_{\text{LMS}} \to 0$ as $N \to \infty$.

LMS-Like Procedure for Data with Missing Values

Now, in addition to the conditions of the first section, let information be lost with a probability of λ ($0 \leq \lambda < 1$), for any observation in the columns of argument of the IM (i.e., columns of $\phi_1(x), \phi_2(x), \ldots \phi_n(x)$, but not $f(x,\omega)$). For different observations the missing data appear differently. Henceforth, we shall use an IM with $N/(1-\lambda)$ rows; therefore, under the conditions given above, we have, in each column of argument, N data and $\lambda N/(1-\lambda)$ missing values.

‡$\|p_{ij}\|$ is the Grammian matrix for a linear independent system of functions $\{\phi_j\}$, so $d > 0$.
§It is evident that $f(x)$ may be written as $f(x) = \sum_{i=1}^{n} d_i/d\,\phi_i(x)$.

Let us now consider the approximation function of the form of (3), but, instead of (5) and (6), the elements of the determinants $\hat{d}$ and $\hat{d}_i$ ($i = 1, 2, \ldots n$) are as follows:

$$\hat{c}_i = \frac{1}{N} \sum_{r=1}^{N} f(x^r, \omega^r) \phi_i(x^r), \tag{12}$$

$$\hat{p}_{ii} = \frac{1}{N} \sum_{r=1}^{N} \phi_i^r(x^r), \tag{13}$$

$$\hat{p}_{ij} = \frac{1}{N(1-\lambda)} \sum_{r=1}^{N(1-\lambda)} \phi_i(x^r) \phi_j(x^r), \quad i \neq j. \tag{14}$$

The sums in ((12) and (13)) [(14)] are extended over the N [$N(1 - \lambda)$] rows of the IM, where column(s) ϕ_i [ϕ_i and ϕ_j] contain(s) no missing values.

We shall call this procedure the LMS-like procedure; it is based on the use of all available observations to estimate the unknown parameters p_{ij} (8) and c_i (9).

First, let us note that the estimates above (12–14) are unbiased ones, i.e., relations similar to (8) and (9) may be written for them as well. Second, in Reference 10, it was proved that var $\hat{c}_i$, var $\hat{p}_{ii}$, var $\hat{p}_{ij}$, and their covariances are on the order of $1/N$. So, as was true under the condition of the first section, when $N \to \infty$ the following estimates converge in probability to their respective values: $\hat{c}_i \to c_i$, $\hat{p}_{ii} \to p_{ii}$, $\hat{p}_{ij} \to p_{ij} (i \neq j)$, $\hat{d}_i \to d_i$, $\hat{d} \to d$, $\hat{f}(Q, x) \to f(x)$ for any $x \in R$.

As in the case mentioned in the first section, the set of estimates $\{\hat{a}_i\}$ is asymptotically normal. The covariance matrix of its multivariate normal density may be estimated, although it differs from its respective covariance matrix in the case of the first section. With the help of this matrix, the asymptotic expression for (7) may be obtained, through the use of (10). In Reference 10, all these calculations are performed (in a slightly different way) and, as a result, the asymptotic expression for the criterion function (7) for the LMS-like approximation may be written as

$$D_1 = D_{\text{LMS}} + \Delta D = \frac{n\sigma^2}{N} + \frac{\lambda d^2}{N} + \frac{\lambda^2 \omega^2}{N} + O\left(\frac{1}{N}\right), \tag{15}$$

where d^2 and ω^2 are quantities that may be expressed through moments of different products of functions ϕ_i, genuine values of coefficients a_i, and so on ($d^2, \omega^2 \geq 0$). These quantities and their properties are discussed in Reference 10. It is clear that $\Delta D \geq 0$ and that $D_1 \to 0$ as $N \to \infty$.

REMARK CONCERNING THE PROCEDURE BASED ON DISCARDING ALL INCOMPLETE OBSERVATIONS

As follows from the description of the model used above, the probability that a row of the IM will contain no missing values equals $(1 - \lambda)^n$, which means that, after discarding all rows with incomplete observations, we obtain an IM with $(1 - \lambda)^n (N/1 - \lambda) = (1 - \lambda)^{n-1} N$ rows.

After the LMS procedure, we obtain a certain result, the quality of which may be asymptotically estimated according to (11) as follows:

$$D_2 = \frac{n\sigma^2}{(1-\lambda)^{n-1}N} + O\left(\frac{1}{N}\right). \qquad (16)$$

In Reference 10, we compare this approach with the LMS-like one and discuss situations when the former is better than the latter and vice versa.

The Procedure Based on Inserting the Mean Value of the Respective Variable Instead of Missing Values

Let us consider the IM with missing values introduced above. For any column of argument let us calculate the mean value

$$\overline{\phi}_j = \frac{1}{N} \sum_{r=1}^{N} \phi_j(x^r). \qquad (17)$$

The sum in (17) is extended over N lines in the IM, where column $\phi_j(x)$ contains no missing values.

Let us use the quantity $\overline{\phi}_j$ instead of each of the $\lambda N/(1-\lambda)$ missing values in column $\phi_j(x)$, $j = 1, 2, \ldots n$. After this procedure, we have an IM without missing values and the function $\hat{f}(Q, x)$ may be obtained with the help of the LMS procedure of the first section, where we replace N in (5) and (6) by $N/(1-\lambda)$; for now, we consider the IM with $N/(1-\lambda)$ rows.

But, taking into account the fact that missing values in the IM were replaced by (17), the formulae for estimates $\hat{c}_j$, $\hat{p}_{jj}$, and $\hat{p}_{ij}$ ($i \neq j$) may be transformed in the following way:

$$\hat{c}_j = \frac{1-\lambda}{N} \sum_{r=1}^{N/(1-\lambda)} f(x^r, \omega^r)\phi_j(x^r) = \frac{1-\lambda}{N}\left[\sum_{r=1}^{N} f(x^r, \omega^r)\phi_j(x^r)\right.$$
$$\left. + \overline{\phi}_j \sum_{r=N+1}^{N/(1-\lambda)} f(x^r, \omega^r)\right], \qquad (18)$$

$$\hat{p}_{jj} = \frac{1-\lambda}{N} \sum_{r=1}^{N/(1-\lambda)} \phi_j^r(x^r) = \frac{1-\lambda}{N}\left[\sum_{r=1}^{N} \phi_j^r(x^r) + N\frac{\lambda}{1-\lambda}(\overline{\phi}_j)^r\right]. \qquad (19)$$

In (18) and (19), the first sums in square brackets are extended over N rows of the IM, where column ϕ_j contained no missing data before insertion.

$$\hat{p}_{ij} = \frac{1-\lambda}{N} \sum_{r=1}^{N/(1-\lambda)} \phi_i(x^r)\phi_j(x^r)$$
$$= \frac{1-\lambda}{N}\left[\sum_{r=1}^{N(1-\lambda)} \phi_i(x^r)\phi_j(x^r)\right.$$
$$\left. + \overline{\phi}_j \sum_{r=N(1-\lambda)+1}^{N(1-\lambda)+\lambda N} \phi_i(x^r) + \overline{\phi}_i \sum_{r=N+1}^{N+\lambda N} \phi_j(x^r) + \lambda^r \frac{N}{1-\lambda}\overline{\phi}_i\overline{\phi}_j\right], \quad i \neq j. \quad (20)$$

In (20), the first sum in square brackets is extended over $N(1-\lambda)$ rows of IM, where columns ϕ_i and ϕ_j contained no missing value before insertion; the second [third] sum is that where column $\phi_i[\phi_j]$ contained no missing values and, at the same time, column $\phi_j[\phi_i]$ contained missing values only.

Let us now calculate the average values of the estimates (18–20).

After some simple transformations, we obtain‖

$$E\hat{c}_j = (1 - \lambda)c_j + \lambda[(\int \phi_j \, d\mu) - (\int f \, d\mu)]$$
$$= c_j - \lambda[\int f\phi_j \, d\mu - (\int f \, d\mu)(\int \phi_j \, d\mu)], \quad (21)$$

$$E\hat{p}_{jj} = p_{jj} - \lambda\left(1 - \frac{1}{N}\right)[(\int \phi_j^2 \, d\mu) - (\int \phi_j \, d\mu)^2], \quad (22)$$

$$E\hat{p}_{ij} = p_{ij} - \left[2\lambda - \lambda^2 - \frac{\lambda^2(1-\lambda)}{N}\right] \quad (23)$$
$$\times [\int \phi_i \phi_j \, d\mu - (\int \phi_i \, d\mu)(\int \phi_j \, d\mu)], \quad i \neq j.$$

Therefore, the estimates (18–20) are biased and the biases do not converge to zero as $N \to \infty$. It does not seem possible to suggest a reliable correction to eliminate the biases.

It is easy to prove that var $\hat{c}_j$, var $\hat{p}_{jj}$, and var $\hat{p}_{ij}$ are on the order of $1/N$. From this fact and (21–23), it follows that the estimates (18–20) converge in probability to the respective values one obtains from (21–23) when $N \to \infty$.

So, if $N \to \infty$, then $\hat{f}(Q, x) \to \hat{f}_\infty(x)$ for all $x \in R$. Here

$$\hat{f}_\infty(x) = \sum_{i=1}^{N} \frac{\hat{d}_i^\infty}{\hat{d}^\infty} \phi_i(x), \quad (24)$$

$$\hat{d}^\infty = \det \|\hat{p}_{ij}^\infty\|, \quad (25)$$

$$\hat{p}_{jj}^\infty = p_{jj} - \lambda[(\int \phi_j^2 \, d\mu) - (\int \phi_j \, d\mu)^2], \quad (26)$$

$$\hat{p}_{ij}^\infty = p_{ij} - (2\lambda - \lambda^2)[\int \phi_i \phi_j \, d\mu - (\int \phi_i \, d\mu)(\int \phi_j \, d\mu)], \quad i \neq j, \quad (27)$$

where $\hat{d}_i^\infty$ is the determinant obtained from the determinant $\hat{d}^\infty$ after replacing its ith column by a column of values $\hat{c}_i^\infty$, where

$$\hat{c}_i^\infty = c_i - \lambda[(\int f f_i \, d\mu) - (\int f \, d\mu)(\int \phi_i \, d\mu)]. \quad (28)$$

Let us consider the space of parameters c_i ($i = 1, 2, \ldots n$) and p_{ij} (we assume that $i \geq j$ for $p_{ij} = p_{ji}$). The total dimension of the space equals $2n + n(n-1)/2$; let us consider $f(x)$ of the form of (24) as a function on this space. Let (c_i, p_{ij}) be the point corresponding to the genuine values of the respective parameters (see (8) and (9)). Taking (26–28) into account, considering λ a small parameter, and assuming that, in some neighborhood of the point (c_i, p_{ij}) in the space of parameters, there exist limited partial derivatives of the first and second order of the function $f(x)$ with respect to c_i and p_{ij}, we obtain an expansion near the center point (c_i, p_{ij}).

$$\hat{f}_\infty(x) = f(x) + \sum_{i=1}^{n} \frac{\partial \hat{f}}{\partial c_i}(c_i - \hat{c}_i^\infty) + \sum_{i,j=1}^{n} \frac{\partial \hat{f}}{\partial p_{ij}}(p_{ij} - \hat{p}_{ij}^\infty) + O(\lambda), \quad i \geq j. \quad (29)$$

The partial derivatives of the first order in (29) are calculated at the center point. Our assumptions concerning the behavior of partial derivatives in the neighborhood of the

‖ Henceforth, we shall use the notation $\int \phi \, d\mu = \int_R \phi(x) d\mu(x)$.

central point are quite reasonable, for all their singularities are in the region where $d = 0$ and we are considering the case where $d > 0$.

Estimating the limit value of the criterion function (7) with the help of (29), we obtain

$$D_3 \simeq c \lambda^2, \tag{30}$$

where c depends on the probability distribution $\mu(x)$, the variance σ^2, and the genuine values of the coefficients a_j.

So, in our case, the criterion function D_3 does not converge to zero as $N \to \infty$. It makes the situation qualitatively different from the cases considered in the first two sections.

THE PROCEDURE BASED ON USING AN APPROXIMATION, CALCULATED BY OTHER INDEPENDENT VARIABLES, INSTEAD OF MISSING VALUES

Let us consider the IM introduced in the first section and let the missing values be in column ϕ_1 only. That is, let each observation in column ϕ_1 be lost with a probability of λ ($0 \leq \lambda < 1$). For different observations, the missing data appear independently. In this section, we shall consider an IM with $N/(1 - \lambda)$ rows; therefore, in the column of argument ϕ_1, there are N data and $\lambda N/(1 - \lambda)$ missing values. In all other columns, there are $N/(1 - \lambda)$ data and no missing values. Let there be a formula to predict a missing value of ϕ_1 by other independent variables,

$$\hat{\phi}_1(x') = \sum_{j=2}^{r} d_j \phi_j(x'). \tag{31}$$

Let the genuine value $\phi_1(x')$ have the form

$$\phi_1(x') = \hat{\phi}_1(x') + \delta_1(s), \tag{32}$$

$$E\delta_1(S) = 0, \quad E\delta_1^2(S) = \sigma_1^2 = \text{const.} \tag{33}$$

So we assume that, for given values $\phi_2(x'), \phi_3(x'), \ldots \phi_n(x')$, there is a distribution of possible values $\phi_1(x')$ with properties of (32) and (33).¶ S is the variable associated with this distribution.

We shall consider here the estimates $\hat{c}_1$ and $\hat{p}_{1i}$ ($i = 1, 2, \ldots n$), their properties, and the properties of the criterion function (7).

So, let us use the values $\hat{\phi}_1(x')$, which were calculated for each row with a missing value in the column ϕ_1 according to (31), instead of the missing values. After this procedure, one has an IM without missing values and the function $\hat{f}(Q, x)$ may be obtained, with the help of the LMS procedure described above (3–6), by replacing N in (5) and (6) by $N/(1 - N)$.

Taking the insertion procedure into account let us consider the properties of the estimates $\hat{c}_1, \hat{p}_{11}$, and $\hat{p}_{1j}$ ($j = 2, 3, \ldots n$).

¶We use the model with missing values in column ϕ_1 only, instead of the one used above, because the latter model has different sets of missing data in different lines of the IM and, in order to predict $\phi_1(x)$, one must use different formulae of the form of (31) for different lines, which makes the analysis more complicated.

$$E\hat{c}_1 = \frac{1-\lambda}{N} E \sum_{r=1}^{N/(1-\lambda)} f(x^r, \omega') \phi_1(x^r)$$

$$= \frac{1-\lambda}{N} E \left[\sum_{r=1}^{N} f(x^r, \omega') \phi_1(x^r) + \sum_{r=N+1}^{N/(1-\lambda)} f(x^r, \omega') \cdot \left(\sum_{j=2}^{n} d_j \phi_j(x^r) \right) \right]. \quad (34)$$

The first sum in square brackets is extended over N rows of that IM which has no missing values before insertion; the second sum is over the other $\lambda N/(1-\lambda)$ rows.

Averaging the first sum in the square brackets of (34) over all possible sample sets with size N from (1) and (2), we obtain

$$E \sum_{r=1}^{N} f(x^r, \omega') \phi_1(x^r) = N \int_R f(x) \phi_1(x) \, d\mu(x) = N C_1. \quad (35)$$

To do the same operation for the second sum, let us note that it follows from (1), (2), (31), (32), and (34) that

$$f(x^r, \omega') = a_1 \left[\sum_{j=2}^{n} d_j \phi_j(x^r) + \delta_1(s^r) \right] + \sum_{k=2}^{n} a_k \phi_k(x) + \delta(\omega'). \quad (36)$$

Taking (36) into account, let us average the second sum in the square brackets in (34). Provided that $\phi_2, \phi_3, \ldots \phi_n$ are fixed,

$$E_{\phi_2,\ldots,\phi_n}\left[f(x^r, \omega') \left(\sum_{j=2}^{n} d_j \phi_j(x^r) \right) \right] = a_1 \left[\sum_{j=2}^{n} d_j \phi_j(x^r) \right]^2$$

$$+ \left[\sum_{i=2}^{n} a_i \phi_i(x) \right] \left[\sum_{j=2}^{n} d_j \phi_j(x) \right]. \quad (37)$$

After averaging over all possible values of $\phi_2(x), \ldots \phi_n(x)$, we obtain

$$E \sum_{r=N+1}^{N/(1-\lambda)} f(x^r, \omega') \left(\sum_{j=2}^{n} d_j \phi_j(x^r) \right) = \frac{\lambda N}{1-\lambda} \left[a_1 \sum_{i,j=2}^{n} d_i d_j p_{ij} \right.$$

$$\left. + \sum_{i,j=2}^{n} a_i d_j p_{ij} \right]. \quad (38)$$

Let us note that $Ef(x^r, \omega') \phi_1(x^r)$ may be transformed into

$$Ef(x^r, \omega') \phi_1(x^r) = E \left\{ \left[a_1 \sum_{j=2}^{n} d_j \phi_j(x^r) + a_1 \delta_1(s^r) + \sum_{i=2}^{n} a_i \phi_i(x^r) + \delta(\omega') \right] \right.$$

$$\left. \times \left[\sum_{j=2}^{n} d_j \phi_j(x^r) + \delta_1(s^r) \right] \right\}$$

$$= a_1 \sum_{j,k=2}^{n} d_j d_k p_{jk} + \sum_{i,k=2}^{n} a_i d_k p_{ik} + a_1 \sigma_1^2. \quad (39)$$

From (34), (35), (38), and (39), it follows that

$$E\hat{c}_1 = \frac{1-\lambda}{N} \left[NC_1 + \frac{\lambda N}{1-\lambda} (c_1 - a_1 \sigma_1^2) \right] = c_1 - \lambda a_1 \sigma_1^2. \quad (40)$$

Through similar transformations, we obtain

$$E\hat{p}_{11} = E\frac{1-\lambda}{N}\sum_{r=1}^{N/(1-\lambda)}\phi_1^2(x^r)$$

$$= \frac{1-\lambda}{N}E\left\{\sum_{r=1}^{N}\phi_1^2(x^r) + \sum_{r=N+1}^{N/(1-\lambda)}\left[\sum_{j=2}^{n}d_j\phi_j(x^r)\right]^2\right\} = p_{11} - \lambda\sigma_1^2. \quad (41)$$

Here, the first sum in curly brackets is extended over the N rows of the IM in which there were no missing values before insertion and the second sum is over all the other $\lambda N/(1-\lambda)$ rows.

It is easy to see that, in the model considered,

$$E\hat{p}_{1j} = p_{1j}, \quad j = 2, 3, \ldots n. \quad (42)$$

The last equality reflects the fact that there are no missing data in the column ϕ_2, $\phi_3, \ldots \phi_n$.

So, in principal, the method of processing information with missing values considered in this section leads to biases in the estimations (see (40) and (41)) that do not converge to zero as $N \to \infty$. As a result, not only did this convergence take place for the case of the previous section, but also it ensures that $\hat{f}(Q, x)$ converges in probability to a function $\hat{f}_\infty(x)$ (for any $x \in R$), which, generally speaking, differs from $f(x)$. And, similarly to (30), it implies the following relation for criterion function (7):

$$D_4 \simeq c^*\lambda^2. \quad (43)$$

Therefore, the criterion function D_4 does not converge to zero when $N \to \infty$. This fact may be proved just as was done in the previous section.

So, generally speaking, the properties of the method of data processing considered in this section are very similar to those of the method in the previous section. The difference is that the values of biases depend on the known (or easily estimated) values λ and σ_1^2 (in more general models they depend on the characteristics of the quality of approximation formulae like (31) and unknown coefficients $\{a_i\}$. Thus, the biases of estimates $\hat{p}_{ij}$ may be excluded in the same way as the biases of estimates $\hat{c}_i$; one may try to exclude them to a greater extent by an iterative process. Each step of the iterative process consists of calculating new values of coefficients $\{a_i\}$, their previous values having been used for estimating and excluding biases of estimates c_i. But the effectiveness of such an approach must be specially investigated.

References

1. AFIFI, A. A. & R. M. ELASHOFF. 1966. Missing observations in multivariate statistics I. J. Am. Stat. Assoc. **61:** 595–604.
2. BEALE, E. M. L. & R. J. A. LITTLE. 1975. Missing values in multivariate analysis. J. R. Stat. Soc. B. **37:** 129–45.
3. HAITOVSKY, Y. 1968. Missing data in regression analysis, J. R. Stat. Soc. B **30:** 67–82.
4. GLASSER, M. 1964. Linear regression analysis with missing observations among the independent variables. J. Am. Stat. Assoc. **59:** 834–44.

5. ORCHARD, T. & WOODBURY, M. A. A missing information principle: Theory and applications. Proc. 6th Berkeley Symp. Math. Stat. Probab. **1:** 697–715.
6. BUCK, S. H. 1969. A method of estimation of missing values in multivariate data suitable for use with an electronic computer. J. R. Stat. Soc. B **22:** 302–6.
7. VAPNIK, V. N. 1979. Restoration of Dependences by Empirical Data. Nauka, Moscow. (In Russian.)
8. WILKS, S. S. 1962. Mathematical Statistics. John Wiley & Sons, Inc., New York.
9. BRAILOVSKY, V. 1981. Comparative analysis of some procedures of function approximation, based on use of sample data. In press.
10. BRAILOVSKY, V. LMS-like procedure of function approximation based on use of experimental data with missing values. In press.

ON SOME IMPORTANT FEATURES OF EXTRA LOW FREQUENCY AND LOW FREQUENCY ELECTROMAGNETIC WAVES $0 \lesssim \omega \lesssim \omega_L$ IN A MAGNETOPLASMA CONNECTED WITH THE INFLUENCE OF IONS

Yakov L. Al'pert

Moscow, USSR

INTRODUCTION

In some studies of the behavior of electromagnetic waves in a plasma, the influence of the ions is not taken into account. This becomes apparent when one considers the fact that the velocity of ions v_i is omitted in comparison with the velocity of electrons v_e, on the assumption that $v_e \gg v_i$, which is, however, not always correct. Also, formulae without terms that include the influence of the gyrofrequency of ions Ω_H are used when waves with frequencies $\Omega_H \ll \omega \lesssim \omega_L$ are studied. Such assumptions are used for simplicity in many studies. In some cases, they do prove to be correct; nevertheless, they should be made very carefully. In this article, it is shown that, because of such assumptions, erroneous formulae were used for many years and some important properties of the behavior of waves in a magnetoplasma were not discovered. Some results of this study are given briefly below.

THE ELEMENTS OF THE TENSOR; THE REFRACTION INDICES, AND THE ATTENUATION FACTORS IN A MAGNETOPLASMA CONTAINING ONE KIND OF ION

For brevity and simplicity, only results for a magnetoplasma that includes one sort of ion are given in this paper. The results of this study, including results for a plasma containing two kinds of ion, are given in more detail in Reference 1.

To determine the tensor of the dielectric coefficient, we begin, as a rule, with equations of motion of the particles and the total current **J** generated by the monochromatic electric field $\mathbf{E} = \mathbf{E}_0 e^{i\omega t}$, where ω is the angular frequency. These equations are:

$$i\omega m \mathbf{v}_e = -e\mathbf{E} - \frac{e}{c}[\mathbf{v}_e \mathbf{H}_0] - m\nu_{ei}(\mathbf{v}_e - \mathbf{v}_i), \tag{1}$$

$$i\omega M \mathbf{v}_i = e\mathbf{E} + \frac{e}{c}[\mathbf{v}_i \mathbf{H}_0] - m\nu_{ie}(\mathbf{v}_e - \mathbf{v}_i),$$

and

$$\mathbf{J} = -e(N_e \mathbf{v}_e - N_i \mathbf{v}_i) = \begin{Bmatrix} \sigma_{11} & \sigma_{12} & \sigma_{13} \\ \sigma_{21} & \sigma_{22} & \sigma_{23} \\ \sigma_{31} & \sigma_{32} & \sigma_{33} \end{Bmatrix} \mathbf{E}, \tag{2}$$

where $mv_{ei}(\mathbf{v}_e - \mathbf{v}_i)$ is the force of friction on the electron due to its collisions with the ions.

In (1) and (2), $\mathbf{v}_e$ and $\mathbf{v}_i$ are the velocities of the electrons and ions; $\mathbf{H}_0$ is the constant magnetic field; σ_{11}, σ_{12}, ... are the elements of the tensor of complex conductivity linking the current $\mathbf{J}$ with the electric field $\mathbf{E}$; ν_{ie} and ν_{ei} are the numbers of collisions between electrons and ions,

$$\nu_{ei} = 2\sqrt{2\pi}\,\frac{e^4}{m^{1/2}(\kappa T)^{3/2}}\,N_i L(\cdots) \simeq \frac{5.5\,N_i}{T^{3/2}}\ln 220\left(\frac{T}{N^{1/3}}\right), \qquad (3)$$

where $N_i = N_e = N$ are the concentrations of the ions and electrons, respectively; m and M are the masses of the electrons and ions; κ is the Boltzmann constant; $-e$ is the electronic charge; T is temperature; and $L = \ln(0.37\,\kappa T/e^2 N^{1/3})$ is a Coulomb logarithm and $\nu_{ie} = N_e/N_i\,\nu_{ei} = \nu_{ei}$ as $N_e = N_i$.

The complex tensor of the dielectric coefficient is determined by the equation

$$\begin{Bmatrix} \epsilon_{11} & \epsilon_{12} & \epsilon_{13} \\ \epsilon_{21} & \epsilon_{22} & \epsilon_{23} \\ \epsilon_{31} & \epsilon_{32} & \epsilon_{33} \end{Bmatrix} = \begin{Bmatrix} 1 & 0 & 0 \\ 0 & 1 & 0 \\ 0 & 0 & 1 \end{Bmatrix} - i\frac{4\pi}{\omega}\begin{Bmatrix} \sigma_{11} & \sigma_{12} & \sigma_{13} \\ \sigma_{21} & \sigma_{22} & \sigma_{23} \\ \sigma_{31} & \sigma_{32} & \sigma_{33} \end{Bmatrix} \qquad (4)$$

and, in order to determine its elements, (1) must be solved with respect to the velocities $\mathbf{v}_e$ and $\mathbf{v}_i$, i.e., their correlations with the components of the electric field E_x, E_y, and E_z must be established. The corresponding calculations are rather lengthy, especially when there are several kinds of ions.

There follows from (1) and (2) the equation

$$[\mathbf{jH}_0] = i\omega c N\,(m\mathbf{r}_e + M\mathbf{r}_i), \qquad (5)$$

which, after some constructive substitutions into (1), leads to expressions of $\mathbf{v}_e$ and $\mathbf{v}_i$ and then, using (4), we obtain the following elements of the tensor of a magnetoplasma with one kind of ion:

$$\epsilon_1 = 1 - \frac{(\omega_0^2 + \Omega_0^2)\left[\omega^2 - \Omega_H\omega_H - i\omega\left(1 + \frac{m}{M}\right)\nu_{ei}\right]}{p(\omega)}$$

$$\epsilon_2 = \omega\,\frac{\omega_0^2\omega_H - \Omega_0^2\Omega_H}{p(\omega)},\qquad \epsilon_3 = 1 - \frac{\omega_0^2 + \Omega_0^2}{\omega^2\left[1 - i\left(1 + \frac{m}{M}\right)\nu_{ei}\right]}, \qquad (6)$$

where $\epsilon_{11} = \epsilon_{22} = \epsilon_1$, $\epsilon_{12} = -\epsilon_{21} = -i\epsilon_2$, $\epsilon_{33} = \epsilon_3$, $\epsilon_{13} = \epsilon_{31} = \epsilon_{23} = \epsilon_{32} = 0$,

$$p(\omega) = \left[(\omega + \omega_H)(\omega - \Omega_H) - i\left(1 + \frac{m}{M}\right)\nu_{ei}\omega\right]$$
$$\times \left[(\omega - \omega_H)(\omega + \Omega_H) - i\left(1 + \frac{m}{M}\right)\nu_{ei}\omega\right], \qquad (7)$$

$\omega_0^2 = 4\pi N_e^2/m$ and $\Omega_0^2 = 4\pi N_e^2/M$ are the plasma frequencies of the electrons and ions, respectively, and $\omega_H = eH_0/me$ and $\Omega_H = eH_0/Me$ are the gyrofrequencies of the electrons and ions.

However, in order to determine the elements of the tensor of a collisional plasma, two simplifications were formerly made (see, for example, Reference 2). At first, the value v_i was omitted from the bracketed term in (1), i.e., the relative velocities of the particles were not taken into account, though this is necessary in the hydrodynamic theory when the force of friction is introduced in the equations of motion of the particles. Secondly, in the second equation of (1), the term $mv_{ie}v_e$ was replaced by Mv_iv_i — a value v_i, an effective collision frequency of ions, was introduced. In addition, v_{ei} was replaced by v_e — an effective collision frequency of electrons was introduced too. As a result, (1) were simplified, which allowed us to introduce effective masses $M^* = m(1 - iv_e/\omega)$ and $M^* = M(1 - iv_i/\omega)$. In the end, the effect of collisions can be included in different formulae, replacing the masses of particles by effective masses and, instead of (6), the following formulae of the elements of the tensor can be obtained for a plasma containing two kinds of ions:

$$\epsilon_1 = 1 - \frac{\omega_0^2(\omega - iv_e)}{\omega[(\omega - iv_e)^2 - \omega_H^2]} - \frac{\Omega_0^2(\omega - iv_i)}{\omega[(\omega - iv_i)^2 - \Omega_H^2]},$$

$$\epsilon_2 = \frac{\omega_0^2 \omega_H}{\omega[(\omega - iv_e)^2 - \omega_H^2]} - \frac{\Omega_0^2 \Omega_H}{\omega[(\omega - iv_i)^2 - \Omega_H^2]}, \quad (8)$$

$$\epsilon_3 = 1 - \frac{\omega_0^2}{\omega(\omega - iv_e)} - \frac{\Omega_0^2}{\omega(\omega - iv_i)}.$$

Besides, it is assumed that the effective masses in a multicomponent plasma, i.e., one that has $<S>$ sorts of ions, are

$$v_e = v_{ee} + \sum_1^s v_{e,is}, \quad v_i = v_{is,is} + \sum_{k,p}^s v_{ik,ip}, \quad (9)$$

where v_{ee} and $v_{is,is}$ are the collision frequencies between charged particles of the same kind, $K = 1,2,3,\ldots s-1$ and $p = 2,3,\ldots s$. In the case of a plasma with one kind of ion discussed here,

$$v_e = v_{ee} + v_{ei}, \quad v_i = v_{ii}, \quad v_{ee} = \frac{v_{ei}}{\sqrt{2}}, \quad v_{ii} = \frac{v_{ei}}{\sqrt{2}}\sqrt{\frac{m}{M}}. \quad (10)$$

Formulae 8–10 are often used in many papers (see, for example, References 3–7). It is clear that, in a plasma without collision, i.e., when $v_e = v_i = 0$, (6) and (8) coincide.

But, at first, in the hydrodynamic theory, the collisions v_{ee} and $v_{is,is} = v_{ii}$ cannot be taken into account on principle. Terms containing v_{ee} and v_{ii} are absent in (1) because the velocities of their relative motion are equal to zero. The losses of the energy of motion of the charged particles due to v_{ee} and v_{ii}, let us say due to electron and ion viscosity, can be strictly theoretically considered only with the help of the kinetic theory. Secondly, even if it is qualitatively assumed that we can use collisions v_{ee} and v_{ii} in (1) (approximately), it is wrong to replace the term $mv_{ie}v_e$ by Mv_iv_i in the second equation of (1). The substitution $mv_{ei}v_e \simeq -Mv_{ei}v_i$ should be made, as it is well known that the Lorentz force e/c [**vH**] does not contribute to the value of the velocities of the charged particles. Therefore, $m\mathbf{r}_e \simeq -M\mathbf{r}_i$. However,

$$M v_{ii} \mathbf{v}_i = M\left(\frac{m}{M}\mathbf{v}_e\right)\sqrt{\frac{m}{2M}}\,v_{ei} = \sqrt{\frac{m}{2M}}\,(mv_{ei}\mathbf{v}_e) \tag{11}$$

(see **(10)**), i.e., the term $mv_{ei}\mathbf{v}_e$ in the second equation of **(1)** is replaced by a term that is $\sqrt{m/2M}$ times smaller than $mv_{ei}\mathbf{v}_e$.

The above discussion also shows that both erroneous formulae for the elements of the dielectric tensor and erroneous refractive indices n_{12} and attenuation factors κ_{12} of electromagnetic waves in a magnetoplasma were used in many papers. To evaluate the expected values of the appropriate errors, numerical calculations of n_{12} and κ_{12} of the two modes of electromagnetic waves that exist in a magnetoplasma were made, using the well-known formula

$$(n_{12} - i\kappa_{12})^2 = [-B_0 \pm \sqrt{B_0^2 - 4A_0C_0}](2A_0)^{-1}, \tag{12}$$

where

$$A_0 = \epsilon_1 \sin^2\theta + \epsilon_3 \cos^2\theta, \qquad C_0 = \epsilon_3(\epsilon_1^2 - \epsilon_2^2),$$
$$B_0 = -\epsilon_1\epsilon_3(1 + \cos^2\theta) - (\epsilon_1^2 - \epsilon_2^2)\sin^2\theta, \tag{13}$$

and θ is the angle between the wave normal **K** and the vector $\mathbf{H}_0$. The formulae of ϵ_1, ϵ_2, and ϵ_3 determined by **(6)** and **(8)** were used in **(19)**. In **(8)**, both the effective collisions $v_i = v_{ii}$ and $v_i = v_{ei}$ were used. In that way, the following three families of $N_{12}(\omega,\theta)$ and $\kappa_{12}(\omega,\theta)$ dependencies were obtained: (a) elements of tensor $\epsilon_{1,2,3}$ **(6)** with one value v_{ei}, (b) elements of tensor $\epsilon_{1,2,3}$ **(8)** with the values $v_i = v_{ii}$, $v_e = v_{ee} + v_{ei}$, and (c) elements of tensor $\epsilon_{1,2,3}$ **(8)** with the values $v_e = v_{ei}$ and $v_i = v_{ie}$. Some results of these calculations are given in FIGURES 1–3. They were obtained for the following plasma parameters:

$$\omega_0 = 3.75 \times 10^7 \text{ c/s}, \quad \omega_H = 7.5 \times 10^6 \text{ c/s}, \quad \Omega_H(H^+) = 4.1 \times 10^3 \text{ c/s}$$
$$v_{ei} = 3 \times 10^2 \text{ s}^{-1}, \quad v_{ee} = 2.1 \times 10^2 \text{ s}^{-1}, \quad v_{ii} = 5 \text{ s}^{-1}, \quad \frac{m}{M(H^+)} = 5.48 \times 10^{-4}, \tag{14}$$

which characterize a region of the upper part of the ionosphere.

It is seen from FIGURE 1 that the character of the variation of the refractive indices $n_1(\omega,\theta)$ and the attenuation factors $\kappa_1(\omega,\theta)$ of the proton whistler mode, i.e., of the ordinary wave, is the same in all three cases (a, b, and c of FIGURE 1). However, the values of $n_1(\omega,\theta)$ differ appreciably in the region $\omega > \Omega_H$. The values of $\kappa_1(\omega,\theta)$, on the other hand, differ appreciably in the region $\omega < \Omega_H$, especially when $\omega \ll \Omega_H$. However, with the help of **(6)**, new qualitative peculiarities of the behavior of the extraordinary wave, i.e., the tail of the electron whistler mode, were discovered (a in FIGURE 2). In a narrow region of frequencies $\Delta\omega$ in the vicinity of the ion gyrofrequency, the value $n_2(\omega,\theta)$ has a weakly pronounced maximum. But the attenuation factor $\kappa_2(\omega,\theta)$ has a very sharply pronounced maximum. At the same time, the dependencies of $n_2(\omega,\theta)$ and $\kappa_2(\omega,\theta)$ (b and c in FIGURE 2), which were obtained with **(8)** with effective masses, do not have maxima and the values of $\kappa_2(\omega,\theta)$ differ appreciably from the values obtained with **(6)** in the entire frequency range discussed. More detail on the dependencies of $n_2(\omega,\theta)$ and $\kappa_2(\omega,\theta)$ around Ω_H obtained with **(6)** is given in FIGURE 3.

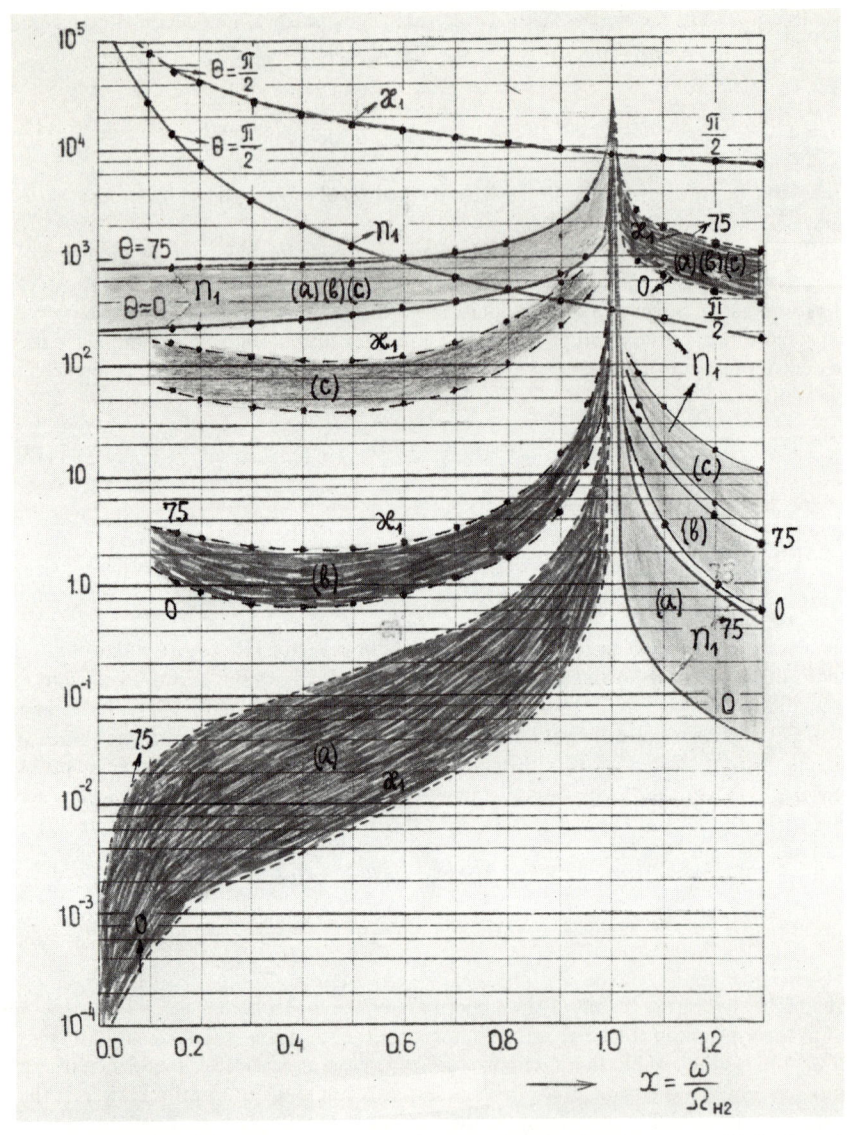

FIGURE 1. Real and imaginary parts of the refractive index for the proton whistler branch (ordinary mode).

In the case discussed, the value $\Delta \omega \simeq 10^{-2} \Omega_H$, but it depends very sensitively on the value of ν_{ie}.

These properties ($n_2(\omega)$, and especially $\kappa_2(\omega)$) of the extraordinary wave around the gyroresonance of ions are new and have not been detected earlier. Their effects can certainly play an important role in different plasma phenomena, but this will not be discussed here. It should be noted that it is interesting to study the behavior of $n_2(\omega, \theta)$ and $\kappa_2(\omega, \theta)$ for different parameters of a magnetoplasma in more detail.

The Amplification of the Electromagnetic Field of the Tail of the Eectron Whistler Mode

In a magnetoplasma, the angle α between the magnetic field $\mathbf{H}_0$ and the ray direction—i.e., the direction of the group velocity $\mathbf{U} = d\omega/d\mathbf{k}$ or of the Poynting vector $\mathbf{S} = c/4\pi\,[\mathbf{EH}]$ of electromagnetic waves—can be determined by the formulae

$$\tan\alpha = -\tan\theta\,\frac{dk_z^2}{dk_x^2} = -\frac{dn_\parallel}{dn_\perp},$$

$$\tan\alpha = \tan\theta\,\frac{\epsilon_2 n^2\cos^2\theta + \dfrac{\epsilon_1^2}{\epsilon_3}\left(n^2 - \dfrac{\epsilon_1^2 - \epsilon_2^2}{\epsilon_1}\right)^2}{\epsilon_2 n^2\cos^2\theta + \epsilon_1(n^2 - \epsilon_1)\left(n^2 - \dfrac{\epsilon_1^2 - \epsilon_2^2}{\epsilon_1}\right)}, \qquad (15)$$

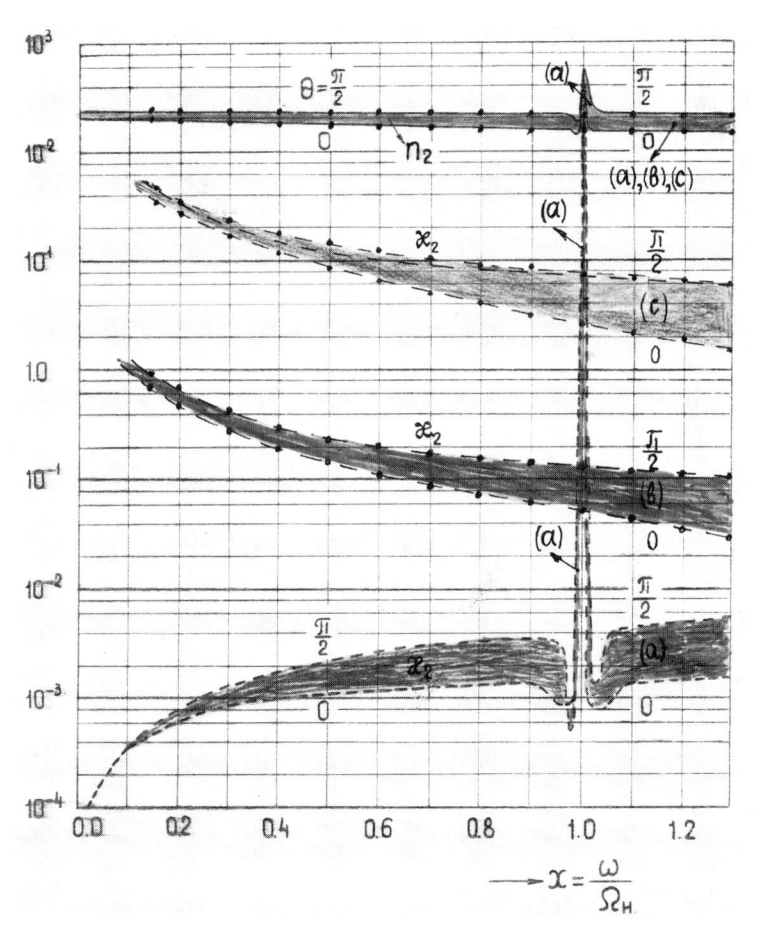

FIGURE 2. Real and imaginary parts of the refractive index for the extraordinary mode (electron whistler branch).

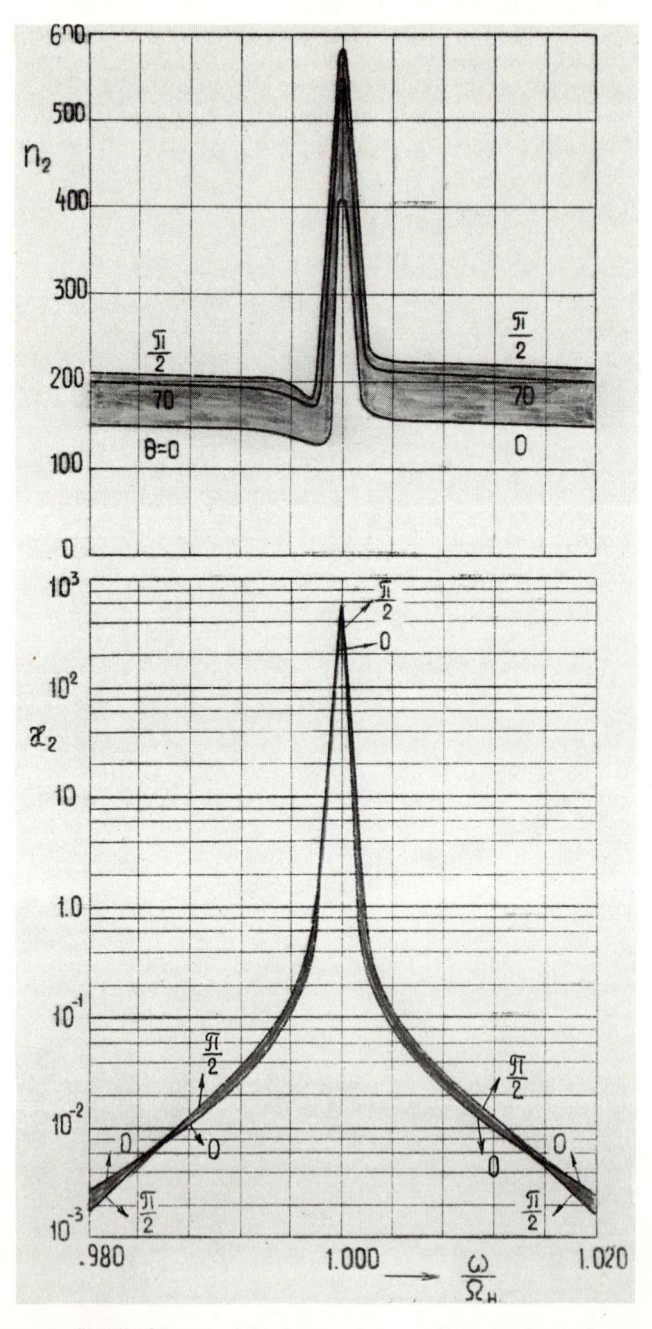

FIGURE 3. Detail of FIGURE 2 for frequencies near the ion cyclotron frequency.

where ϵ_1, ϵ_2, and ϵ_3 are given by (6) or (8) when $\nu_i = \nu_e = 0$, $n_\parallel = n\cos\theta$, $n_\perp = n_{\sin}\theta$, n is the refraction index, and θ is the angle between $\mathbf{H}_0$ and the wave vector $\mathbf{k} = \omega/c\,\mathbf{n}$.

The behavior of $\alpha(\theta, \omega)$ of the electron whistler mode $0 < \omega \lesssim \omega_H$ was studied for the first time in Reference 8 and then in more detail in Reference 9. It was shown that, in part of a frequency range $0 < \omega \lesssim \omega_s$, $\omega_s < \omega_H$, the dependencies $\alpha(\theta)$ have maxima α_M and that the value of α_M, changing with ω, is maximal—$\alpha_{M,\max} = 19.5°$ when $\omega/\omega_H \ll 1$. This value $\alpha_{M,\max}$ was found for the first time by Storey. The values α_M can be determined by the formulae

$$\left(\frac{d^2 n_\parallel}{dn_\perp^2}\right)_{n_\perp M(\omega)} = 0, \quad \tan\alpha_M = -\left(\frac{dn_\parallel}{dn_\perp}\right)_{n_\perp M(\omega)}. \tag{16}$$

The frequency $\omega = \omega_s$ can be determined by the following formula when the maxima of $\alpha(\theta, \omega_s)$ disappear, i.e., $\alpha_M = 0$,

$$\epsilon_1 + \epsilon_3 - \epsilon_2 = 0, \tag{17}$$

which follows from the equation

$$\left(\frac{d^2 n_\parallel}{dn_\perp^2}\right)_{n_\perp = 0} = 0 \tag{18}$$

and the dispersion equation of a magnetoplasma,

$$D = \epsilon_3 n_\parallel^4 + n_\parallel^2[(\epsilon_1 + \epsilon_3)n_\perp^2 - 2\epsilon_1\epsilon_3]$$
$$+ \{\epsilon_1 n_\perp^4 + n_\perp^2[\epsilon_2^2 - \epsilon_1(\epsilon_1 + \epsilon_3)] + \epsilon_3(\epsilon_1^2 - \epsilon_2^2)\} = 0. \tag{19}$$

However, in References 8 and 9 and other papers studying the dependency $\alpha(\theta, \omega)$, the formula $n(\omega, \theta)$ was used without taking the influence of ions into account, that is, the terms depending on the ion gyrofrequency Ω_H were omitted. Therefore, some important properties of $\alpha(\theta, \omega)$ were not found. They were revealed mainly as a result of detailed numerical calculations of $\tan\alpha$ given by the second formula in (15), since the analytical formula of $n(\omega, \theta)$ and its derivatives are very complicated.[10] The following features of $\alpha(\theta, \omega)$ of the electron whistler mode were found.

The dependence $\alpha(\theta)$ has only one extremal point in the frequency range $\omega_s \gtrsim \omega \gtrsim \omega_L$, where ω_L is the lower hybrid frequency. The value $\alpha = \alpha_{M,\max} = 19.5°$ when $\omega = \omega_L$. Then, at frequencies $\omega_L \gtrsim \omega \gtrsim \omega_r$, it has two extremal points: a maximum $\alpha_M > 19.5°$ and a minimum $\alpha_m < 19.5°$. It is clear that α_M and α_m are determined by the same equation

$$\frac{d^2 n_\parallel}{dn_\perp^2} = 0. \tag{20}$$

The minimum disappears, i.e., $\alpha_m = 0$ when $\omega = \omega_L$. At that point $\theta_m = \pi/2$. Both the maximum and the minimum of $\alpha(\theta, \omega)$ disappear when its extremal points and its inflection point coincide, i.e., at that point

$$\frac{d^3(n_\parallel)}{dn_\perp^3} = \frac{d^2(n_\parallel)}{dn_\perp^2} = 0 \tag{21}$$

and $\alpha_{M,\max} = \alpha_{m,\max}$. From **(21)**, the frequency $\omega = \omega_r$ is determined, too. In FIGURE 4 is given a family of curves α $(\theta, \omega/\omega_n = \text{const})$ in the frequency range $0 \le \omega \le \omega_H$ computed with the help of the second formula of **(15)** and **(6)** and **(12)**. The parameters of the plasma,

$$\frac{\omega_0}{\omega_H} = 2, \quad \frac{\omega_L}{\omega_H} = 5.2307 \times 10^{-3}, \quad \frac{m}{M(0^+)} = 3.425 \times 10^{-5}, \quad \nu_{ie} = 0, \quad (22)$$

were used and the following characteristic values of α and ω were obtained:

$$\alpha_{M,\max} = \alpha_{m,\max} \simeq 24.3°, \quad \theta_r = 71.4°, \quad \frac{\omega_r}{\omega_H} \simeq 1.6 \times 10^{-4}. \quad (23)$$

It should be noted that the resonance cutoff angle $\alpha = \alpha_\infty$, i.e., the point where $n^2 \to \infty$, the angle $\theta = \theta_0$, and the $\alpha(\theta)$ curve intersects the axis $\alpha = 0$, is determined by a simple formula

$$\tan \alpha_\infty = \text{ctg}\, \theta_0 = \sqrt{\frac{-\epsilon_1}{\epsilon_3}}. \quad (24)$$

As a result of the properties of the direction of the ray of the whistler mode described above, its electromagnetic field has the following features.[11] In the frequency range $\omega_r \le \omega \le \omega_L$, its field is trapped into two cones inserted one into other with a common symmetry axis along the vector of the magnetic field $\mathbf{H}_0$. At some given value of ω/ω_H, the edge (generatrix) of one cone is determined by the angle $\alpha = \alpha_\infty(\omega/\omega_H)$ — this is the *resonance cone*. The edge of the other cone is determined by the angle $\alpha = \alpha_M$ — this cone was called the Storey cone in Reference 11. In the frequency range $\omega_s \le \omega \le \omega_H$, only one resonance cone is formed by the lines of force of the electromagnetic field. It was shown (see Reference 11) that, in the resonance cone, the amplitude of the field is considerable amplified in a narrow range of angles $\Delta \alpha_\infty$ close to its edge when $\alpha \lesssim \alpha_\infty$. Depending on ω/ω_H, $\Delta \alpha_\infty \simeq 10^{-2} - 1°$ and the amplification factor changes from a value $a \simeq 1$ to $a \simeq 10^5 - 10^6$ and larger. In the Storey cone, the field is weakly amplified close to $\alpha = \alpha_M$ by a factor $a \gtrsim 1$ in an angle range $\Delta \alpha_M \simeq 0.1°$. But it is considerably amplified close to the direction of the magnetic field $\mathbf{H}_0$, i.e., when $\alpha \gtrsim 0$, $\theta \simeq \theta_0$ (see **(24)**). In this region, $\Delta \alpha_0 \simeq 10^{-2°}$ and, depending on ω/ω_H, the amplification factor $a \gtrsim 10^2 - 10^6$.

Recently, some new properties of the whistler mode were found. Due to the behavior of the ray direction described above, the structure of the electromagnetic field is more complicated in the frequency range $\omega_L \gtrsim \omega \gtrsim \omega_r$. At these frequencies it is usually trapped in a cavity formed between two cones. One is the Storey cone with the edge $\alpha = \alpha_M$. The other cone is inserted into the Storey cone. Its edge is formed by a ray with the angle $\alpha = \alpha_m < \alpha_M$. By analogy, this cone may be called the reversed Storey cone. It was shown that, in the frequency range $\omega_r \ll \omega < \omega_L$, the field trapped in this cavity is considerably amplified close to the edge of the second cone, i.e., when $\alpha \lesssim \alpha_m$. Results of some calculations of the modulus of the electric field $|\mathbf{E}(\alpha)|_{\text{rev}}$ at the edge of the reversed Storey cone are given in FIGURE 5. The parameters of the plasma

$$\frac{\omega_0}{\omega_H} = 2, \quad \frac{\omega_L}{\omega_H} = 5.2307 \times 10^{-3}, \quad \frac{\omega_r}{\omega_H} \simeq 1.6 \times 10^{-4}, \quad \omega_H = 7.5 \times 10^6, \quad (25)$$

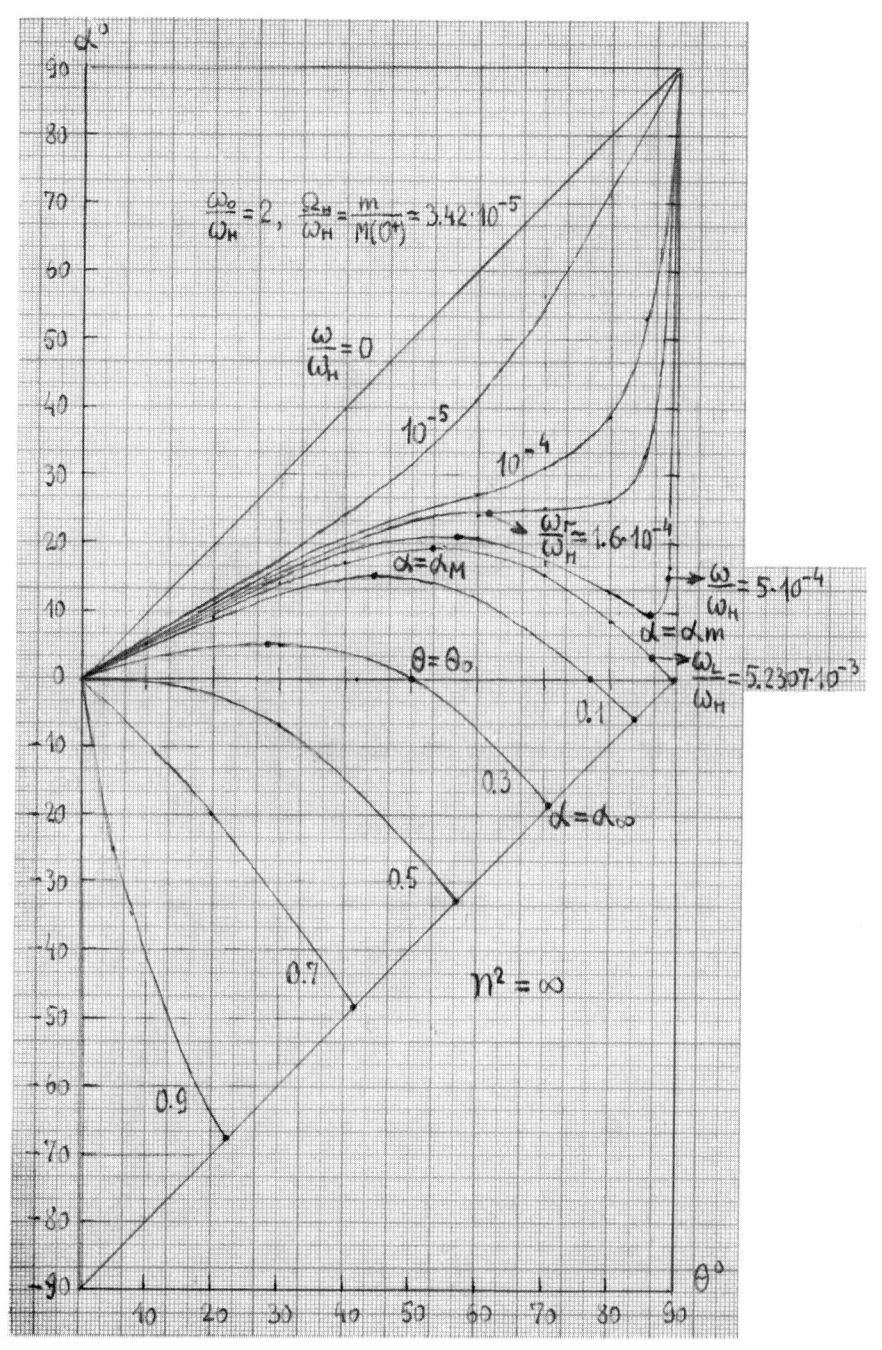

FIGURE 4. The direction of ray propagation for the whistler branch versus the phase-velocity direction.

148　　　　　　　　　Annals New York Academy of Sciences

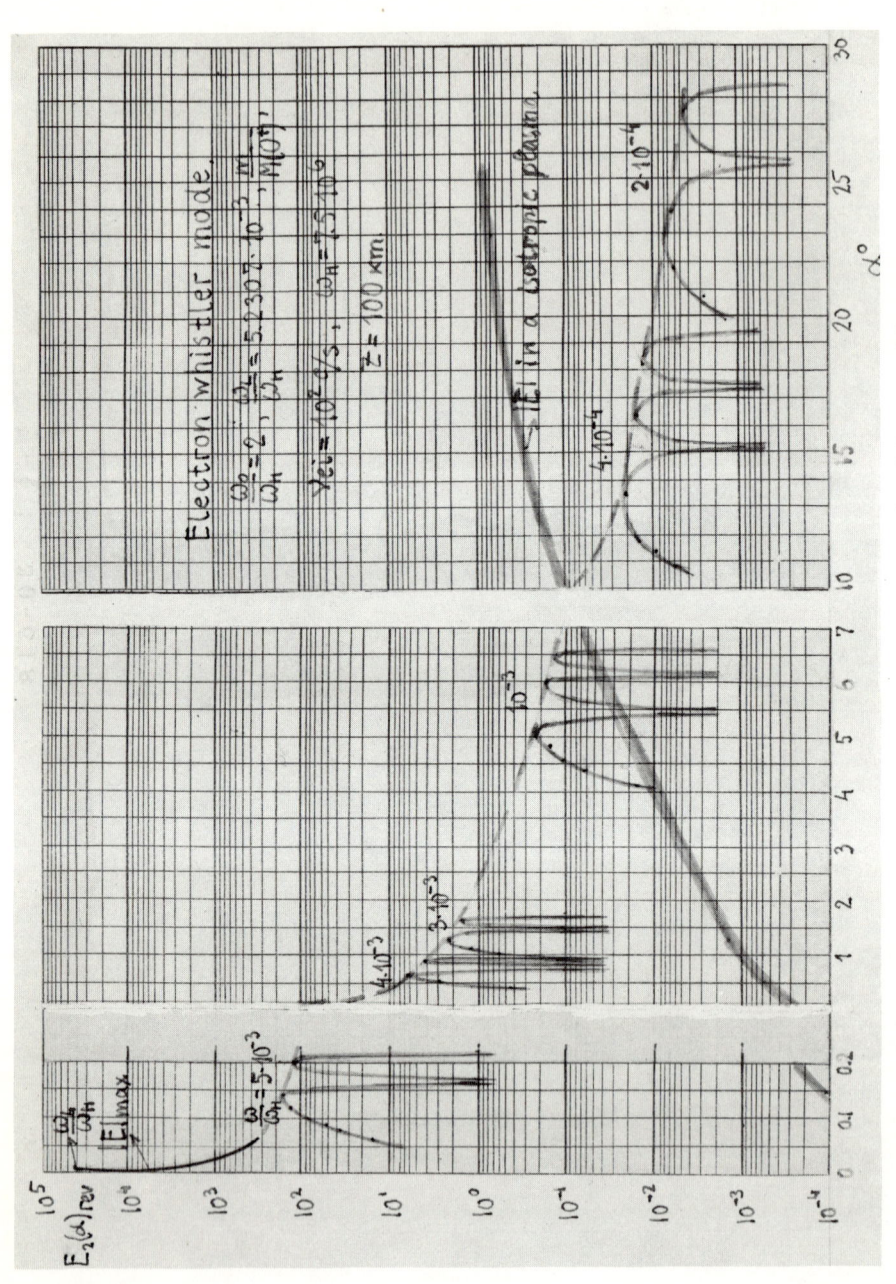

FIGURE 5. The modulus of the wave electric field near the edge of the reversed Storey cone.

and the distance from the point source of the electromagnetic waves $Z = 100$ km were used. It is seen from FIGURE 5 that, in the frequency range $\omega/\omega_H \simeq (1-5) \times 10^{-3}$, the amplification factor $a \simeq 1 - 10^5$ and $\Delta\alpha_m \simeq 1 - 10^{-10}$. This effect can play an important role in different investigations of the behavior of electromagnetic waves in a plasma.

REFERENCES

1. AL'PERT, YA. L. 1980. J. Atmos. Terr. Phys. **49:** 217.
2. SMITH, R. L. & N. N. BRICE 1964. J. Geophys. Res. **69:** 5029.
3. GURNETT, D. A., S. D. SHAWHAN & N. M. BRICE. 1965. J. Geophys. Res. **70:** 1665.
4. JONES, D. 1969. J. Atmos. Terr. Phys. **31:** 971.
5. JONES D. 1972. Planet. Space Sci. **20:** 1173.
6. BOOKER, H. G. 1975. Philos. Trans. R. Soc. London Ser. A **280:** 57.
7. BOOKER, H. G. & R. B. DYCE. 1965. Radio Sci. **69 D:** 463.
8. STOREY, L. R. O. 1953. Philos. Trans. R. Soc. London Ser. A **246:** 113.
9. SMITH, R. L. 1958. Ann. Geophys. **14:** 144.
10. AL'PERT, YA. L. 1980. J. Atmos. Terr. Phys. **42:** 205.
11. AL'PERT, YA. L. & B. S. MOISEYEV. 1980. J. Atmos. Terr. Phys. **42:** 521.

THE EFFECT OF THE EARTH'S ROTATION ON THE PROPAGATION OF WEAK NONLINEAR SURFACE AND INTERNAL LONG OCEANIC WAVES*

A. I. Leonov

*No. 1 Academician Il'yushin Street, Apt. 54
Moscow, 125319, USSR*

Introduction

A number of papers have dealt with the weak nonlinear theory of long internal waves.[1-4] However, those results can be applied to the propagation of long internal and surface waves in the horizontally homogeneous ocean only in an intermediate asymptotic case corresponding, on one hand, to a wavelength, λ, that is much greater than the oceanic depth, H, and, on the other hand, to a phase velocity of the wave, C_f, that is much greater than λf, where f is the local parameter of the Earth's rotation. This paper considers effects of the earth's rotation on the propagation of weak nonlinear surface and internal oceanic waves within the framework of a long wave approximation. An amplitude equation is derived and its stationary one-dimensional solutions are considered. It is shown that soliton-like solutions of the equation do not exist, but solutions in the form of weak nonlinear periodic waves do exist and are stable with respect to both longitudinal and transverse disturbances. In the last section of the paper, the amplitude equation is derived directly from the hydrodynamic equation of a stratified fluid using an asymptotic procedure.

Derivation of an Equation Describing Longitudinal Wave Motions

We shall start with the following linear dispersion relation, which holds for a rotating medium with negative dispersion,

$$\omega = (f_*^2 + C_f^{0\,2}\kappa^2 - \beta^2 \kappa^4 + \cdots)^{1/2}, \tag{1}$$

and which is characteristic of internal waves. Here f_* is a parameter of the Earth's rotation, ω is the wave frequency, κ is the wave number, $\beta^2 > 0$ is the dispersion parameter, and C_f^0 is the phase velocity of a long wave disturbance in the absence of the earth's rotation.

Let us further consider the case of very slow rotation when

$$f_*/(C_f^0 \kappa) = \mathcal{O}(\kappa^2). \tag{2}$$

---

*Due to "communication difficulties," this paper was corrected by Martin Kruskal (Princeton University) without the author's knowledge or consent.

Then, taking (2) into account, we can reduce (1) to the form

$$\omega = C_f^0 \kappa - \frac{\beta^2}{2C_f^0} \kappa^3 + \frac{f_*^2}{2C_f^0 \kappa} + \cdots . \tag{3}$$

Letting

$$\omega \to -i\frac{\partial}{\partial t}, \qquad \kappa \to i\frac{\partial}{\partial x},$$

and considering (3) an operator equation, we obtain

$$u_t + C_f^0 u_x + \frac{\beta^2}{2C_f^0} u_{xxx} = \frac{f_*^2}{2C_f^0} \int u(t, x')dx'.$$

If one adds a "typical" nonlinear term $\gamma u u_x$ to the equation and passes to the moving frame of reference $\xi = x - C_f t$, where C_f is the phase velocity of the wave, we will have, finally,

$$u_t - C_f^{(1)} u_\xi + \gamma u u_\xi + \frac{\beta^2}{2C_f^0} u_{\xi\xi\xi} = \frac{f_*^2}{2C_f^0} \int u(t, \xi')d\xi', \tag{4}$$

where $C_f^{(1)} = C_f - C_f^0$.

The left-hand side of (4) corresponds to the familiar Korteweg–de Vries (KdV) equation, while its right-hand side includes an additional term, which is a result of the rotation effect in (2).

If we also consider another effect, viz. very long scale transverse motions, then, instead of (4), it is possible to obtain the following equation by analogy with Reference 5:

$$u_t - C_f^{(1)} u_\xi + \gamma u u_\xi + \frac{\beta^2}{2C_f^0} u_{\xi\xi\xi} + \frac{C_f^0}{2} \int u_{\eta\eta}(t, \xi', \eta)d\xi'$$

$$= \frac{f_*^2}{2C_f^0} \int u(t, \xi', \eta)d\xi', \tag{5}$$

the left-hand side of which is identical with the Kadomtsev-Petviashvili equation employed in References 5 and 6 to investigate the stability of KdV stationary solutions with respect to transverse disturbances.

Equations 4 and 5 were derived in References 7 and 8, respectively, directly from the hydrodynamic equations with the aid of different asymptotic methods. Moreover, the stationary solutions of (4) at $\beta^2 = 0$ were also studied in Reference 7. Another procedure for deriving (5), one different from the procedures used in References 7 and 8, is presented below.

Equation 5 is invariant with respect to the following transformations:

$$t \to -t, \qquad \xi \to -\xi, \qquad \eta \to -\eta \tag{6}$$

$$\gamma \to -\gamma, \qquad u \to -u \tag{7}$$

Then, because of (7), it is sufficient to assume that the parameter γ is positive, although, in general, its sign is arbitrary.

On Stationary Solutions of Equation 4

After differentiation with respect to ξ, the stationary equation (4) describing longitudinal propagation of waves can be written as

$$u'''' - au'' + \mathcal{C}(uu'' + u'^2) - Cu = 0 \tag{8}$$

$$a = 2C_f^{(1)} C_f^0/\beta^2, \quad \mathcal{C} = 2\gamma C_f^0/\beta^2, \quad C = f_*^2/\beta^2, \tag{9}$$

where $a \sim C_f^{(1)}$ is the eigenvalue of a certain nonlinear boundary value problem for (8). Equation 8 can be rewritten in the following form:

$$u'' + \frac{\mathcal{C}}{2} u^2 - au - Cv = 0, \quad v'' = u. \tag{10}$$

Proceeding from the fact that (10) is analogous to the set of equations that describes the motion of a material point in the (u, v) plane, it is easy to obtain a first integral (Hamiltonian function) for (10):

$$H = \frac{1}{2}(u'^2 + Cv'^2) + \frac{\mathcal{C}}{6} u^3 - \frac{a}{2} u^2 - Cuv = \text{const.} \tag{11}$$

No other independent first integral of (10) has been found.

Let us suppose that there exists a solution for (10) that satisfies certain given initial conditions. Then it is simple to find the equation for a trajectory $u = U(v)$ of the material point in the (u, v) plane,

$$[g''(v)]^2 g - H = \frac{a}{2} g'^2 - \frac{\mathcal{C}}{6} g'^3 + C(g'v - g), \tag{12}$$

where

$$g(v) = \int U(v) dv + \mathcal{A} \tag{13}$$

and $\mathcal{A}$ is a new integration constant. Moreover, the second equation in (10) yields the first integral

$$\frac{1}{2}[v'(\xi)]^2 = g(v). \tag{14}$$

Now we shall establish the nonexistence of soliton-like solutions of (10), i.e., solutions for which $u^{(n)}, v^{(n)} \to 0$ as $\xi \to \pm\infty$ ($n = 0, 1$). Let us assume the opposite, that is, the existence of a soliton-like solution of (10). Then it immediately follows from (11) that $H = 0$ for this solution and (12) takes the form

$$[g''(v)]^2 g = \frac{a}{2} g'^2 - \frac{\mathcal{C}}{6} g'^3 + C(g'v - g). \tag{15}$$

Indeed, with appropriate choice of $\mathcal{A}$, the function $g(v)$ satisfies the following initial conditions in this case:

$$g(0) = g'(0) = 0. \tag{16}$$

Equation 15 satisfies the conditions of the uniqueness and existence theorem along the entire v-axis, with the exception of the vicinity of the branch point $v = 0$. If one seeks the solution of (15) in the form of the power series

$$g(v) = \sum_{n=1}^{\infty} \frac{\lambda_n}{n+1} v^{n+1}, \tag{17}$$

then (15) yields as recurrence relation for the coefficients of the series

$$-\lambda_n[\lambda_1(n^2+1) - a] = \sum_{m=1}^{n-2} \lambda_{m+1}\lambda_{n-m}(m+1)\frac{m+n+1}{n-m+1} + \frac{e}{2}\sum_{m=1}^{n-1}\lambda_m\lambda_{n-m} \tag{18}$$

for $n \geq 2$, where λ_1 is a root of the quadratic equation

$$\lambda_1^2 = a\lambda_1 + C. \tag{19}$$

For sufficiently large n, the following rough estimate for λ_n holds true:

$$|\lambda_n| \leq \frac{|\lambda_1|}{2n \ln (3/2)}, \tag{20}$$

which proves the convergence of the series.

The first three coefficients λ_n have the form

$$\lambda_1 = \frac{a}{2} \pm \frac{1}{2}\sqrt{a^2 + 4C}, \quad \lambda_2 = -\frac{1}{2}\frac{e\lambda_1^2}{5\lambda_1 - a}, \quad \lambda_3 = \frac{2}{3}\frac{\lambda_2^2}{\lambda_1}\frac{10\lambda_1 - 3a}{10\lambda_1 - a}. \tag{21}$$

As follows from (21), each of the two solutions of (15) is defined by the choice of the sign of λ_1. According to (14), the value of $g(v)$ is necessarily positive for small values of v, therefore one has to adopt the plus sign in the formula for λ_1 in (21); thereby, one of the two solutions of (15) is chosen, and is represented in the form of a power series converging within a certain neighborhood of the point $v = 0$. Furthermore, owing to the existence and uniqueness theorem, this positive solution can be continued on the entire finite v-axis. Now we shall note that the solitary wave solutions, as follows from (14), can exist only if the function $g(v)$, whose behavior is defined by (17–21), vanishes at a certain value v_0 ($|v_0| > 0$). This is impossible, however, because the function $g(v)$ increases (decreases) monotonically for $v > 0$ ($v < 0$) in the vicinity of the point $v = 0$ and is monotonic outside that vicinity, too. Indeed, in order to break the monotonicity, it would be necessary that $g'(v_*) = 0$ at a certain point $v = v_*$. ($|v_*| < |v_0|$). But this equality is inconsistent with (15), which proves the nonexistence of soliton-like solutions of (10).

Now let us consider the problem of finding a strictly periodic solution of (8) (or (10)) at sufficiently small values of amplitude parameter α. If one seeks, for instance, a solution of (8) in the expansion form

$$u = \alpha u_1(\theta) + \alpha^2 u_2(\theta) + \alpha^3 u_3(\theta) + \mathcal{O}(\alpha^4)$$

$$a = a_1 + \alpha^2 a_2 + \mathcal{O}(\alpha^4), \quad (\theta = \kappa\xi + \theta_0), \tag{22}$$

then one readily obtains

$$u = \alpha \sin\theta - \alpha^2 \frac{\mathcal{C}\kappa^2 \cos 2\theta}{3(C + 4\kappa^4)} - \alpha^3 \frac{3\mathcal{C}^2 \kappa^4 \sin 3\theta}{16(C + 4\kappa^4)(C + 9\kappa^4)} + \mathcal{O}(\alpha^4)$$

$$a = \frac{C}{\kappa^2} - \kappa^2 + \alpha^2 \frac{\mathcal{C}^2 \kappa^2}{6(C + 4\kappa^4)} + \mathcal{O}(\alpha^4). \tag{23}$$

The second formula, (23), taking account of the expression for the parameter a in (9), yields a so-called nonlinear dispersion relation,

$$\omega(\kappa, \alpha^2) = C_f^0 \kappa + \frac{\beta^2}{2C_f^0}\left[\frac{C}{\kappa} - \kappa^3 + \alpha^2 \frac{\mathcal{C}\kappa^3}{6(C + 4\kappa^4)} + \mathcal{O}(\alpha^4)\right]. \tag{24}$$

Using the well-known stability criterion for a weak nonlinear stationary wave against longitudinal disturbances,[9]

$$\left(\frac{\partial^2 \omega}{\partial \kappa^2}\right)_{\alpha=0}^{-1} \left(\frac{\partial \omega}{\partial \alpha^2}\right)_{\alpha=0} > 0, \tag{25}$$

and against transverse disturbances,[10]

$$\left(\frac{\partial \omega}{\partial \alpha^2}\right)_{\alpha=0} > 0, \tag{26}$$

it is easy to obtain, from (24),

$$\left(\frac{\partial \omega}{\partial \alpha^2}\right)_{\alpha=0} = \frac{\beta^2 \mathcal{C} \kappa^3}{12 C_f^0 (C + 4\kappa^4)} > 0, \quad \left(\frac{\partial^2 \omega}{\partial \kappa^2}\right)_{\alpha=0} = \frac{\beta^2}{C_f^0}\left(\frac{C}{\kappa^3} - 3\kappa\right). \tag{27}$$

As we can see from (26) and (27), the long weak nonlinear wave (23) is always stable with respect to transverse disturbances. From (25) and (27), it follows that it is stable with respect to longitudinal disturbances only if

$$\kappa < \left(\frac{f_*^2}{3C_f^0 \beta}\right)^{1/4}. \tag{28}$$

Thus, the effect of the Earth's rotation on the long oceanic wave propagation results, as has been demonstrated here, in two conclusions diametrically opposed to those usual for KdV equations: (1) nonexistence of soliton-like solutions and (2) stability of those solutions which are like periodic weak nonlinear long waves.

AN ASYMPTOTIC PROCEDURE FOR DERIVING THE REDUCED EQUATION 5 DIRECTLY FROM THE THEORY OF INTERNAL WAVES

Despite its generality, the derivation of (5) given above does not elucidate some subtle questions: What is the relation between the "longitudinal" function u in (4) and (5) and the dynamical variables occuring in the initial equations for internal waves? What is the asymptotic procedure within the framework of which the reduced equation 5 can be obtained from the initial set of equations? How is it possible to

Leonov: Effect of Earth's Rotation on Wave Propagation 155

calculate the coefficients of (5)? In order to answer these questions, we shall consider an asymptotic procedure that allows us to derive (5) directly from the hydrodynamic equations of a stratified rotating fluid.

In a local frame of reference, the dimensionless set of equations and boundary conditions for wave motion of a stratified fluid can be written as follows:

$$(\rho_0 + \rho)(\partial_t \mathbf{v} + (\mathbf{v} \cdot \nabla)\mathbf{v}) + \nabla p + (\rho_0 + \rho)\boldsymbol{\kappa} + 2(\rho_0 + \rho)\boldsymbol{\Omega} \times \mathbf{v} = 0$$

$$\nabla \cdot \mathbf{v} = 0, \quad \partial_t \rho + (\mathbf{v} \cdot \nabla)\rho + \rho_0' \mathbf{v} \cdot \boldsymbol{\kappa} = 0,$$

$$[\partial_t p + (\mathbf{v} \cdot \nabla)p]_{z=\zeta} = 0, \quad [\partial_t \zeta + \mathbf{v}_\parallel \cdot \nabla_\parallel \zeta - \mathbf{v} \cdot \boldsymbol{\kappa}]_{z=\zeta} = 0, \quad \mathbf{v} \cdot \boldsymbol{\kappa}|_{z=-1} = 0 \quad (29)$$

$$(\boldsymbol{\Omega} = f_* \{\cos \phi \sin \psi, \cos \phi \cos \psi, \sin \phi\}),$$

where $\mathbf{v} = \{u, v, w\}$ is the velocity vector, $\mathbf{v}_\parallel = \{u, v\}$ represents its horizontal component, p is pressure, ρ is the deviation of density from the distribution $\rho_0(z)$ in the absence of wave motion, ζ is the displacement of the fluid free surface, the z axis (with unit vector $\boldsymbol{\kappa}$) is directed opposite to the gravity force, $\boldsymbol{\Omega}$ is the vector of the angular velocity of the earth's rotation corresponding to the local frame considered, the x axis of the system makes an angle ψ with the "geophysical" axis x' directed towards the east, and φ is latitude. Wave motion of the fluid within a horizontal homogeneous layer of depth H is considered. All the variables in (29) have been made dimensionless by means of the constants H, g, and ρ_*, where g is the acceleration due to gravity and ρ_* is the density of the fluid at the surface.

The following asymptotic expansions may be proposed for (29):[11]

$$\xi = \epsilon(x - C_f^0 t) - \epsilon^3 C_1 t + \cdots, \quad \eta = \epsilon^2 y, \quad z \equiv z, \quad \tau = \epsilon^3 t,$$

$$u = \epsilon^2 u_1 + \epsilon^4 u_2 + \cdots, \quad v = \epsilon^3 v_1 + \epsilon^5 v_2 + \cdots,$$

$$w = \epsilon^3 w_1 + \epsilon^5 w_2 + \cdots, \quad (30)$$

$$p = -\int \rho_0 dz + \epsilon^2 p_1 + \epsilon^4 p_2 + \cdots, \quad \rho = \epsilon^2 \rho_1 + \epsilon^4 \rho_2 + \cdots,$$

$$\zeta = \epsilon^2 \zeta_1 + \epsilon^4 \zeta_2 + \cdots, \quad f_* = \epsilon^2 f/2.$$

The last relation in (30) shows that the small positive parameter ϵ is related to the dimensionless modulus of angular velocity of the earth's rotation, $f_* = |\Omega| \sqrt{H/g}$, where $|\Omega| \simeq 0.7 \times 10^{-5}$ s^{-1} is the velocity of the earth's rotation.

Substituting (30) into (29) and equating all the terms of the same order, we obtain, in the lowest order, the following set of first order equations and corresponding boundary conditions:

$$-\rho_0 C_f^0 \partial_\xi u_1 + \partial_\xi p_1 = 0, \quad -\rho_0 C_f^0 \partial_\xi v_1 + \partial_\eta p_1 + \rho_0 f u_1 \sin \phi = 0,$$

$$\rho_1 + \partial_z p_1 = 0, \quad \partial_\xi u_1 + \partial_z w_1 = 0, \quad -C_f^0 \partial_\xi \rho_1 + \rho_0' w_1 = 0, \quad (31)$$

$$[C_f^0 \partial_\xi p_1 + w_1]_{z=0} = 0, \quad C_f^0 \partial_\xi \zeta_1 + w_1|_{z=0} = 0, \quad w_1|_{z=-1} = 0.$$

The set of second order equations and boundary conditions is of the form:

$$-\rho_0 C_f^0 \partial_\xi u_2 + \partial_\xi p_2 = f_1, \qquad -\rho_0 C_f^0 \partial_\xi v_2 + \partial_\eta p_2 + \rho_0 f u_2 \sin \phi = f_2,$$

$$p_2 + \partial_z p_2 = f_3, \qquad \partial_\xi u_2 + \partial_z w_2 = f_4, \qquad -C_f^0 \partial_\xi \rho_2 + \rho_0' w_2 = f_5, \qquad (32)$$

$$[C_f^0 \partial_\xi p_2 + w_2]_{z=0} = f_6, \qquad C_f^0 \partial_\xi \zeta_2 + w_2|_{z=0} = f_7, \qquad w_2|_{z=-1} = 0,$$

where the f_i are

$$f_1 = -\rho_0(\partial_\tau u_1 - C_1 \partial_\xi u_1 + u_1 \partial_\xi u_1 + w_1 \partial_z u_1) + \rho_1 C_f^0 \partial_\xi u_1$$
$$- f\rho_0(w_1 \cos\phi \cos\psi - v_1 \sin\phi),$$

$$f_2 = -\rho_0(\partial_\tau v_1 - C_1 \partial_\xi v_1 + u_1 \partial_\xi v_1 + w_1 \partial_z v_1)$$
$$+ \rho_1 C_f^0 \partial_\xi v_1 - f\rho_1 u_1 \sin\phi + f\rho_0 w_1 \cos\phi \frac{\sin\psi}{\epsilon}, \qquad (33)$$

$$f_3 = \rho_0 C_f^0 \partial_\xi w_1 + \rho_0 f u_1 \cos\phi \cos\psi, \qquad f_4 = -\partial_\eta v_1,$$

$$f_5 = -(\partial_\tau \rho_1 - C_1 \partial_\xi \rho_1 + u_1 \partial_\xi \rho_1 + w_1 \partial_z \rho_1),$$

$$f_6 = [(\partial_\tau p_1 - C_1 \partial_\xi p_1 + u_1 \partial_\xi p_1 + w_1 \partial_z p_1) - (C_f^0 \zeta_1 \partial_\xi \partial_z p_1 + \zeta_1 \rho_0' w_1$$
$$+ \zeta_1 \partial_z w_1)]_{z=0},$$

$$f_7 = [\partial_\tau \zeta_1 - C_1 \partial_\xi \zeta_1 + u_1 \partial_\xi \zeta_1 - \zeta_1 \partial_z u_1]_{z=0}.$$

Let us note that, owing to the expressions for f_2 in (33), the expansions (30) are applicable to waves propagating outside the equatorial zone (cos $\phi \neq 0$) only if the waves move almost parallel to the geophysical axis x', or, to be more precise, if sin $\psi \simeq \psi \simeq \epsilon\psi_0$. Near the equatorial zone (cos $\phi \sim \epsilon$), the waves may have arbitrary directions of propagation. In order to be more definite, we shall discuss below only the first case: sin $\psi \sim \epsilon$, cos $\psi \simeq 1$.

Now let us consider (31) for the first order approximation. The second equation of the set and the second boundary condition are split off from the other ones and just serve to determine the variables v_1, and ζ_1, respectively. One can also notice that it follows that

$$\partial_\xi v_1 - \partial_\eta u_1 = \frac{f \sin\phi}{C_f^0} u_1.$$

Thus, this approximation exhibits some vortex motion in the horizontal plane, which is due to the earth's rotation.

Equations 31 can be easily reduced to the boundary value problem for the function w_1:

$$\partial_z(\rho_0 \partial_z w_1) - (C_f^0)^{-2} \rho_0' w_1 = 0, \qquad \left[\partial_z w_1 - \frac{w_1}{C_f^{0\,2}}\right]_{z=0} = 0, \qquad w_1|_{z=-1} = 0, \quad (34)$$

which is separable, that is, it is possible to separate the variables: $w_1 = \mathcal{A}(\tau, \xi, \eta) W(z)$. This is the basic condition for choosing asymptotic expansions in the form of (30). It may be shown that such asymptotic expansions are unique within a class of regular (with respect to parameter ϵ) asymptotic expansions, resulting in the separable boundary value problem (34).

Leonov: Effect of Earth's Rotation on Wave Propagation 157

All the dynamical variables of the first order approximation can be expressed through the "longitudinal" function $U(\tau, \xi, \eta)$ and the vertical function $W(z)$ as follows

$$u_1 = UW', \qquad v_1 = W' \int \left(\partial_\eta U + \frac{f \sin \phi}{C_f^0} U\right) d\xi, \qquad w_1 = -W \partial_\xi U$$

$$p_1 = C_f^0 \rho_0 UW', \qquad \rho_1 = -\frac{\rho_0' UW}{C_f^0}, \qquad \zeta_1 = \frac{UW(0)}{C_f^0}. \tag{35}$$

From (34), we have the following boundary value problem for $W(z)$:

$$(\rho_0 W')' + \frac{\rho_0 \Omega}{C_f^{0\,2}} W = 0, \qquad \left[W' - \frac{W}{C_f^{0\,2}}\right]_{z=0} = 0, \qquad W(-1) = 0, \tag{36}$$

where $\Omega = -\rho_0'/\rho_0 > 0$ is the square of the Väisäla-Brunt dimensionless frequency.

This boundary value problem is self-conjugate; it generates a complete set of denumerable eigenfunctions $\{W_n(z)\}$ corresponding to the set of eigenvalues $(C_{f,n}^0)^{-2}$; moreover, the set $\{W_n(z)\}$ may be chosen to be orthonormal:

$$(W_n, W_j) \equiv W_n(0) W_j(0) + \int_{-1}^{0} \rho_0 \Omega W_n W_j \, dz = \delta_{nj}. \tag{37}$$

Furthermore, for the inhomogeneous boundary value problem associated with (36),

$$(\rho_0 W')' + \frac{\rho_0 \Omega}{C_f^{0\,2}} W = F_1(z), \qquad \left[W' - \frac{W}{C_f^{0\,2}}\right]_{z=0} = F_2, \qquad W|_{z=-1} = 0, \tag{38}$$

the following solvability condition must be fulfilled:

$$-F_2 W(0) + \int_{-1}^{0} F_1 W \, dZ = 0. \tag{39}$$

Now let us consider the second-order approximation in (32). In that set of equations, the second equation and second boundary condition are, again, split off from the other ones and, again, just serve to determine the variables v_2 and ζ_2, respectively. This set of equations, as well as the first-approximation set, can be reduced to the following inhomogeneous boundary value problem for w_2:

$$\partial_z(\rho_0 \partial_z w_2) - \frac{\rho_0' w_2}{C_f^{0\,2}} = F_1, \qquad \left[\partial_z w_2 - \frac{w_2}{C_f^{0\,2}}\right]_{z=0} = F_2, \qquad w_2|_{z=-1} = 0, \tag{40}$$

where the right-hand sides (the functions F_1 and F_2) can, because of (33), (35), and (36), be represented as follows:

$$F_1 = \frac{\partial_z f_1 - \partial_\xi f_3}{C_f^0} + \partial_z(\rho_0 f_4) - \frac{f_5}{C_f^{0\,2}}$$

$$= \frac{2\rho_0 \Omega W}{C_f^{0\,3}} (U_\tau - C_1 U_\xi)$$

$$- \frac{f \cos \phi \rho_0 \Omega W}{C_f^0} U_\xi + \frac{UU_\xi}{C_f^0} \left\{[(WW'' - W'^2)\rho_0 - \rho_0' W' W]' + \frac{W^2 \rho_0''}{C_f^{0\,2}}\right\}$$

$$+ \rho_0 W U_{\xi\xi\xi} + \frac{\rho_0 \Omega W}{C_f^{0\,2}} \int U_m d\xi - \frac{f^2 \sin^2\phi\, \rho_0 \Omega W}{C_f^{0\,4}} \int U d\xi,$$

$$F_2 = \left[\frac{f_1}{C_f^{0\,4}} + f_4 + \frac{f_6}{C_f^{0\,2}} \right]_{z=0}$$

$$= -\frac{2W(0)}{C_f^{0\,3}} (U_\tau - C_1 U_\xi) - \frac{f \cos\phi}{C_f^0} W(0) U_\xi$$

$$+ \frac{UU_\xi}{C_f^{0\,3}} \left(\Omega(0) W^2(0) - \frac{W^2(0)}{C_f^{0\,2}} \right) - \frac{W(0)}{C_f^{0\,2}} \int U_m d\xi \quad (41)$$

$$+ \frac{f^2 \sin^2\phi}{C_f^{0\,4}} W(0) \int U d\xi.$$

Let us seek the solution $w_2(\tau, \xi, \eta, z)$ of the linear boundary value problem (40) in the form

$$w_2 = B(\tau, \xi, \eta) W(z) + w_2^*(\tau, \xi, \eta, z), \quad (42)$$

where $B(\tau, \xi, \eta)$ is a "longitudinal" function of the second-order approximation, which is to be determined from the third-order approximation, and w_2^* is a solution of the inhomogeneous boundary value problem (40). Then the function w_2^* will exist only if the solvabilty condition (39) is fulfilled, which, taking account of (41), (36), and (37), yields the amplitude equation (5) for the second-order approximation:

$$\partial_\tau U - C_f^{(1)} \partial_\xi U + \gamma U \partial_\xi U + \frac{\beta^2}{2C_f^0} \partial_{\xi\xi\xi}^3 U + \frac{C_f^0}{2} \int \partial_m^2 U d\xi = \frac{f_*^2}{2C_f^0} \int U d\xi,$$

where

$$f_* = f \sin\phi, \quad C_f^{(1)} = C_1 + \frac{f \cos\phi}{2} C_f^{02}, \quad \beta^2 = C_f^{04} \int_{-1}^0 \rho_0 W^2 dz,$$

$$\gamma = \frac{C_f^{0\,2}}{2} \left[W^2(0) \left(\frac{1}{C_f^{0\,2}} - \Omega(0) \right) \right. \quad (43)$$

$$\left. + \int_{-1}^0 \left(-2\rho_0' W'^2 W + \rho_0 W'^3 - 4\rho_0 \frac{W' W^2}{C_f^{0\,2}} \right) dz \right]$$

Equations 43 allow us to express the coefficients of (5) through functionals of the eigenfunction $W(z)$ of the linear boundary value problem (36), if a certain mode is excited. It should be noted, besides, that the difference between $C_f^{(1)}$ and C_1 in (43) appears because of the additional anisotropy of wave motions due to the Coriolis force. Equations 35 give the relation in question between the dynamical variables of the first-order approximation and the "longitudinal" function U satisfying (5).

Summary

Stationary solutions of an amplitude equation that describes the propagation of long surface and internal oceanic waves, taking account of the earth's rotation, have been considered. The equation was derived both from the dispersion relation and directly from the hydrodynamic equations. It was shown that a solitary wave solution of the equation does not exist; however, solutions in the form of sufficiently long periodic waves with small finite amplitude are stable with respect to longitudinal and transverse disturbances.

References

1. TER-KRIKOROV, A. M. 1965. On the theory of stationary waves in nonuniform fluid. Prikl. Mat. Mekh. **28**(3): 440–52.
2. BENJAMIN, T. B. 1966. Internal waves of finite amplitude and permanent form. J. Fluid Mech. **25**(2): 241–55.
3. LEONOV, A. I. & YU. Z. MIROPOL'SKII. 1975. On the theory of nonlinear internal gravitational waves of permanent form. Izv. Akad. Nauk SSSR Fiz. Atmos. Okeana. **11**: 491-502.
4. LEONOV, A. I., YU. Z. MIROPOL'SKII & R. E. TAMSALU. 1979. Nonlinear stationary internal and surface waves in shallow seas. Tellus **32**(2): 150–60.
5. KADOMTSEV, B. B. & V. I. PETVIASHVILI. 1970. On the stability of solitary waves in weak dispersive media. Dokl. Akad. Nauk SSSR **192**(4): 753–56.
6. ZAKHAROV, V. YE. 1975. Instability and nonlinear oscillations of solitons. Zh. Eksp. Teor. Fiz. Pis'ma Red. **22**(7): 364–67.
7. OSTROVSKII, L. A. 1978. Nonlinear internal waves in rotating ocean. Oceanology (USSR) **18**(2).
8. ODULO, A. B. 1978. On the equations of long nonlinear waves in the ocean. Oceanology (USSR) **18**(6): 965–71.
9. LIGHTHILL, M. J. 1965. Inst. Math. Appl. **1**: 269.
10. KARPMAN, V. I. 1973. Nonlinear Waves in Dispersive Media. Nauka. Moscow. (In Russian.)
11. LEONOV, A. I. 1976. On two-dimensional KdV equation in nonlinear theory of surface and internal waves. Dokl. Akad. Nauk SSSR **229**(4): 820–23.

RECENT DEVELOPMENTS IN CONTOUR DYNAMICS FOR THE EULER EQUATIONS*†

Norman J. Zabusky

Institute for Computational Mathematics and Applications
Department of Mathematics and Statistics
University of Pittsburgh
Pittsburgh, Pennsylvania 15261

INTRODUCTION

The evolution of inviscid flows in two dimensions often elucidates realistic nearly inviscid physical situations. The method of contour dynamics, a generalization of the "water bag" model, is ideally suited to treating the dynamics of incompressible, nondissipative fluids in two dimensions. For example, Longuet-Higgins and Cokelet have studied incompressible shallow and deep water waves on boundaries between regions where the density is piecewise-constant.[1] Baker et al.[2] and Meiron et al.[3] have investigated the Rayleigh-Taylor instability and free-surface flows described by the Euler equations. Zabusky et al.,[4] Deem and Zabusky,[5] and Landau and Zabusky[6] have investigated the Euler equations with piecewise-constant finite-area vortex regions (FAVRs). Finally, Overman and Zabusky have studied the evolution of a piecewise-constant, weakly ionized, and strongly magnetized plasma (or deformable dielectric) in an electric field.[7] There are no existence theorems for these flows and we have no assurance that solutions exist for all times. Computational evidence suggests that the Rayleigh-Taylor and magnetized plasma problems are poorly posed and that interfacial corners often arise after a finite time.

In this article, I review recent progress with the Euler equations, including stationary singly-connected "V-states" and their stability and the dynamical "breaking" of an unstable elliptical vortex.

EULER EQUATIONS OF MOTION AND CONTOUR DYNAMICS ALGORITHM

The Euler equations in two space dimensions can be written in vorticity-stream function form as

$$\omega_t + u\omega_x + v\omega_y = 0, \tag{1a}$$

$$\Delta\psi \equiv \psi_{xx} + \psi_{yy} = -\omega, \tag{1b}$$

where

$$u = \psi_y, \quad v = -\psi_x. \tag{1c}$$

---

*Dedicated to Irina Brailovsky, loving mother and steadfast wife.
†This research was supported by contracts from the Office of Naval Research, nos. N00014-77-C-0520 (Task NR 062 583) and N00013-78-C-0074.

Zabusky: Developments in Contour Dynamics for Euler Equations

If the vorticity is composed of a set of piecewise-constant finite area vortex regions (FAVRs), that is, if each member of the set is a characteristic function χ_i of magnitude ω_i and boundary Γ_i, or $\omega(x, y, t) = \Sigma_i \, \omega_i \, \chi_i \, (x, y, t)$, then

$$\psi(x, y) = -(2\pi)^{-1} \sum_i \int\!\!\int_{\mathbf{R}^2} \omega_i \chi_i \, G(x - \xi, y - \eta) \, d\xi d\eta, \tag{2}$$

where we use the two-dimensional Green's function

$$G = (\tfrac{1}{2}) \log [(x - \xi)^2 + (y - \eta)^2] = (\tfrac{1}{2}) \log r^2 \tag{3}$$

for flow in an unbounded domain. Equation 1 says that every point of the fluid including the boundary is convected with the flow. The evolution equation for boundary points is the *area-preserving* mapping

$$(x_t, y_t) \equiv (u(x, y, t), v(x, y, t)) = (2\pi)^{-1} \sum_j \int_{\Gamma_j} \log r (d\xi, d\eta), \tag{4}$$

where $(x, y) \in \Gamma_i$ and $(\xi, \eta) \in \Gamma_j$. We have used Green's theorem to replace the area integral over the domain of χ_i by the line integral over its boundary Γ_i, thus reducing the dimensionality by one.

STATIONARY VORTEX STATES (V-STATES) OF THE EULER EQUATIONS

If we add a small harmonic perturbation to a circular FAVR of radius a and density ω_1, we obtain a unidirectional wave on the surface. That is, if

$$r = a[1 + \epsilon \cos (m\theta - \Omega_m t)], \tag{5}$$

then the linear dispersion relation is

$$\Omega_m = (\omega_1/2)(m - 1)/m. \tag{6}$$

In the late nineteenth century, Kirchoff found that an elliptical FAVR is stationary in a frame of reference rotating with angular velocity $\Omega^{(2)} = \omega_1 \, ab/(a + b)^2$, where a and b are the semimajor and semiminor axes.

Recently, Deem and Zabusky found several examples of rotating and translating stationary solutions, which they called "V-states."[5] The set of uniformly rotating V-states, $V(m, \Omega^{(m)})$, are single FAVRs of constant vorticity density, having m-fold symmetry and angular velocity $\Omega^{(m)}$. They are natural extensions of the Kirchoff ellipitical vortex and represent bifurcations from the dispersion relation, (6). The set of uniformly translating V-states are the "desingularized" representations of two oppositely signed point vortices.

Recently, using an improved first-order relaxation algorithm, we have found more completely filled sets of singly-connected rotating V-states, translating V-states, and doubly-connected rotating states.[6] Saffman and colleagues have used a Newton-Raphson procedure to find doubly-connected symmetric rotating V-states,[8] a single-row periodic array of FAVRs, analogous to a free-shear layer,[9] and a double-row periodic array of FAVRs, analogous to an asymmetric wake.[10]

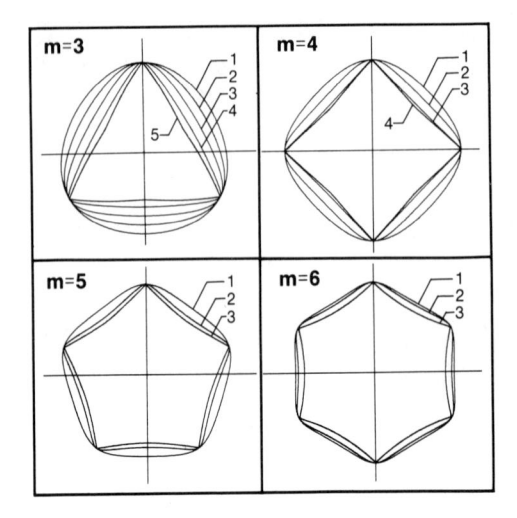

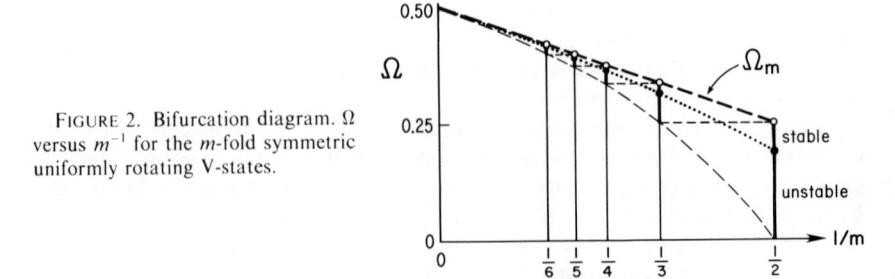

FIGURE 1. Uniformly rotating V-states: (a) $V(3,\Omega^{(3)})$, (b) $V(4,\Omega^{(4)})$, (c) $V(5,\Omega^{(5)})$, and (d) $V(6,\Omega^{(6)})$. (The contour labels correspond to the case numbers in Table 1 of Reference 6, where details are given.)

These states were obtained by using the boundary condition $\mathbf{n} \cdot \mathbf{v}_{particle} = \mathbf{n} \cdot \mathbf{v}_{boundary}$. For a singly connected rotating region of vorticity, this can be written as

$$\partial_s\psi + \Omega r(dr/ds) = 0, \quad (x, y) \in \Gamma. \tag{7}$$

Integrating once, we obtain

$$\psi(x, y) + \Omega r^2/2 = c, \quad (x, y) \in \Gamma. \tag{8}$$

Numerical results for $m = 3$ through $m = 6$ are shown in FIGURE 1. There is strong evidence that, for $m > 2$, the region of each bifurcation parameter $\Omega^{(m)}$ is finite (the vertical dark line segments in FIGURE 2), or

$$\frac{(m-2)}{(m-1)} \leq \Omega^{(m)} \leq \frac{(m-1)}{m}, \quad m > 2, \tag{9}$$

where the upper termination point (open circles) is associated with the circle, (6). At the lower termination point (light dashed line), the boundaries become nonanalytic.

FIGURES 3, 4, and 5 show doubly-connected regions. FIGURES 3 and 4 correspond

FIGURE 2. Bifurcation diagram. Ω versus m^{-1} for the m-fold symmetric uniformly rotating V-states.

FIGURE 3. Rotating doubly-connected symmetric V-states with $\omega_1 = \omega_2 = 1.0$. (The contour labels correspond to the case numbers of Table 3 of Reference 6.)

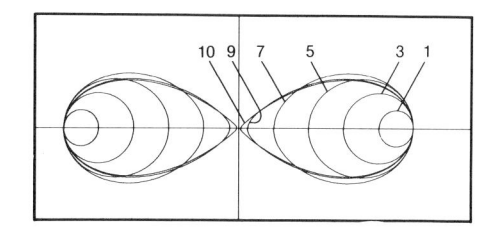

to fixed contour rotation about the origin and various positions along the x-axis to the right of the origin. FIGURE 5 shows the upper half plane for the states that translate uniformly along the x-axis obtained from the boundary relation

$$\psi(x, y) + Uy = c_i, \quad (x, y) \in \Gamma_i, \quad i = 1, 2. \tag{10}$$

Our work confirms the main body of the calculations of Pierrehumbert.[11] However, we are presently uncertain about his limiting case, where parts of the boundaries of the positive and negative regions nearly coincide.

Stability of m-Fold Symmetric Uniformly Rotating V-States[12]

Love investigated the stability of Kirchoff's elliptical vortex ($m = 2$) by expressing the inner and the outer stream function in elliptic coordinates and adjusting them on the slightly disturbed elliptical boundary to satisfy the conditions of continuity of the tangential and normal velocity and the condition that the surface of the vortex always contain the same particles.[13] Recently, Burbea developed a new method that uses the so-called Grunsky, Toeplitz, and Hankel operators to form a certain self-adjoint operator (Burbea's operator) in the space of complex sequences l_2.[14,15] He proved the theorem that relates the stability of m-fold symmetric states to the spectrum of this operator. The major advantage of this approach is that it is independent of a particular coordinate system. Using the results of Burbea's theortical work, Landau developed an efficient numerical algorithm that found the conformal mapping g of the exterior of

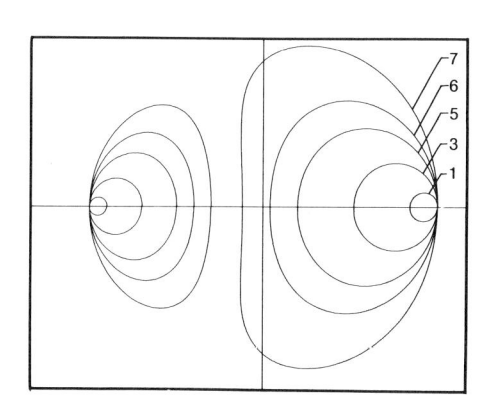

FIGURE 4. Rotating doubly-connected asymmetric V-states with $\omega_1 = -0.8$ (left) and $\omega_2 = 1.0$ (right). (The contour labels correspond to the case numbers in Table 5 of Reference 6.)

164   Annals New York Academy of Sciences

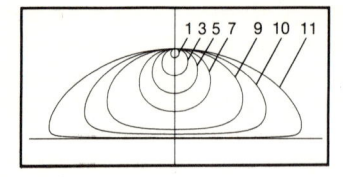

FIGURE 5. Translating symmetric V-states. Only the upper half plane contours are shown. (The contour labels correspond to the case numbers in Table 7 of Reference 6).

the unit disk onto the exterior of a FAVR by employing Theodorsen's method.[12] Next, he calculated the coefficients of the Fourier expansion of the above determined g and the truncated Grunsky operator. The truncated Toeplitz and Hankel operators are easily computed from the Fourier expansion of the so-called vortex-function.[14,15] Finally, he obtained a dispersion relation for perturbations to $V(m, \Omega^{(m)})$ proportional to $\exp[i\sigma_n^{(m)}t]$.

FIGURE 6 shows the dispersion diagram $(\sigma_n^{(m)})^2$ versus $\Omega^{(m)}$ for the V-states $V(2, \Omega^{(2)})$ (Kirchoff elliptic vortex), $V(3, \Omega^{(3)})$, and $V(4, \Omega^{(4)})$. The index n, $1 \leq n \leq \infty$, within each region labels the eigenvalues corresponding to the eigenfunctions of the associated perturbation. These eigenfunctions are obtained from the mapping, which takes a circle in the transform plane to the V-state in the physical plane. In the transform plane, the eigenfunctions are $\cos n\theta'$ and $\sin n\theta'$.

The solid lines in the $m = 2$ region correspond to eigenvalues of the perturbed Kirchoff elliptic vortex

$$(\sigma_n^{(2)})^2 = \frac{1}{4}\left\{\left[\frac{2n\alpha}{(1+\alpha)^2} - 1\right]^2 - \left(\frac{1-\alpha}{1+\alpha}\right)^{2n}\right\},$$

where $\alpha < 1$ is the ratio of the minor axis to the major axis and the vorticity density is

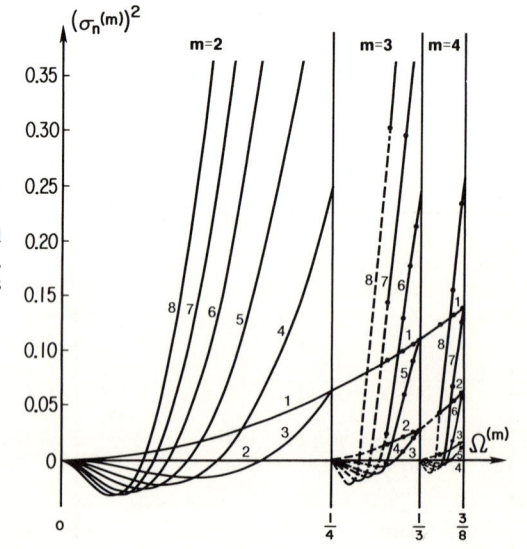

FIGURE 6. Eigenvalues for m-fold symmetric V-states: $V(2,\Omega^{(2)})$, $V(3,\Omega^{(3)})$, and $V(4,\Omega^{(4)})$. $(\sigma_n^{(m)})^2$ versus $\Omega^{(m)}$.

unity. This result was first obtained by Love[13] and repeated in a terse and elegant calculation by Burbea and Landau (Sec. 2.6 of Reference 12). By numerical calculation they also found many of the points on these curves to three significant figures, thereby validating the algorithm for eigenvalue and eigenfunction determination. The solid lines in the $m = 3$ and $m = 4$ regions connect points calculated by the algorithm and the dashed lines are qualitative predictions of how the curves approach zero as one approaches the region terminator (at the left of each region).

A remarkable result is that one-fourth of each region corresponds to stable perturbations; that is, the range of stability of $V(m, \Omega^{(m)})$ is

$$\Omega_c^{(m)} < \Omega^{(m)} \leq \frac{(m-1)}{2m}, \tag{11a}$$

and the range of instability is

$$\frac{(m-2)}{2(m-1)} < \Omega^{(m)} < \Omega_c^{(m)}, \tag{11b}$$

TABLE 1
CRITICAL VALUES OF ASPECT RATIOS AND ANGULAR VELOCITIES
FOR m-FOLD SYMMETRIC ROTATING V-STATES

m	$\alpha_c^{(m)}$	$\Omega_c^{(m)}$
2	0.3333	0.1875
3	0.6654	0.3122
4	0.7854	0.3641
5	0.8532	0.3933
6	0.8911	0.4120

where

$$\Omega_c^{(m)} = \frac{1}{8}\left[\frac{3(m-1)}{m} + \frac{m-2}{m-1}\right]. \tag{11c}$$

That is, the point $(\sigma_{m+1}^{(m)})^2 = 0$ occurs at $\Omega_c^{(m)}$. The values of the critical aspect ratio (minimum radius/maximum radius) or rotational velocity $\Omega_c^{(m)}$ are shown in TABLE 1. Also,

1. $\sigma_m^{(m)} = 0$
2. $[\sigma_{m-n}^{(m)}|_{\alpha=1}]^2 = [\sigma_{m+n}^{(m)}|_{\alpha=1}]^2$, for $n = 1, 2 \ldots (m-1)$,

where $\alpha = 1$ corresponds to the right-most point of the region. Thus, at $\Omega = 3/8$, $(\sigma_1^{(4)})^2 = (\sigma_7^{(4)})^2$, $(\sigma_2^{(4)})^2 = (\sigma_6^{(4)})^2$ and $(\sigma_3^{(4)})^2 = (\sigma_5^{(4)})^2$.

3. $(\sigma_{m+1}^{(m)})^2 \neq (\sigma_{m+1}^{(m+1)})^2$ as $\Omega \to \Omega_m = (m-1)/2m$. That is, there is an evident discontinuity.

Desingularization

The existence of the doubly-connected V-states allows us to make the following "desingularization" conjecture:

> For the two-dimensional inviscid Euler equations, if one can find stationary (uniformly rotating or uniformly translating) arrays of *point* vorticies each with circulation γ_i, then one can find stationary arrays of sets of *nested* contours with corresponding circulations in the same spatial region.[6]

That is, each point vortex γ_i can be desingularized into J_i nested contours Γ_{ij} ($j = 1$, ... J_i) of area A_{ij} and vorticity increment $[\omega_{ij}]$, such that

$$\gamma_i = \sum_{j=1}^{J_i} A_{ij} [\omega_{ij}]. \qquad (12)$$

If the FAVRs in each group are small in "diameter" compared to the interset distance, then probably neither J_i, A_{ij}, nor $[\omega_{ij}]$ are required to be identical as long as (12) is preserved. That is, we conjecture that it is possible to find asymmetric translating V-states. Also, it is possible that we can desingularize J. J. Thomson's vortex "atom," that is, identical point vortices spaced uniformly on a circle and a point vortex at the center.

DYNAMICAL EVOLUTION OF STABLE AND UNSTABLE ELLIPTICAL V-STATES

We discretized the integral of (4) by assuming that each contour is composed of N_j nodes that are connected by linear segments; we then carried out the resulting integrals exactly.[4] The resulting $2N_j$ ordinary differential equations per contour are solved by predictor-corrector methods. We used an ad hoc node insertion algorithm if the perimeter grew to the point that the spacing between nodes exceeded an a priori magnitude h_{max} or if the ratio of adjacent segments fell outside a range

$$r_{min} < \left(\frac{h_{n+1}}{h_n}\right) < r_{max}.$$

If either condition was met, we inserted a node and used a quadratic interpolation procedure to obtain nearly equal spacings among the nodes involved. Recently, Zabusky and Overman rationalized the node-insertion and removal process.[16] Their algorithm maintains an a priori accuracy in regions where corners "tend" to form on the contour; that is, the interval (Δs) between nodes in the interval

$$(\Delta s)_{min} \leq s \leq \frac{c_1}{\max(|\kappa|, c_2)},$$

or the density of nodes, is bounded above and proportional to the local curvature κ.

We studied the influence of perturbations of stable and unstable ellipses.‡ To

---

‡Reference 12, Chapter III. A primitive node-insertion algorithm was used.

Zabusky: Developments in Contour Dynamics for Euler Equations 167

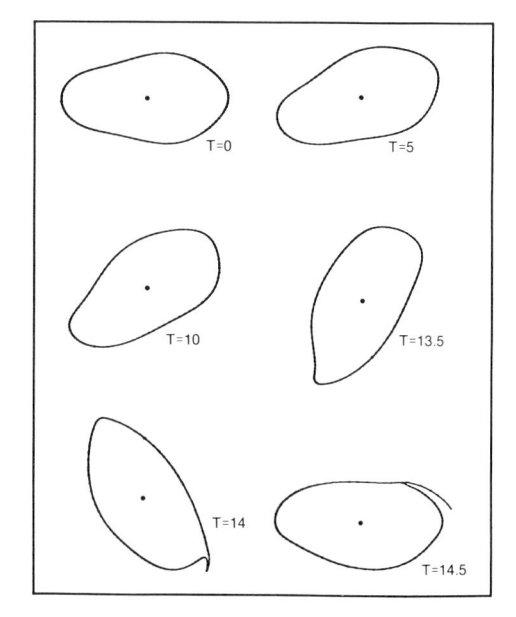

FIGURE 7. Motion of 2:1 ellipse subjected to a ($M = 3$, $\epsilon = 0.05$) perturbation ((13) and (14)).

mitigate the magnitude of the computation, we chose the asymmetrical perturbation

$$\xi = \xi_0 - \epsilon \cos(M\theta)[(\alpha \sin\theta)^2 + (\cos\theta)^2], \qquad (13)$$

with

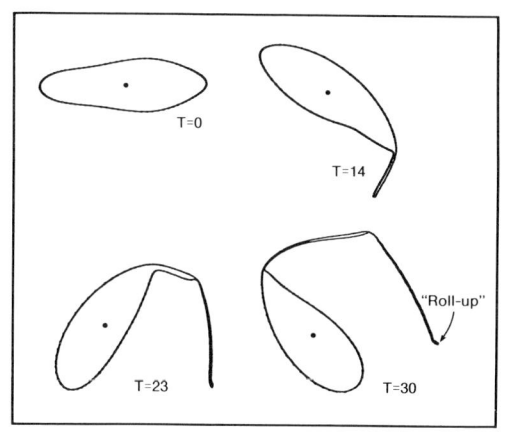

$$x = \cosh\xi \, \cos\theta, \qquad (14a)$$

$$y = \sinh\xi \, \sin\theta. \qquad (14b)$$

FIGURE 8. Motion of 3.5:1 ellipse subjected to a ($M = 3$, $\epsilon = 0.05$) perturbation ((13) and (14)) with an ad hoc node-insertion algorithm.

Here, ξ is an elliptic radius of the perturbed ellipse, ξ_0 is an elliptic radius of the unperturbed ellipse, and α is the aspect ratio. We performed experiments with $M = 3,4$ and different values of ϵ.

Experiments were first conducted with a *stable* (2:1 or $\alpha = \frac{1}{2}$) ellipse. We took $\epsilon = 0.01$, $M = 3,4$ and ran this case five full rotations. The area oscillated only in the sixth significant figure. The circumference oscillated in the third significant figure, but the circumference averaged over the period oscillated only in the sixth significant figure. Thus, we have agreement with the linear stability analysis described above.

We tried $\epsilon = 0.03$ and $\epsilon = 0.05$. The circumference began to grow rapidly after a short period of oscillatory behavior and soon corners developed, as shown in FIGURE 7. There seems to be a tendency for the area of the ejected filaments to decrease to zero. Results for an unstable ellipse (3.5:1 or $\alpha = 2/7$) for $M = 3$ and $\epsilon = 0.05$ are shown in FIGURE 8. Note that the area of the vortex filament being ejected is proportional to ϵ (see TABLE 2) and remains nearly constant for the remainder of the calculation. As the filament becomes longer and narrower, a second eruption takes place. Again, the area of this second region does not change during the calculation. After some time, the FAVR starts to look like a spiral with the dominant part approximating to a perturbed

TABLE 2

AREAS OF PERTURBED 3.5:1 ELLIPSES AND EJECTED FILAMENTS FOR $M = 3$ AND VARIOUS ϵ

ϵ	Area of Initial Perturbed Ellipse	Area of Filament After 1st Ejection	Area of Filament After 2nd Ejection
0.01	0.8969	0.00204	0.04150
0.03	0.8972	0.01020	0.04192
0.05	0.8978	0.02051	0.04271

stable ellipse (3:1). Note, too, the incipient clockwise "roll-up" of the end-region of the filament.

REGULARIZATION

Our regularization algorithm is a two-part procedure for modeling weak dissipation.[16] Thus, we are able to compute realistic motions of isolated piecewise-constant regions for "intermediate" times. The two parts are performed sequentially and approximate the small-scale and large-scale effects of "true" viscous dissipation. The small-scale procedure uses a second derivative with respect to arc length (which is proportional to the local curvature). In the large-scale procedure, we model global spreading and decay by allowing the boundary Γ to expand and by allowing the magnitude Ω to decrease monotonically, all in a self-consistent manner, as described below.

For the small-scale procedure, we add a term $(X, Y) = \mu(x_{ss}, y_{ss})$ to the evolution algorithm, or

$$(x_t, y_t) = (\hat{x}_t, \hat{y}_t) + \mu(x_{ss}, y_{ss}), \quad (x, y) \in \Gamma, \tag{15}$$

where the circumflex designates the advective part of the mapping and s is the arc length. This algorithm models the two-dimensional diffusion equation $\omega_t = \mu(\omega_{xx} + \omega_{yy})$ to the first order. One finds that the area and perimeter obey the evolution equations

$$A_t = \hat{A}_t - \mu t \quad \text{and} \quad P_t = \hat{P}_t - \mu \oint \kappa^2 ds. \tag{16}$$

The Euler and plasma-cloud equations are area-preserving mappings and $\hat{A}_t = 0$.

The decrease in area observed above contradicts the physically expected slow increase resulting from diffusive "spreading." To obtain global spreading, we include the large-scale procedure for increasing A and decreasing the magnitude $\Omega = \Omega(t)$ of piecewise-constant functions. These are based on the integral expressions

$$\partial_t \iint_{\mathrm{IR}^2} \omega \, da = 0 \implies d(\Omega A)/dt = 0, \tag{17}$$

$$\partial_t \iint_{\mathrm{IR}^2} \omega^2 \, da = -2\mu \iint_{\mathrm{IR}^2} |\nabla \omega|^2 \, da \implies$$

$$d(\Omega^2 A)/dt \simeq -\mu \Omega^2 \, P/[(2\pi)^{1/2} \, (\delta_0^2 + 4\mu t)^{1/2}], \tag{18}$$

where P is the perimeter and δ_0 is a length-scale smaller than any wavelength on the contour.

Acknowledgments

The work above was accomplished in collaboration with Dr. M. Landau (stationary V-states and dynamical evolution) and Prof. E. A. Overman II (regularization and dynamical evolution). The work on linear stability was performed by Dr. M. Landau and Prof. J. Burbea.

References

1. Longuet-Higgins, M. S. & E. D. Cokelet. 1976. The deformation of steep surface waves on water. I. A numerical method of computation. Proc. R. Soc. London Ser. A **350**: 1–26; 1978. The deformation of steep surface waves on water. II. Growth of normal-mode instabilities. Proc. R. Soc. London Ser. A. **364**: 1–28.
2. Baker, G. R., D. I. Meiron & S. A. Orszag. 1980. Phys. Fluids **23**: 1485–1490.
3. Meiron, D. I., S. A. Orszag & M. Israeli. 1981. Applications of numerical conformal mapping. J. Comput. Phys. **40**: 345–60.
4. Zabusky, N. J., M. H. Hughes & K. V. Roberts. 1979. Contour dynamics for the Euler equations in two dimensions. J. Comput. Phys. **30**: 96–106.
5. Deem, G. S. & N. J. Zabusky. 1978. Vortex waves: Stationary V-states, interactions, recurrence and breaking. Phys. Rev. Lett. **40**: 859–62. Also see: 1978. Stationary V-states, interactions, recurrence and breaking. *In* Solitons in Action. K. Lonngren and A. Scott, Eds.: 277–93. Academic Press. New York.
6. Landau, M. & N. J. Zabusky. 1981. Stationary solutions of the Euler equations in two dimensions. Singly- and doubly-connected V-states. Submitted for publication.

7. OVERMAN, E. A. & N. J. ZABUSKY. 1980. Stability and nonlinear evolution of plasma clouds via regularized contour dynamics. Phys. Rev. Lett. **45:** 1693–96.
8. SAFFMAN, P. G. & R. SZETO. 1980. Equilibrium shapes of a pair of equal uniform vortices. Phys. Fluids. **23:** 2339–42.
9. SAFFMAN, P. G. & R. SZETO. 1981. Structure of a linear array of uniform vortices. Stud. Appl. Math. To be published.
10. SAFFMAN, P. G. & J. C. SCHATZMAN. 1981. Properties of a vortex street of finite vortices. SIAM J. Sci. Stat. Comput.; 1981. Stability of a vortex street of finite vortices. To be published.
11. PIERREHUMBERT. R. T. 1980. A family of steady translating vortex pairs with distributed vorticity. J. Fluid Mech. **99:** 129.
12. LANDAU, M. 1981. The structure and stability of finite area vortex regions of the two-dimensional Euler equations. Ph.D. Thesis, University of Pittsburgh. Pittsburgh, Pennsylvania. The detailed numerical investigation of the linear stability is given in Chapter II and the dynamical evolution studies are given in Chapter III.
13. LOVE, A. E. H. 1893. On the stability of certain vortex motions. Proc. London Math. Soc. **35**(1): 18–42.
14. BURBEA, J. 1981. On the stability of vortex motions in the plane. To be published.
15. BURBEA, J. Vortex motions and their stability. Proc. Nonlinear Phenomena in Math. Sci., Arlington 1980, Academic Press. New York.
16. ZABUSKY, N. J. & E. A. OVERMAN II. 1981. Regularization of contour dynamical algorithms. J. Comput. Phys. In press.

CONFINEMENT AND PHASE TRANSITIONS IN GAUGE THEORIES

Laurence Jacobs

Instituto de Física, UNAM
Apdo. Postal 20-364
México 20, D.F.

INTRODUCTION

In this article I shall describe work done recently at the Brookhaven National Laboratory in collaboration with Michael Creutz and Claudio Rebbi.[1] Time limitations force me to be brief; however, both the method of analysis and a more detailed presentation of results will be soon available elsewhere.[2]

Gauge theory is generally believed to be the most promising framework for the description of all elementary interactions. Indeed, preliminary steps in the construction of a true unified field theory are already in hand: The electroweak $SU(2) \times U(1)$ model of Weinberg and Salam and the $SU(3)$ Quantum Chromodynamics are notably successful. Moreover, schemes based on larger gauge groups, of which $SU(5)$ is the simplest example, provide a highly satisfactory machinery for the unification of strong and electroweak interactions. Among the most spectacular new physics which this type of "grand unification" predicts is the previously unsuspected decay of the proton, implying the ultimate instability of all matter as well as providing profound new insight into the basic structure and history of the universe.

This impressive advance in our understanding of high energy physics has, nonetheless, encountered a very serious obstacle. Although we believe that QCD is as good a theory of the strong interaction as QED is of the electromagnetic interaction, the predictive power of the former is by no means comparable to that of the latter. The central reason for this deficiency is well understood: While perturbation theory is superb for QED, it is, in most cases, virtually useless in QCD.

A notable example to which perturbative techniques are not applicable is the low-energy, long-distance structure of quark dynamics. The demonstration of quark confinement and the extraction of hadron masses from QCD, important goals for a field theory, are outside the reach of conventional perturbative methods.

To overcome this failing and to progress further in our ability to extract information from field theory, a great deal of work has recently been done to develop effective nonperturbative techniques. Several directions have been explored in this connection. The most promising—in my belief—is based on the formulation of gauge theory proposed independently by Wilson and Polyakov.[3] This formalism, now known generally as lattice gauge theory, is a nonperturbative, gauge-invariant ultraviolet cutoff. The space-time manifold is replaced by a discrete hypercubical lattice of spacing a, and the gauge theory is formulated on this lattice in such a way that the conventional theory is regained in the limit as $a \rightarrow 0$. Besides defining the theory well, this regularization scheme has an added advantage: The field theory, when defined on the lattice, becomes a theory of nonlocal spin variables. Thus, the emerging spin system becomes, formally, a statistical system in which the powerful methods of statistical mechanics can be fruitfully applied. Moreover, methods devised for the study of ordinary spin systems can be generalized and applied to the case of gauge theories.

The generalized method that I will now describe is based on the so-called Monte Carlo technique,[4] which has been extensively used in the study of spin systems. I do not have time here to describe the method in any detail, but the interested reader can consult a number of good reviews on the subject.[2,5]

The basic idea of the method is quite simple. One wishes to compute the expectation values of quantum field operators, which, in the language of Feynman path integrals, are given by

$$\langle O[\phi] \rangle = \int (D\phi)\, O[\phi] e^{-S[\phi]} \Big/ \int (D\phi) e^{-S[\phi]}. \tag{1}$$

Starting with a trial configuration, ϕ_0 (i.e., a given collection of values for the dynamical variables), the algorithm consists of the generation of a new trial configuration ϕ according to a given probability distribution $P(\phi_0 \to \phi)$. P must satisfy a detailed balance condition:

$$P(\phi_0 \to \phi) \exp(-S[\phi_0]) = P(\phi \to \phi_0) \exp(-S[\phi]). \tag{2}$$

If P satisfies (2), after iterating the procedure described below, the probability of encountering any given configuration in the sequence will be given by the normalized Boltzmann distribution and, hence, will correspond to a configuration in "thermal equilibrium" with a "heat bath" at some appropriately defined "temperature."

The updating procedure, that is, the algorithm by which trial configurations are either kept in the sequence or rejected is as follows. If the new configuration ϕ is such that $S[\phi_0] > S[\phi]$, it is accepted; if the opposite is true, a random number, r, is generated uniformly in the unit interval and ϕ is accepted only if

$$\exp\{-(S[\phi] - S[\phi_0])\} > r. \tag{3}$$

A sequence of configurations $\phi^{(i)}$ is thus generated with the property that (1) can be evaluated as a finite sum

$$\langle O[\phi] \rangle \approx \frac{1}{n} \sum_{i=1}^{n} O[\phi^{(i)}]. \tag{4}$$

Equation 4 is valid independently of the choice of P (as long as it satisfies (2) and the distribution is chosen for computational and convergence advantages.[1,2]

In our analysis, we have used, for S, the Wilson action. Given a gauge theory for a group G, the spin variables are defined on the link joining two lattice sites i and j as

$$U_{ij} = \exp\left[i \int_i^j A_\mu dx^\mu\right] \in G, \tag{5}$$

where A_μ corresponds to the gauge potential as a matrix representation of the generators of G.

The gauge-invariant action is then constructed as a sum over the lattice of the product of four U's around elementary squares or plaquettes (p),

$$S = \frac{1}{8} \sum_{\{p\}} (1 - \omega). \tag{6}$$

where

$$\omega = U_{ij} U_{jk} U_{kl} U_{li}. \tag{7}$$

Inserting this action into a path integral and calling the inverse gauge coupling $\beta = 1/g^2$ defines a partition function for an equivalent statistical system at inverse temperature β,

$$Z = \sum_{U \in G} \exp[-\beta S]. \tag{8}$$

In practice, one deals with finite systems consisting of, say, N_s lattice sites. One can then define a free energy as

$$F = \lim_{N_s \to \infty} \ln Z. \tag{9}$$

A phase transition is signaled by a singularity in F or its derivatives with respect to β. The quantity that we use most in our analysis is the average action per plaquette, defined by (9) as

$$E = \langle 1 - \operatorname{Re} \omega \rangle = -\frac{1}{6} \frac{\partial F}{\partial \beta}. \tag{10}$$

Although E is not strictly an order parameter, its behavior as a function of β can serve to identify the phase structure of a dynamical system.

Critical Behavior

Once a gauge theory has been defined on the lattice, a study of its phase structure is of primary interest. Since the invention of the lattice regulator, it was shown that, in the strong coupling regime of both the U(1) theory and the SU(2) theory, elementary excitations are confined.[3] In fact, little difference can be seen between Abelian and non-Abelian theories in this regime. Moreover, conventional perturbation theory for the weak coupling regime indicates the existence of a massless phase in both cases. It is crucial, then, to see whether or not these two pairs of phases are connected through a phase transition. In the case of QED, where electrons and photons represent physical states, the Wilson model should have a phase transition from the confined, disordered phase observed for large coupling to the ordered, unconfined phase seen for weak coupling. If the lattice model is to be trusted in the study of non-Abelian theories, it is crucial that it work in the prototype gauge theory—the U(1) model.

Our first goal in applying the Monte Carlo technique to the study of QCD was, therefore, to map out the phase structure of the U(1) model. We found it convenient, in the pursuit of this goal, to first study the U(1) theory indirectly, as the large N limit of a sequence of theories corresponding to the discrete subgroups Z_N. Of course, gauge theories for finite groups are only sensible in discrete manifolds, and the considerably simpler structure of these theories is another unique advantage provided by the lattice formulation.

Once the theory has been defined, one has, in fact, created a small—but statistically significant—crystal of the gauge theory in the computer. With this setup, one can then perform a variety of "experiments" with this crystal. I shall describe some of the most useful simulations we have performed in our analysis.

Starting with the system in some definite configuration, say at zero bare coupling, one performs a single Monte Carlo iteration (that is, one updating of all the U_{ij} in the lattice) at some small coupling. One then increases the coupling by a small amount and performs another Monte Carlo iteration. This procedure is continued until $\beta = 0$ (infinite coupling) and then reversed until the initial value of β is reached. In this way, the system is never really in equilibrium but is generally close to it. However, if the sequence of values of β goes through a critical point β_c, the relaxation time increases and the system will remain in that region, relatively far from equilibrium. If, at the end of each iteration, one computes the average action per plaquette, E, and plots it as a function of β, one will observe the following. Since $E(\beta = \infty) = 0$ and $E(\beta = 0) = 1$, a phase transition far from these points will not affect the values of $E(\beta)$ derived from the "heating" and "cooling" ends of the cycle, and both halves of these cycles will therefore coincide at the ends of the cycle. Near a phase transition, however, the values of $E(\beta)$ coming from either end of the half-cycle will differ substantially. If the system undergoes a phase transition, this plot will have the typical shape of a hysteresis loop. The precise value of β_c cannot be obtained in this way and further analysis is needed to determine it,[1,2] but this "thermal cycle" provides a global overview of the phase structure of the model. Examples of thermal cycles for the groups $Z(3)$, $Z(5)$, and $Z(8)$ are shown in FIGURE 1. Notice that, while $Z(3)$ has a single transition, $Z(5)$ shows further structure, which becomes evident for $Z(8)$. On the basis of further analysis,[1,2] we determined that $Z(N)$ has a single first-order transition for $N \leq 4$ and two continuous transitions for $N > 4$. The critical temperature of the first transition (smaller β) becomes $\beta_c^I \simeq 1$ as $N \to \infty$, whereas $B_c^{II} \simeq \alpha/[1 - \cos(2\pi/N)]$ with α a constant ($\simeq 0.78$). As $N \to \infty$, $\beta_c^{II} \to \alpha N^2$ and one concludes that $U(1)$ has a single continuous transition at finite bare coupling. Further direct analysis of the $U(1)$ theory confirms these results. In FIGURE 2 the curve $E(\beta)$ for $U(1)$ is displayed. The continuous transition at $\beta_c \simeq 1$ is seen as an inflection point in the graph.

Another powerful tool of analysis for the confinement problem is provided by the Wilson function, defined as

$$W \equiv \langle \text{Re}(U_{i_1 i_2} U_{i_2 i_3} \ldots U_{i_n i_1}) \rangle. \tag{11}$$

W measures the average value of a closed loop of U's in a given configuration. For a plaquette, $W = 1 - E$. As the size of the loop is increased, W always tends to zero, as the spin correlations diminish with separation. The speed with which $W \to 0$, however, is a property of the phase in which it is measured. It has been argued by Wilson that, in a confining phase, W decreases exponentially with the minimal area enclosed by the loop; in a deconfining (ordered) phase, the decay is slower, with W falling off exponentially with the length of the loop.[3] The results of comparing the predictions (high and low temperature Padé approximants) for side one and side two square loops

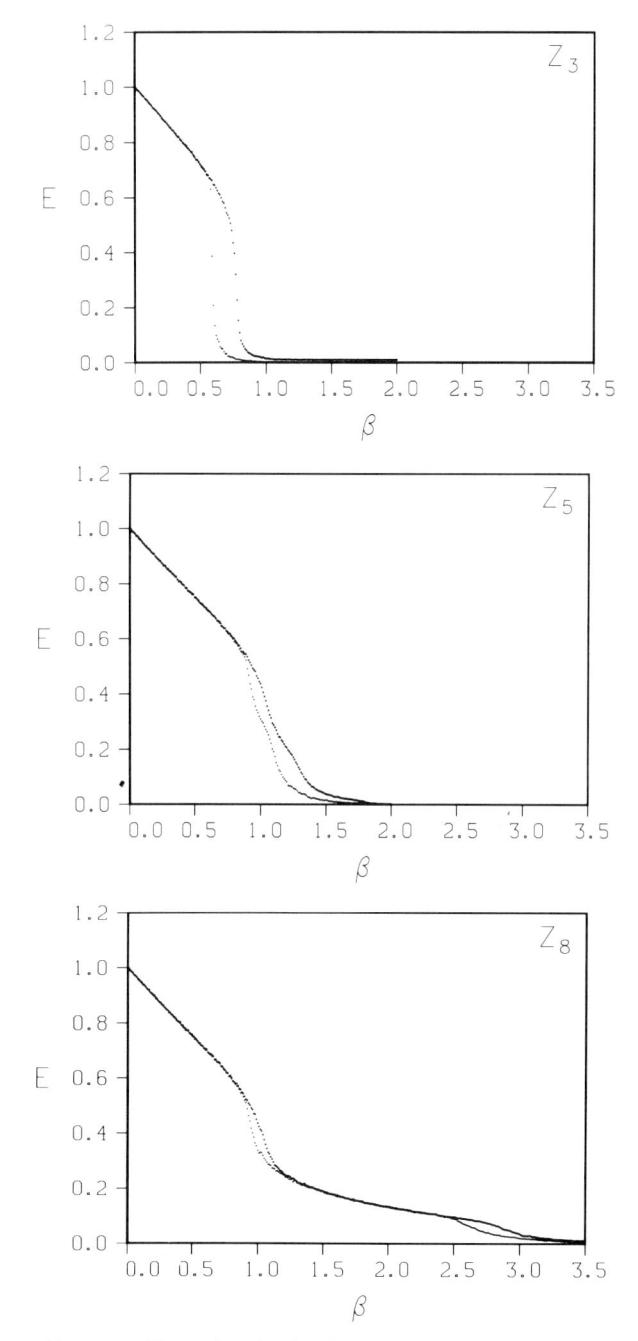

Jacobs: Confinement and Phase Transitions in Gauge Theories 175

FIGURE 1. Thermal cycles for the groups Z(3), Z(5), and Z(8).

with Monte Carlo simulations of the group Z(2) are shown in FIGURE 3. The impressive agreement with the area law in the high temperature phase and the perimeter law in the low temperature phase is clear evidence of a deconfining transition. The behavior is shown in the larger groups as well as in the U(1) model.

These are some important examples of the analysis that has proven the validity of the Wilson model for QED and encouraged its use for QCD.

Of course, the major goal of these analyses is not to prove that electrons and photons are physical particles but rather to apply the technique to the case of quarks and gluons.

This analysis has not been completed and I will only mention some of the major results that have been obtained so far.

The first attempts were to follow the lessons of the Abelian case and carry out the analysis of discrete subgroups of SU(2). Of course, there is only a finite number of subgroups that are truly non-Abelian (the symmetry groups of the regular polyhedra) and the sequence of groups is never really close to SU(2). Nevertheless, this has proven to be a remarkably good approximation to the physics of SU(2). The first case studied was the eight element group of quaternions.[6] This model was shown to have a single, first order transition at $\beta \simeq 1.3$. The analysis was then generalized to larger (24, 48, and eventually 120 element) subgroups of SU(2).[7] All models show first order transitions at finite β, with the transition temperature moving to zero as the order of the group is increased. Parallel direct analysis of the SU(2) model showed no phase transition at finite β,[8] in agreement with the results obtained with the discrete groups. Moreover, these simulations were shown to be compatible with both the hypothesis of confinement and the property of asymptotic freedom. Moreover, in the region of most interest (in β), the results of the relatively small $\tilde{O}$ group (the 48 element product group of Z(2) and the rotation group of the octahedron) and SU(2) are essentially identical.[7] This is an unexpected and extremely useful result, since it allows the analysis of quark dynamics with the much simpler structure of the $\tilde{O}$ group. Calculations with a 120 element subgroup of SU(2) are under way; they should

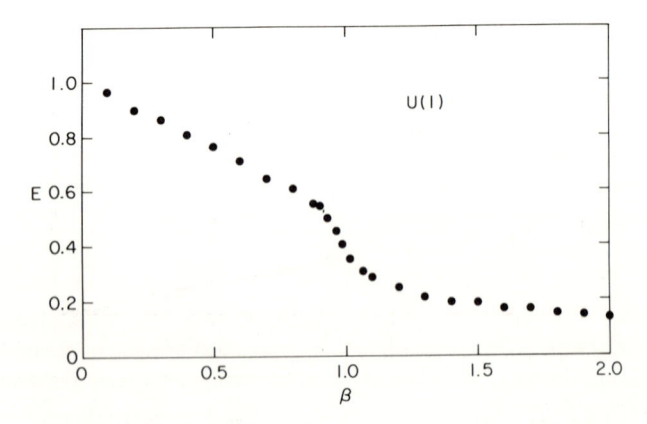

FIGURE 2. The average action per plaquette for the group U(1).

Jacobs: Confinement and Phase Transitions in Gauge Theories 177

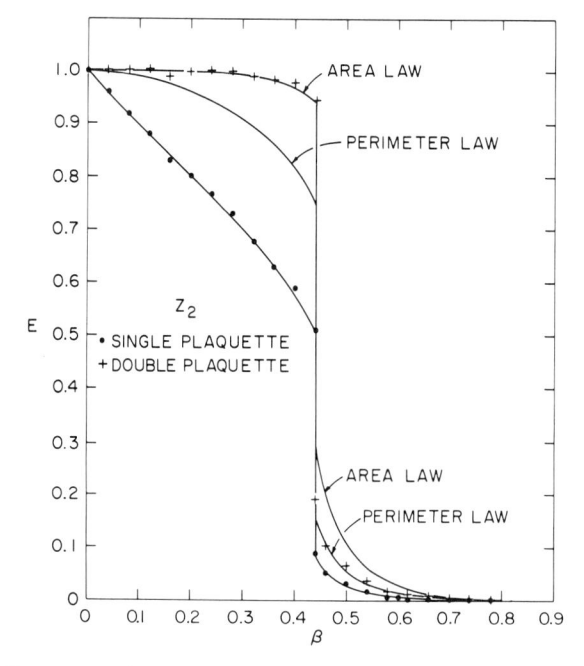

FIGURE 3. Comparison of the area and perimeter law predictions with the simulation of the group Z(2).

greatly increase the statistics obtained from the direct simulation of the SU(2) model.[8]

Conclusions

The results I have presented constitute a major step in our understanding of the dynamics of gauge fields. They lead to the long-conjectured permanent confinement of quarks and gluons. Most important, for the first time we have a truly nonperturbative tool for the analysis of quantum field theory; this will lead, in the near future, to quantitative predictions of QCD and to the possibility of trial by experiment—the ultimate test of any physical theory.

Summary

A survey of recent work on lattice gauge theories that have produced strong evidence for the permanent confinement of quarks was presented. As a primary step, the U(1) gauge theory in the Wilson-Polyakov formulation was analyzed and shown to be consistent with our understanding of Quantum Electrodynamics. The extension of this analysis to SU(2) was briefly discussed.

Acknowledgments

I am most grateful to Dr. Viktor Brailovsky and the rest of the organizing committee of the Fourth International Conference on Collective Phenomena for their warm hospitality in Moscow and for the opportunity to present this material. I also wish to thank the Comité pro Derechos de la Minoría Judía en la URSS for their support as well as the New York Academy of Sciences and the other bodies that have made these conferences possible.

References

1. CREUTZ, M., L. JACOBS & C. REBBI. 1979. Phys. Rev. Lett. **42:** 1390; 1979. Phys. Rev. D **20:** 1915; CREUTZ, M. 1979. Phys. Rev. Lett. **43:** 553.
2. A comprehensive review of these results will appear in CREUTZ, M., L. JACOBS & C. REBBI. 1980. Phys. Rep. C. In press.
3. WILSON, K. 1974. Phys. Rev. D **10:** 2445; POLYAKOV, A. M. 1975. Phys. Lett. B **59:** 82.
4. METROPOLIS, N., A. W. ROSENBLUTH, A. H. TELLER & E. TELLER. 1953. J. Chem. Phys. **21:** 1087.
5. For a recent review see BINDER, K., Ed. 1979. Monte Carlo Methods. Springer-Verlag.
6. JACOBS, L. & C. REBBI. 1979. Unpublished.
7. REBBI, C. 1980. Phys. Rev. D. **21:** 3350.
8. CREUTZ, M. 1979. Phys. Rev. Lett. **43:** 553; BNL Report 26847, Sep. 1979; WILSON, K. 1979. Cargese Lectures.

SOME CONSIDERATIONS OF STABILITY IN SOLIDIFICATION OF LAMELLAR EUTECTICS*

J. S. Langer

*Physics Department
and
Center for the Joining of Materials
Carnegie-Mellon University
Pittsburgh, Pennsylvania 15213*

The directional solidification of lamellar eutectics is a relatively simple but nontrivial pattern-forming process in which the mechanism for pattern selection may provide useful clues toward the understanding of a much broader range of collective phenomena.[1] Eutectic patterns seem to have much in common with nonequilibrium phenomena in hydrodynamics, chemistry, and perhaps even biology. A useful first step toward understanding the pattern-selection mechanism is a study of the stability of the growth forms predicted by steady-state analysis. The emphasis in the following presentation will be on certain fundamental considerations regarding the stability of these systems. More detailed calculations are presented in other publications.[2,3]

The lamellar solidification process of interest here is illustrated schematically in FIGURE 1, where a periodic array of solid phases α and β is shown growing upward into the fluid. These lamellae may be visualized as semi-infinite plates perpendicular to the plane of the paper; however, for our present purposes, it will be convenient to consider primarily the thin-film geometry in which the thickness of the sample is small compared to the lamellar spacing. One version of this situation that we shall study is that in which the velocity of the solidification front is fixed by the imposition of a moving temperature gradient, as in a zone-refining process. In this case, it is known experimentally that the lamellar spacing and the temperature (i.e., location) of the solidification front are uniquely determined by the imposed growth velocity. In a second version of the problem, the eutectic fluid is simply quenched uniformly, in which case it turns out that both the spacing and the velocity are unique functions of the quench temperature.

As is becoming increasingly common in situations of this kind, steady-state analysis fails to produce the experimentally observed relations between temperature, spacing, and growth velocity. The definitive steady-state calculations have been performed by Jackson and Hunt,[4] who obtained a relation of the form

$$\Delta T = T_E - T_S = \frac{1}{2} \Delta T_{\min} \left(\frac{\lambda}{\lambda_{\min}} + \frac{\lambda_{\min}}{\lambda} \right). \tag{1}$$

Here, ΔT is the undercooling at the solidification front, that is, the difference between the actual temperature at the front T_S and the eutectic temperature T_E. The lamellar spacing is λ; the quantity denoted by $\lambda_{\min}$ is a length of the form

---

*This research was supported by a grant from AFOSR, no. 80-0034.

$$\lambda_{\min} \propto \left(\frac{D\, d_0}{v}\right)^{1/2}, \tag{2}$$

where D is the diffusion constant in the fluid, d_0 is a capillary length, and v is the growth velocity. The temperature $\Delta T_{\min}$ is proportional to the dimensionless group of parameters $(d_0 v/D)^{1/2}$. Note that ΔT, as a function of λ, passes through a minimum at $\Delta T_{\min}$, $\lambda = \lambda_{\min}$.

The two parts of the right-hand side of (1) can be understood qualitatively as follows. The first, proportional to $\lambda v/D$, is a diffusion effect. Each advancing α region rejects β molecules and vice versa; thus, the fluid ahead of each solid region is supersaturated and the interfacial temperature is correspondingly depressed. The larger λ or v becomes, the larger must be the concentration of, say, β molecules in front of each α region in order to drive the required diffusion flux. Therefore, this term in (1) increases with the product λv. The second term in (1), proportional to d_0/λ, is a capillary term associated with the Gibbs-Thomson effect, that is, the depression of the melting point at a curved interface. The forward bulges shown in FIGURE 1 are necessary in order that the capillary forces balance at the α-β-liquid triple points; in

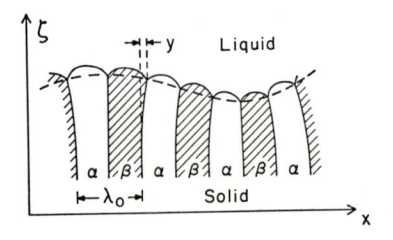

FIGURE 1. Schematic illustration of a lamellar eutectic growing up the page with a deformed solidification front.

fact, the opening angles at these points are completely determined by the force-balance conditions. Thus, this term is the product of the capillary length d_0 and the curvature, which is proportional to λ^{-1}.

Obviously, (1) provides us with only a single relation of the form $\Delta T(\lambda, v)$ instead of the pair of relations, say $\lambda(v)$, $\Delta T(v)$, required by directional-solidification experiments. In the metallurgical literature, the conventional way out of this dilemma has been to assume, essentially without derivation, that the directionally solidifying system operates at its point of minimum undercooling, that is, at $\lambda = \lambda_{\min}$. The resulting prediction that $\lambda^2 v$ is a constant has been shown to be consistent with most experimental data, but there has been no successful attempt to check whether theory and experiment are actually in quantitative agreement. Similarly, in the quenched, constant-temperature version of this problem, (1) implies that v passes through a maximum as a function of λ at constant ΔT; the metallurgical assumption has been that the system operates at this point of maximum velocity.

The interesting result of the stability analysis to be described below is that both the minimum-undercooling and maximum-velocity conditions locate points of marginal stability; that is, they identify values of λ at which the system is just on the verge of becoming unstable against some mode of deformation. To see this, suppose that the solidification front is slowly deformed on a length scale much greater than λ. Let this deformation be described by the function $\zeta(x, t)$, the dashed curve in FIGURE 1, which measures the forward displacement of the interface away from its undeformed position in the frame of reference moving with velocity v. Next, define $y(x, t)$ to be the sidewise displacement of lamellae originally at position x, as shown in the figure. That is, the local lamellar spacing is

$$\lambda(x, t) \simeq \lambda_0(1 + \partial y/\partial x), \tag{3}$$

where λ_0 is the original spacing of the undeformed system.

A crucial assumption—one which permits the stability problem to be solved without a much more elaborate calculation—is that ζ and y are coupled by the condition that each lamella must grow in a direction that is locally perpendicular to the solidification front. Thus

$$\frac{\partial y}{\partial t} \simeq -v \frac{\partial \zeta}{\partial x}. \tag{4}$$

Taking a derivative of (4) with respect to x and using (3), we find

$$\frac{\partial \lambda}{\partial t} \simeq -v\lambda_0 \frac{\partial^2 \zeta}{\partial x^2}. \tag{5}$$

This is a purely geometrical relationship that should be valid at least to linear order in the deformations.

To apply (5) in the case of the directionally solidifying system, assume that a temperature gradient G is fixed in the frame of reference moving with velocity v, so that the undercooling ΔT given by (1) is related to the position ζ by

$$\Delta T(\lambda) = -G\zeta. \tag{6}$$

Inserting this expression for ζ on the right-hand side of (5), we find

$$\frac{\partial \lambda}{\partial t} \simeq \frac{v\lambda_0}{G} \frac{\partial^2}{\partial x^2} \Delta T(\lambda), \tag{7}$$

or, to linear order,

$$\frac{\partial \lambda}{\partial t} \simeq \mathcal{D}(\lambda_0) \frac{\partial^2 \lambda}{\partial x^2}, \tag{8}$$

where

$$\mathcal{D}(\lambda_0) = \frac{\lambda_0 v}{G} \frac{\partial}{\partial \lambda_0} \Delta T(\lambda_0) \propto 1 - \left(\frac{\lambda_{min}}{\lambda_0}\right)^2 \tag{9}$$

plays the role of a diffusion constant in (8). Thus, all deformations decay diffusively as long as $\mathcal{D}(\lambda_0)$ is positive. However, $\mathcal{D}(\lambda_0)$ is greater than zero only for $\lambda_0 > \lambda_{min}$; that is, λ_{min} locates the point of marginal stability, as advertised.

In the case of the quenched system, the thermal gradient G vanishes and ΔT is fixed, so that (1) produces a relationship $v(\lambda)$. We have no analogue of (6), but we can define a local growth velocity[5]

$$v(x, t) = v_0 + \frac{\partial \zeta}{\partial t}, \qquad (10)$$

where v_0 is the unperturbed steady-state velocity of the system whose stability is being examined. Differentiating (5) with respect to t and keeping only the linear order on the right-hand side, we find

$$\frac{\partial^2 \lambda}{\partial t^2} \simeq v_0 \lambda_0 \frac{\partial^2 v}{\partial x^2} \simeq -v_0 \lambda_0 \frac{\partial v}{\partial \lambda_0} \frac{\partial^2 \lambda}{\partial x^2}. \qquad (11)$$

On the large-λ_0 side of the maximum velocity, where $\partial v/\partial \lambda_0$ is negative, (11) is a wave equation with stable solutions. On the small-λ_0 side of the maximum, however, (11) admits indefinitely growing modes of deformation. As advertised, the maximum velocity identifies a point of marginal stability.

In our recent papers on the theory of dendritic crystal growth,[6,7] Müller-Krumbhaar and I have suggested that there may be a general tendency for pattern-forming systems, when driven by thermal fluctuations or other noise sources, to drift toward a point of marginal stability. The argument as applied to the directionally solidifying eutectic, for example, is that $\mathcal{D}(\lambda)$ in (8) decreases with decreasing λ; thus, a fluctuation that drives λ downwards persists for a longer time than one which goes in the other direction. If there exists somewhere in the system a mechanism for creating new lamellae (defects, edges, etc.), then the time-averaged effect of noise on a stable system with $\lambda > \lambda_{min}$ must be to drive the system into new stable configurations with diminishing values of λ. At λ_{min}, of course, the system becomes unstable, but one can show that the effect of this instability is the occasional termination of a lamella with the resulting restoration of the local spacing to a more stable value $\lambda > \lambda_{min}$. The result is that the system operates, with fluctuations, at or near the spacing λ_{min}. The idea that, under some circumstances, eutectic solidification patterns are governed by this marginal-stability mechanism seems attractive. As yet, however, there is no compelling reason to believe that this mechanism is the dominant one in eutectics, nor do we know whether or not marginal stability is relevant to a broader range of pattern-forming phenomena.

References and Notes

1. There is a large metallurgical literature on the subject of eutectic solidification. Useful reviews include HOGAN, L. M., R. W. KRAFT & F. D. LEAKEY. 1971. Eutectic grains. *In* Advances in Materials Research, Vol. 5. H. Herman, Ed. Wiley-Interscience. New York; LESOULT, G. 1980. Ann. Chim. (Paris) **5**: 154.
2. LANGER, J. S. 1980. Phys. Rev. Lett. **44**: 1023.
3. DATYE, V. & J. S. LANGER. 1981. To be published.
4. JACKSON, K. A. & J. D. HUNT. 1966. Trans. Metall. Soc. AIME **236**: 1129.
5. This analysis was originally suggested by J. Kirkaldy.
6. LANGER, J. S. & H. MÜLLER-KRUMBHAAR. 1978. Acta Metall. **26**: 1681; 1689; 1697.
7. LANGER, J. S. 1980. Rev. Mod. Phys. **52**: 1.

ON TWO-ELEMENT SUBSETS IN GROUPS

L. V. Brailovsky and G. A. Freiman

Moscow, USSR

Introduction

Let $E = (a, b)$ be a subset of a group. In the set E^2 there are no more than four different elements—a^2, ab, ba, b^2—and the multiplication table is

$$\begin{array}{c|cc} & a & b \\ \hline a & a^2 & ab \\ b & ba & b^2 \end{array} \qquad (1)$$

Such a table is said to be a square.

Using capital letters and denoting equal elements by the same letters, we get four possible types of squares:

$$\begin{array}{cccc} \text{I} & \text{II} & \text{III} & \text{IV} \\ \begin{array}{cc} A & B \\ B & A \end{array} & \begin{array}{cc} A & B \\ B & C \end{array} & \begin{array}{cc} A & B \\ C & A \end{array} & \begin{array}{cc} A & B \\ C & D \end{array} \end{array} \qquad (2)$$

Here A, B, C, and D are distinct elements.

A description of all types of third order squares in groups was given in Reference 1. We will call (1) a second stage multiplication table; the products it consists of each contains two factors. Let us now consider multiplication tables at the third stage.

In case IV of (2), this table will look like

$$\begin{array}{c|cccc} & a^2 & ab & ba & b^2 \\ \hline a & a^3 & a^2b & aba & ab^2 \\ b & ba^2 & bab & b^2a & b^3 \end{array} \qquad (3)$$

Again, we shall denote equal elements by the same letters and distinct elements by different letters. For example, a third stage multiplication table of the set $M = ((1234), (132))$ will take the following form:

$$\begin{array}{cccc} A' & B' & C' & D' \\ E' & A' & F' & G' \end{array} \qquad (4)$$

In case I of (2), the third stage multiplication table contains two columns, while, in cases II and III, it contains three columns.

Having changed the order of the elements of the set E, we may, generally speaking, change (3) as well. Thus, changing the order of elements in set M, we obtain

183

the following multiplication table at the third stage,

$$\begin{array}{cccc} A' & B' & C' & D' \\ E' & F' & G' & C' \end{array} \quad (5)$$

instead of (4).

According to the definition in Reference 2, two sets are called "isomorphic up to the third stage" if the second and third stage multiplication tables of one of the sets may be obtained from those of the other set by the proper change of rows and columns with lexicographical redesignations. We shall call the tables of such sets "isomorphic" (for example, (4) and (5)).

The purpose of the present work is to describe all classes of two-element sets isomorphic up to the third stage, i.e., having isomorphic tables at the third stage, existing in groups.

In the last section, there are some results that illustrate the research directions arising from introducing the notion of the class of isomorphic sets in a group.

A Third Stage Multiplication Table for Case I of (2)

In case I of (2), we have

$$A = a^2 = b^2, \quad (6)$$
$$B = ab = ba.$$

Let us construct a multiplication table at the third stage. It can have any of these four forms:

$$\begin{array}{cccc} A' \; B' & A' \; B' & A' \; B' & A' \; B' \\ B' \; A' & B' \; C' & C' \; A' & C' \; D' \end{array} \quad (7)$$

Actually, not all these types are found in groups. From (6) we get the following equations: $ab = ba$, $ab^2 = bab$, $aA = bB$, $aba = ba^2$, and $aB = bA$. This means that only the first case in (7) may be realized in a group. For example, it exists in a C_2-cyclic group of the second order.

Third Stage Multiplication Tables for Case II of (2)

Considering case II of (2), we have

$$ab = ba, \quad ab^2 = bab, \quad aC = bB$$
$$aba = ba^2, \quad aB = bA,$$

so only two possible types of third stage multiplication tables remain:

$$\begin{array}{cc} 1 & 2 \\ \begin{array}{ccc} A' & B' & C' \\ B' & C' & A' \end{array} & \begin{array}{ccc} A' & B' & C' \\ B' & C' & D' \end{array} \end{array} \quad (8)$$

Case 1 of (8) may be realized in group C_3, and Case 2 of (8) may be realized as the table of the set (a, a^2), where $a \in C_4$.

THIRD STAGE MULTIPLICATION TABLES FOR CASE III OF (2)

Let us consider case III of (2). Here the following equations hold:

$$a^2 = b^2, \qquad a^2b = b^3, \qquad aB = bA$$
$$a^3 = b^2a, \qquad aA = bC,$$

i.e., there are only two possible types of tables:

$$
\begin{array}{ccc}
1 & & 2 \\
A' \;\; B' \;\; C' & & A' \;\; B' \;\; C' \\
B' \;\; C' \;\; A' & & B' \;\; D' \;\; A'
\end{array}
\qquad (9)
$$

Case 1 of (9) has its realization in S_3 as a table of the set $((12), (13))$. Case 2 of (9) has its realization in S_4 as a table of the set $((12), (13)(24))$.

THIRD STAGE MULTIPLICATION TABLES FOR CASE IV OF (2)

The Case IV of (2) implies the following inequalities:

$$a^2 \neq b^2, \qquad ab \neq ba.$$

Here we shall point out some types of third stage multiplication tables that cannot be realized in groups.

These types are

$$
\begin{array}{cccc}
a & b & c & d \\
A' \;\; B' \;\; C' \;\; D' & A' \;\; B' \;\; C' \;\; D' & A' \;\; B' \;\; C' \;\; D' & A' \;\; B' \;\; C' \;\; D' \\
-\;-\;A'\;- & -\;-\;-\;B' & C'\;-\;-\;- & -\;D'\;-\;- \\
\\
e & f & g & h \\
A' \;\; B' \;\; C' \;\; D' & A' \;\; B' \;\; C' \;\; D' & A' \;\; B' \;\; C' \;\; D' & A' \;\; B' \;\; C' \;\; D' \\
B'\;-\;-\;A' & B' \;\; C'\;-\;- & -\;-\;D'\;A' & -\;C'\;D'\;-
\end{array}
$$

For case a, we have $aA = BC$, i.e., $aa^2 = bba$ and $a^2 = b^2$, which is a contradiction.

For case e, we have

$$
\begin{array}{cc}
aA = bD & aa^2 = bb^2 \\
\text{i.e.} & \\
aB = bA & aab = ba^2
\end{array}
$$

so $a^3b = aba^2$, $ba^3 = aba^2$, and $ab = ba$, which is a contradiction.

Other cases may be proved impossible in the same way.

Given a multiplication table, in order to find one isomorphic to it, one must switch rows 1 and 2 and columns 1 and 4 and 2 and 3 and make lexicographical redesignations. We shall now enumerate isomorphic pairs of tables, giving only their second lines.

$B'A'D'E'—B'E'D'C'$, $B'A'E'C'—E'A'D'C'$, $B'A'E'F'—E'F'D'C'$,
$B'E'F'C'—E'A'D'F'$, $B'E'F'G'—E'F'D'G'$, $D'A'B'E'—D'E'B'C'$,
$D'A'E'F'—D'E'F'C'$, $E'A'B'F'—E'F'B'C'$, $E'A'F'G'—E'F'G'C'$.

Let us write all the types of tables that remain after deleting the impossible ones and picking one of every pair of isomorphic tables (see above).

1. $B'A'D'C'$	9. $D'A'B'E'$	17. $D'E'B'F'$	25. $E'C'B'F'$
2. $B'A'D'E'$	10. $D'A'E'C'$	18. $D'E'F'A'$	26. $E'C'F'A'$
3. $B'A'E'C'$	11. $D'A'E'F'$	19. $D'E'F'G'$	27. $E'C'F'G'$
4. $B'A'E'F'$	12. $D'C'B'A'$	20. $E'A'B'C'$	28. $E'F'B'A'$
5. $B'E'D'F'$	13. $D'C'B'E'$	21. $E'A'B'F'$	29. $E'F'B'G'$
6. $B'E'F'C'$	14. $D'C'E'A'$	22. $E'A'F'C'$	30. $E'F'G'A'$
7. $B'E'F'G'$	15. $D'C'E'F'$	23. $E'A'F'G'$	31. $E'F'G'H'$
8. $D'A'B'C'$	16. $D'E'B'A'$	24. $E'C'B'A'$	

Many types of tables in (**10**) still cannot be realized in groups.
In the case 2, we have

$$aA = bB \qquad aa^2 = bab$$
$$aB = bA \quad \text{or} \quad aab = ba^2$$
$$aD = bC \qquad ab^2 = bba$$

so $ba^2 = a^2b$, $ba^3 = a^2ba$, $b^2ab = a^2ba$, $ab^2b = a^2ba$, $b^3 = aba$, and $bD = aC$, which is a contradiction.

Similar contradictions can be found in cases 3, 8, 9, 11, 13–16, 18, 21, 24, 25, and 28.

Three Cases with Nontrivial Realizations

The other cases in (**10**) turned out to be possible in groups. Now we shall consider three cases with nontrivial realizations.

Case 10 gives us the following form of the multiplication table:

	A'	B'	C'	D'	E'
a	A''	B''	C''	D''	E''
b	E'''	C'''	A'''	B'''	D'''

It is easy to demonstrate that set $I = a^{-3} \{A', B', \ldots E'\}$ is a cyclic group of the fifth order. Having noticed that the tables of the multiplication of set (a, b) by sets I

and a^4I have the same form, we shall try to find the realization in the semidirect product of cyclic group C_5 by the cyclic group of the fourth order, C_4, i.e., in the group with relations $g^5 = e$, $k^4 = e$, and $kg = g^2k$. We find elements k and kg to fit case 10 (**10**) at the third stage. Representing group L as the permutation group of degree 5, we obtain the following realizations, k equal to (12345) and g equal to (2354), $a = (2354)$ and $b = (1243)$.

The investigation of case 20, if made in a similiar way, would lead to the same group L. Elements to be sought are here k^3 and k^3g, or, realizing them as permutations, $a = (2453)$ and $b = (1254)$.

Let us consider case 26 of (**10**). The fourth stage multiplication table takes the following form:

	A'	B'	C'	D'	E'	F'
a	A''	B''	C''	D''	E''	F''
b	B''	G''	E''	C''	H''	A''

The fifth stage multiplication table is as follows.

	A''	B''	C''	D''	E''	F''	G''	H''
a	A'''	B'''	C'''	D'''	E'''	F'''	G'''	H'''
b	C'''	D'''	G'''	F'''	B'''	E'''	H'''	A'''

We shall show only that $bG'' = aH''$, i.e., $ab^2a^2 = b^2a^2b$. Indeed, $ba^3 = a^3b$, $abaa^2 = a^4b$, $ababa^2 = a^5b$, $bab^2a^2 = b^3a^2b$, and $ab^2a^2 = b^2a^2b$.

It can be shown, in the same way, that $aF'' = bD''$. Set $Q = a^{-5}(A''', \ldots H''')$, which is the quaternion group with the following designations:

$$a^{-1}ba^{-1}b = -1, \qquad a^{-2}ba = i, \qquad ab^{-1} = j, \qquad b^{-1}a = k.$$

If we set $a^3 = 1$, then the set (Q, aQ, a^2Q) is the desired group of the twenty-fourth order. It is the semidirect product of Q by the cyclic group of the third order, C_3. Representing this group by permutations of the eighth degree, we obtain $a = (164)(253)$ and $b = (138)(247)$.

The complete list of all types of third stage multiplication tables existing in groups is given in TABLE 1.

THE STRUCTURES OF GROUPS AND THE PROPERTIES OF THEIR SUBSETS*

The description of the classes of isomorphic sets is not the ultimate goal. Having got the list of such classes, one is able to begin the study of the connection between the structure of a group and the properties of its subsets. Let us point out some possible directions of the study.

The following theorem is true.[1]

THEOREM 1. If G is a finite non-Abelian group and $P_4 = 0$ (P_4 is the number of

---

*This section is written in cooperation with Ya. G. Berkovich.

TABLE 1

N	Type at the Second Stage	Second line of Table at the Third Stage	Relations at the Third Stage	Example of Realization
1	I	$B'\ A'$		C_2
2	II	$B'\ C'\ A'$	$a^3 = b^3$	C_3
3	II	$B'\ C'\ D'$		$g, g^2, g \in C_4$
4	III	$B'\ C'\ A'$	$aba = bab$	(12), (13)
5	III	$B'\ D'\ A'$		(12), (13) (24)
6	IV	$B'\ A'\ D'\ C'$	$a^3 = bab$, $ab^2 = b^2a$ $a^2b = ba^2$, $aba = b^3$	(12345678) (12785634)
7	IV	$B'\ A'\ E'\ F'$	$a^3 = bab$, $a^2b = ba^2$	$\begin{pmatrix} -1 & 2 \\ 0 & 1 \end{pmatrix}, \begin{pmatrix} 1 & -2 \\ 0 & 1 \end{pmatrix}$
8	IV	$B'\ E'\ D'\ F'$	$a^2b = ba^2$, $ab^2 = b^2a$	$\begin{pmatrix} 1 & 0 \\ 0 & -1 \end{pmatrix}, \begin{pmatrix} 1 & 2 \\ 3 & -1 \end{pmatrix}$
9	IV	$B'\ E'\ F'\ C'$	$a^2b = ba^2$, $aba = b^3$	$\begin{pmatrix} 1 & -i\sqrt{2} \\ 0 & -1 \end{pmatrix}, \begin{pmatrix} i\sqrt{2} & 1 \\ 1 & 0 \end{pmatrix}$
10	IV	$B'\ E'\ F'\ G'$	$a^2b = ba^2$	(12) (34), (15342)
11	IV	$D'\ A'\ E'\ C'$	$ab^2 = ba^2$, $a^3 = bab$, $aba = b^3$	(2354), (1243)
12	IV	$D'\ C'\ B'\ A'$	$ab^2 = ba^2$, $a^2b = b^2a$ $aba = bab$, $a^3 = b^3$	(123456) (126453)
13	IV	$D'\ E'\ B'\ F'$	$ab^2 = ba^2$, $a^2b = b^2a$	$\begin{pmatrix} 1 & 1 \\ 0 & w \end{pmatrix}, \begin{pmatrix} w^2 & -w \\ 1 & 0 \end{pmatrix}$ $w = \frac{1}{2} + i\sqrt{3}/2$
14	IV	$D'\ E'\ F'\ G'$	$ab^2 = ba^2$	$\begin{pmatrix} 1 & 1 \\ 0 & c \end{pmatrix}, \begin{pmatrix} 1 & 7 \\ 0 & c \end{pmatrix}$ $c = (1 + \sqrt{5})/2$
15	IV	$E'\ A'\ B'\ C'$	$a^2b = b^2a$, $a^3 = bab$, $aba = b^3$	(2453), (1254)
16	IV	$E'\ A'\ F'\ C'$	$a^3 = bab$, $aba = b^3$	$\begin{pmatrix} 1 & -1-i \\ 0 & i \end{pmatrix}, \begin{pmatrix} 1+i & -i \\ 1 & 0 \end{pmatrix}$
17	IV	$E'\ A'\ F'\ G'$	$a^3 = bab$	(1234), (132)
18	IV	$E'\ C'\ F'\ A'$	$a^3 = b^3$, $aba = bab$	(164)(253),(138)(247)
19	IV	$E'\ C'\ F'\ G'$	$aba = bab$	(1234), (1342)
20	IV	$E'\ F'\ B'\ G'$	$a^2b = b^2a$	$\begin{pmatrix} 1 & 1 \\ 0 & r \end{pmatrix}, \begin{pmatrix} 1 & 3 \\ 0 & r \end{pmatrix}$ $r = (-1 + \sqrt{5})/2$
21	IV	$E'\ F'\ G'\ A'$	$a^3 = b^3$	(123456789), (189423756)
22	IV	$E'\ F'\ G'\ H'$		(12345), (134)

two-element subsets of type IV of (2)), then $G = Q \oplus E$, where Q is a quaternion group and $e^E = 2$.

It would be natural to set a general problem concerning the description of all the groups, which, given the list of k-element sets up to the fifth stage, have no sets from a certain part of this list.

Let $k = 2$ and $s = 2$. Except for the case $P_4 = 0$, there is only one more nontrivial case $P_3 = 0$ (P_3 is the number of two-element sets of type III of (2)).

THEOREM 2 (*Ya. G. Berkovich*). Let $H \in \mathrm{Syl}_2(G)$ and G be finite; then

$$P_3(G) = 0 \tag{18}$$

if and only if

$$H \triangleleft G \tag{19}$$

and

$$P_3(H) = 0. \tag{20}$$

Proof. Equation 20 is obviously implied by (18). We shall show that (19) is implied by (18) as well. Let us suppose that $H \ntriangleleft G$. Then (see Reference 3) G contains Schmidt subgroup $H°$ of an even order, and $P° \subset H$, $P° = \mathrm{Syl}_2(H)$, $P°$—cyclic. (A group U is called a Schmidt group if it is not nilpotent but all of its subgroups are nilpotent. Then $|U| = p^\alpha q^\beta$, $p \neq q$, are primes, the subgroup of order p^α is cyclic, and the subgroup of order q^β is normal in U.) Let X generate a subgroup of order $\frac{1}{2}|P°|$ in H, and let y and z generate two different subgroups $\mathrm{Syl}_2(H)$. H contains only one subgroup of order $\frac{1}{2}|P°|$, that is, $\langle x \rangle$. So one may find y and z such that $y^2 = z^2 = x$. But y and z do not commute (as generators of the different $\mathrm{Syl}_2(H)$), which is a contradiction showing that $P \triangleleft G$.

Let us prove now that (18) follows from (19) and (20).

Let $P_3(H) = 0$, $H \in \mathrm{Syl}_2(G)$, and $x^2 = y^2$. We shall show that $xy = yx$. We have $x = x_o x_e$, $y = y_o y_e$ (here $|x_o| \equiv 1$ (2), $|x_e| = 2^h$, $|y_e| = 2^{h_1}$; the same for y_o, y_e, i.e., such a representation of x and y is unique). But then $x_o^{2^h} x_e^{2^h} = y_o^{2^h} y_e^{2^h}$ (if $h \geq h_1$), $x_o^{2^h} = y_o^{2^h}$ and $x_o = y_o$. From $x_e^2 = y_e^2$ and the fact that $x_e, y_e \in H$ (because $H \triangleleft G$), it follows that $x_e y_e = y_e x_e$. Together with $x_o = y_o$, it gives $xy = x_o x_e y_o y_e = x_o x_o x_e y_e = x_o x_o y_e x_e = x_o y_o y_e x_e = x_o y_e y_o x_e = y_o y_e x_o x_e = yx$.

Thus, the problem is reduced to the description of the two-groups outside case III of (2). This problem seems to be difficult.

Let $R(G) = k/g$, where k is the number of conjugate classes G and $g = G$. The study of groups with large $R(G)$ was begun by K. S. Joseph[4] and D. Rusin.[5] In fact, the questions they considered are a part of our general scheme. Actually, $R(G) = P/g^2$, where P is the number of ordered pairs $\{x, y\}$, $x, y \in G$, $xy = yx$, and $P = P_1 + P_2$. The relevant results are given in Reference 6.

Let $R_4(G) = P_4/g^2$. As in the example given above, one may study groups with small $R_4(G)$.

THEOREM 3 (*K. Nekrasov*). Let G be a finite group. If $R_4(G) \neq 0$, then $R_4(G) \geq \frac{5}{32}$ and this bound is attainable.

A general problem for $k = 2$ and $S = 3$ has been studied in the course of a seminar on group theory at Kalinin University during 1979–80. Starting from Theorem 1, one could not only move in the direction of $R_4(G) = P_4/g^2$, but also pose the following problem: What is the structure of a finite group G if any of its two-element subsets of type IV of (2) at the second stage can have types at the third stage only from a certain subset of the set of all 17 types given in TABLE 1?

So, the following theorem is true.

THEOREM 4 (*K. Nekrasov*). Let G be a finite group such that each of its two-element subsets of type IV of (2) at the second stage is a set of type 6 or 7 of TABLE 1.

This is true iff G satisfies one of the following conditions:

1. $G = C_2 \wedge H$ is a semi-direct product of C_2 and an Abelian group H, the following being true:

$$a \in G, \quad a \neq e, \quad \phi_a(x) = x^{-1}, \quad \forall x \in H.$$

2. $p(G) = 2$, where $p(G)$ is the number of equivalence classes with respect to the equality of the squares of the elements.

Multiplication tables for finite subsets of G may be classified according to the number of different elements of the tables.

THEOREM 5. (*Ya. G. Berkovich and G. A. Freiman*). Let $M^2 < 9$ for any three-element subset M of a finite group G. Then all the Sylov subgroups $G/\phi\,(G)$ are elementary ($\phi(G)$ is Frattini's group), and G has an invariant Abelian two-complement. If G is a non-Abelian nilpotent group, then G is a two-group.

Results of this type have been summarized in Reference 6 and presented in Reference 7.

REFERENCES

1. FREIMAN, G. A. 1981. On two- and three-element subsets of groups. Aequationes mathematicae. In press.
2. FREIMAN, G. A. 1973. Foundations of a Structural Theory of Set Addition. Translations of Mathematical Monographs, Vol. 37. Am. Math. Soc., Providence, R. I.
3. BERKOVICH, YA. G. 1965. Cb. Konechniye gruppy. Minsk.
4. JOSEPH, K. S. 1969. Commutativity in non-Abelian groups. Ph.D. Thesis. University of California, Los Angeles.
5. RUSIN, D. 1979. What is the probability that two elements of a finite group commute? Pac. J. Math. 2(1): 237–47.
6. BERKOVICH, YA. G. & G. A. FREIMAN. 1981. On the connection between some numeric characteristics of a finite group and the structure of the group. In press.
7. BERKOWICH, YA. G. & G. A. FREIMAN. 1981. On the connection between the structure of a group and the maximal cardinality of the square of its two- and three-element subsets. In press.

INTEGER PROGRAMMING AND NUMBER THEORY

P. L. Buzytsky and G. A. Freiman

Moscow, USSR

INTRODUCTION

This article consists of the results established by the authors during the past year and a half. The research work conducted was aimed at developing efficiently solvable classes of integer programs (IP) and creating new methods, based on analytic number theory concepts, to solve IP.

Integer programming was first studied in the early 1950's. During that period, this field of scientific activity involved many specific approaches and some perspectives and directions were realized. It is worth noting that the development of this field was greatly affected by the theory of continuous linear programming (LP), the basic method of which (simplex-method) proved to be quite efficient. At the beginning, it was mistakenly assumed that there was a similarity between the two problems. This is witnessed by cutting plane methods, which were very popular (i.e., specialists paid much attention to them) during the 1960's. Once, just after Gomory's methods had appeared, it was thought that all the difficulties had been overcome. However, attempts at numerical implementation of cutting plane algorithms have shown that they are impractical even for problems of medium size. The reason for this seems to be that the combinatorial nature of IP, which is the crucial distinction between IP and LP, was neglected.

As numerical experience was accumulated, application specialists changed their attitude to these methods, which had seemed so promising at the beginning. The only scheme that proved to be viable was the scheme of implicit enumeration, which is like the branch-and-bounds method in its various modifications. Still, in spite of its practical utility, it was considered theoretically inconsistent, since it did not guarantee efficient solution and might involve complete enumeration.

The possibility for escaping complete enumeration in IP, as well as in many other combinatorial problems, proved to be one of the main theoretical questions. About ten years ago, the first papers forming the foundation of the modern concept of the complexity of combinatorial problems appeared. The concept of NP-complete problems has become central for the entire field in question. Without giving any further details, let us note that even such a "simple" problem as the 0,1 knapsack problem has been found to be in the list of NP-complete problems, i.e., it appears to be hopeless. However, awareness that IP is principally complicated in nature does not release one from the necessity of solving these problems. In a sense, a paradoxical situation developed: great theoretical efforts were aimed at proving the "impossibility" of solving certain general problems, while specific problems were solved by means of some presumably "bad" methods.

Thus, at the beginning, IP did not attract the serious attention it was due, while, later, the attitude became too serious.

In both cases an important detail was neglected, namely that IP, in nature, is one of the directions of number theory. Number theory dealt with many problems like IP, differing from the latter only in emphasis.

IP may be easily reduced to the problem of the existence of natural or 0,1 solutions for linear simultaneous equations. In analytic number theory, a question of the number of solutions of such equations in terms of a function of right-hand sides is quite natural, the behavior of such a function being studied by means of various asymptotic formulae. In IP theory, it is important to study such equations for fixed and possibly small number of variables, which leads to the necessity of deriving formulae with possibly accurate error estimation. The other feature of IP is the computational aspect. It is necessary to compute the derived formulae efficiently.

The considerations above provide a basis for the development of a number theory approach to IP. At the beginning of our research work, our efforts were concentrated on the derivation of asymptotic formulae for the number of solutions of linear equations with 0,1 variables.[1] Having obtained such formulae and investigated the limits of their applicability,[2] the authors managed to realize further advantages of this approach.[3] The established formulae provided some solvability conditions for one equation and two simultaneous equations. The study of these conditions enabled us to formulate the concept of stability regions (see below), which, in turn, provided a nontrivial estimation for the optimum of the knapsack problem. This led to a further direction: the calculation of an intermediate solution giving this estimation. Subsequent analysis, which gave rise to this report, completely supported the fundamental correctness of those suppositions. The concept of stability regions, together with the concept of the core of a problem (both of which are described below), led us to an efficient solution scheme for IP with many variables and few constraints.

The approach under consideration involves a broad class of problems with the following properties. The coefficients are small; for instance, we assume them to be bounded by a polynomial of the number of variables. The number of constraints is small in comparison with the number of variables. The variables are Boolean. For this class of IP, reduced to simultaneous equations, effective formulae for the number of solutions will be established, firstly for one equation and secondly for two simultaneous equations; then, the method of efficient solution of the whole problem will be given, followed by a comparison with some known approaches. An important aspect of the approach in question is its statistical nature. Using such an approach, we are able to solve almost all the problems in the aforementioned class. One of our previous papers contains a rigorous statistical analysis of the case of one equation,[2] while, at the present time, similar work is being conducted for many simultaneous equations. Thus, one more aspect of the number theory approach should be pointed out; it enables us to describe broad classes of problems that are statistically easy to solve, which provides a possibility for the creation of a new complexity classification differing from the scheme of NP-completeness.

An initial version of the main results has appeared in Reference 4. The detailed representation that follows is given in References 5–8.

On the Number of Solutions of One Linear 0,1 Equation

Let us consider the equation

$$a_1 x_1 + \cdots a_n x_n = b, \tag{1}$$

where a_j and b are positive integers, $x_j = 0,1$, $j = 1, \ldots n$, and

$$b \leq \sum_{j=1}^{n} a_j/2.$$

The number of solutions of (1) may be expressed as follows:

$$I_n = \exp(\sigma b) \prod_{j=1}^{n} (1 + \exp(-a_j\sigma))$$

$$\times \int_0^1 \prod_{j=1}^{n} (p_{1j} + p_{2j} \exp(2\pi i \alpha a_j)) \exp(-2\pi i \alpha b) d\alpha, \quad (2)$$

where

$$p_{1j} = (1 + \exp(-\sigma a_j))^{-1}, \qquad p_{2j} = 1 - p_{1j}.$$

Equation 2 is true for any real σ.

Equation 2 may be interpreted in terms of probability theory. The integral in this equation may be considered the probability $P(\xi = b)$ that a random variable $\xi = \xi_1 + \cdots \xi_n$ will equal b. Here ξ_j, $j = 1, \ldots n$, are independent random variables taking values 0 and a_j with probabilities p_{1j} and p_{2j}, respectively.

Conducting the proof along a scheme resembling that of local limit theorems, we can establish the following result.

Let σ be chosen as a solution of the equation

$$\sum \frac{a_j}{(1 + \exp(\sigma a_j))} = b.$$

Introduce the following notation:

$$D = \sum_j p_{1j} p_{2j} a_j^2,$$

$$h = \max_j \frac{p_{1j} p_{2j} a_j^2}{D},$$

$$\rho_3 = \sum_i a_j^3 p_{1j} p_{2j},$$

$$t = \frac{1}{cD^{1/2}}, \qquad c \text{ is a positive constant,}$$

$$\nu = \sqrt{2\pi D} \int_t^{1/2} \left| \prod_{j=1}^{n} (p_{1j} + p_{2j} \exp(2\pi i \alpha a_j)) \right| d\alpha,$$

$$\Phi(x) = \left(\frac{1}{\sqrt{2\pi}}\right) \int_0^x \exp\left(\frac{-z^2}{2}\right) dz,$$

$$q = \frac{\exp\left(\exp\left(\frac{2\pi^2 h}{c^2}\right) 8\pi^3 \frac{\rho_3}{3c^3 D^{3/2}}\right) - 1}{\exp\left(2\pi^2 \frac{h}{c^2}\right) 8\pi^3 \frac{\rho_3}{3c^3 D^{3/2}}},$$

$$u = \frac{\sqrt{2/9\pi} q \rho_3}{(1-h)^2 D^{3/2}} \left(1 - \exp\left(-2\pi^2 \frac{(1-h)}{c^2}\right)\right) \left(1 - 2\pi^2 \frac{(1-h)}{c^2}\right).$$

Then

$$I_n = \exp(\sigma b) \prod_{j=1}^{n} (1 + \exp(-\sigma a_j)) \frac{1}{\sqrt{2\pi D}} \left(1 + 2\theta \left(\left(\frac{1}{2} - \phi\left(\frac{2\pi}{c}\right)\right) + u + v\right)\right), \tag{3}$$

where $|\theta| \leq 1$.

Equation 3 makes sense only if the values h, $\rho_3/D^{3/2}$, and v are sufficiently small. The analysis in Reference 2 has shown that if b in (**1**) does not grow too fast with respect to n, then the condition on these values holds true for almost all (**1**). Also notice that, by appropriately letting c tend to zero, we may transform (**3**) into an asymptotic equation.

ON THE NUMBER OF SOLUTIONS OF TWO SIMULTANEOUS EQUATIONS

Consider the following system of equations:

$$\begin{aligned} a_{11}x_1 + \cdots + a_{1n}x_n &= b_1, \\ a_{21}x_1 + \cdots + a_{2n}x_n &= b_2, \end{aligned} \tag{4}$$

where all the quotients are positive integers and the variables take the values 0 and 1.

The number of solutions of (**4**) is given by the formula

$$I_n = \exp(\sigma_1 b_1 + \sigma_2 b_2) \prod_{j=1}^{n} (1 + \exp(-\sigma_1 a_{1j} - \sigma_2 a_{2j}))$$

$$\times \int_0^1 \int_0^1 \prod_{j=1}^{n} (p_{1j} + p_{2j} \exp(2\pi i (\alpha_1 a_{1j} + \alpha_2 a_{2j}))) \tag{5}$$

$$\times \exp(-2\pi i (\alpha_1 b_1 + \alpha_2 b_2)) \, d\alpha_1 \, d\alpha_2,$$

where σ_1 and σ_2 are any real numbers and

$$p_{1j} = (1 + \exp(-\sigma_1 a_{1j} - \sigma_2 a_{2j}))^{-1}, \qquad p_{2j} = 1 - p_{1j}.$$

Just as in the case of one equation, the following result can be established.

Let σ_1 and σ_2 be determined by the system

$$\begin{cases} \sum_{j=1}^{n} a_{1j} p_{2j} = b_1, \\ \sum_{j=1}^{n} a_{2j} p_{2j} = b_2. \end{cases} \quad (6)$$

Such σ_1 and σ_2 are shown to exist under some reasonable conditions on b_1 and b_2.
Introduce the following notation:

$$D_1 = \sum_j a_{1j}^2 p_{1j} p_{2j},$$

$$D_2 = \sum_j a_{2j}^2 p_{1j} p_{2j},$$

$$D_{12} = \sum_j a_{1j} a_{2j} p_{1j} p_{2j},$$

$$\rho_{13} = \sum_j a_{1j}^3 p_{1j} p_{2j},$$

$$\rho_{23} = \sum_j a_{2j}^3 p_{1j} p_{2j},$$

$$h_1 = \max_j \frac{a_{1j}^2 p_{1j} p_{2j}}{D_1}$$

$$h_2 = \max_j \frac{a_{2j}^2 p_{1j} p_{2j}}{D_2}$$

$$\Delta = D_1 D_2 - D_{12}^2,$$

$$G(\lambda_0) = \{\alpha_1, \alpha_2 / D_1 \alpha_1^2 + 2 D_{12} \alpha_1 \alpha_2 + D_2 \alpha_2^2 \leq \lambda_0^2\},$$

where λ_0 is a positive constant,

$$T = \{\alpha_1, \alpha_2 / |\alpha_1| \leq 1/2, \ |\alpha_2| \leq 1/2\},$$

$$\nu = 2\pi \sqrt{\Delta} \iint_{T \setminus G(\lambda_0)} \left| \prod (p_{1j} + p_{2j} \exp(2\pi i (\alpha_1 a_{1j} + \alpha_2 a_{2j})) \right| d\alpha_1 d\alpha_2,$$

$$\beta_0 = \lambda_0 \left(\frac{D_1 D_2}{\Delta} \right)^{1/2},$$

$$P = \frac{\rho_{13}}{(D_1 - D_{12}^2/D_2)^{3/2}} + \frac{\rho_{23}}{(D_2 - D_{12}^2/D_1)^{3/2}},$$

$$\delta_0 = \exp(4\pi^2 \beta_0^2 (h_1 + h_2)) \frac{32\pi^3 \lambda_0^3 P}{3},$$

$$u = (\exp(\delta_0) - 1) 2(3\sqrt{2\pi} \Phi(2\pi\lambda_0) - \frac{2\pi\lambda_0 (4\pi^2 \lambda_0^2 - 3) \exp(-2\pi^2 \lambda_0^2))}{(2\pi)^4 \lambda_0^3}.$$

Then, if the condition

$$\Delta > 4\lambda_0^2 \max(D_1, D_2) \tag{7}$$

holds, the following formula is valid.

$$I_n = \exp(\sigma_1 b_1 + \sigma_2 b_2) \prod_{j=1}^{n} (1 + \exp(-\sigma_1 a_{1j} - \sigma_2 a_{2j})) \frac{1}{2\pi\sqrt{\Delta}}$$

$$\times (1 + \theta (\exp(-2\pi^2\lambda_0^2) + u + v)), \tag{8}$$

where $|\theta| \leq 1$.

Notice that (8) makes sense only if the values $\beta_0^2(h_1 + h_2)$, P, and ν are sufficiently small and Δ is sufficiently large (see Reference 7).

ON AN EFFICIENT METHOD OF SOLVING IP WITH MANY VARIABLES AND A FEW CONSTANTS

This method is based on the results presented above. The number of constraints has not yet been studied, but the principal nature of the restriction is clear: the number of constraints should be not too large in comparison with the number of variables. Some aspects of this method have to be examined further; however, this does not obstruct its practical implementation.

Consider an IP, such as the following.

$$\sum_{j=1}^{n} c_j x_j \to \max,$$

$$\sum_{j=1}^{n} a_{ij} x_j \leq b_i, \quad i = 1, \ldots, m, \tag{9}$$

$$x_j = 0, 1.$$

We assume that c_j, a_{ij}, and b_i are nonnegative integers. Let us denote by I the number of solutions of the system

$$\begin{cases} \sum_{j=1}^{n} c_j x_j = c, \\ \sum_{j=1}^{n} a_{ij} x_j = \bar{b}_i, \quad i = 1, \ldots, m, \\ x_j = 0, 1. \end{cases} \tag{10}$$

Problem 9 may be reduced to the solvability question of (10) for various right-hand sides. Naturally, we are interested in $0 \leq \bar{b}_i \leq b_i$ and possibly large c. Thus, we have to investigate I.

Introduce the following notation:

$$w_j = \rho c_j + \sum_{i=1}^{m} \sigma_i a_{ij},$$

$$p_{1j} = (1 + \exp(-w_j))^{-1},$$

$$p_{2j} = \exp(-w_j)/(1 + \exp(-w_j)),$$

$$D_{kl} = D_{lk} = \sum_{j=1}^{n} p_{1j} p_{2j} a_{kj} a_{lj}, \qquad k, l = 1, \ldots m,$$

$$D_{k,m+1} = D_{m+1,k} = \sum_{j=1}^{n} p_{1j} p_{2j} a_{kj} c_j, \qquad k = 1, \ldots m,$$

$$D_{m+1,m+1} = \sum_{j=1}^{n} p_{1j} p_{2j} c_j^2,$$

$$\Delta = \det D = \det(D_{kl}).$$

If σ_i and ρ are determined by the system

$$\begin{cases} \sum_{j=1}^{n} a_{ij} p_{2j} = \bar{b}_i, & i = 1, \ldots, m, \\ \sum_{j=1}^{n} c_j p_{2j} = c, \end{cases}$$

then the following approximate formula is valid:

$$I = \exp\left(\rho c + \sum_{i=1}^{m} \sigma_i \bar{b}_i\right) \frac{1}{(2\pi)^{(m+1)/2} \sqrt{\Delta}} (1 + R). \qquad (11)$$

A detailed proof of this formula will be published soon.

Let us now consider the polyhedral region $E = \{y/y \in R^{m+1}, y_i = \Sigma_j a_{ij} x_j, i = 1, \ldots m; y_{m+1} = \Sigma_j c_j x_j, x_j \in [0,1]\}$. For $m = 1$, this region is shown in FIGURE 1.

Analysis of (11) has revealed that the so-called stability region exists in the case where n is sufficiently large as compared with m and the coefficients in (9) are sufficiently small. This convex stability region $E' \subset E$ is characterized by the following property: each integer point of it, used as the right-hand side of (10), provides a large number of solutions of this system. Setting various values of the right-hand sides and computing (11) ($I \geq 2$, $R < 1$), we may estimate the boundary of E'.

The above considerations enable us to solve (9). Let $\bar{b}_i = b_i = b_i$ in (10), and let $Y_0 = (b_1, \ldots b_m, c_0)$ be the highest intersection point of the straight line $L = \{y/y_1 = b_1, \ldots y_m = b_m\}$ with E' (FIGURE 1). The main aspect of the approach in question is solving (10) with the right-hand side equal to $[Y_0]$; we are able to find Y_0 by using (10). Finding Y_0 has two important implications.

1. Y_0 provides a nontrivial estimate from below of the optimum of (9), $c \geq [c_0]$, and so it leads to an efficient solution of the problem.

2. $[c_0]$ is, in a sense, the "true" or "stable" optimum of (9).

Let us consider these two items in detail. Denote by Y_c a point corresponding to a continuous basic solution of (9), where the constraint $x_j = 0,1$ is relaxed by $0 \leq x_j \leq 1$.

The continuous solution has some variables equal to 1. Denote the set of such variables by X_1. We include all the variables taking continuous optimal values equal to 0 in the set X_0. All the other variables are included in the set X_f. Obviously, $|X_f| \le m$.

Let us take a variable from X_1 (X_0) and set it equal to 0(1). By an appropriate ordering, we may assume the variable to be $x_1(x_n)$. As a result, we would have a reduced problem with $n - 1$ variables. Denote by $\overline{Y}_c$ the point corresponding to continuous optimum of the reduced problem. If $\overline{Y}_c$ is situated lower than Y_0, then the variable x_1 must be equal to 1 in an optimum solution of (9) (if $\overline{Y}_c$ is higher than Y_0, then x_n must be equal to 0). Such variables are said to be unessential, while the others compose the core of the problem. Clearly, the computational complexity is determined by the cardinality of the core.

Of course, the core may be approximated by solving the appropriate reduced

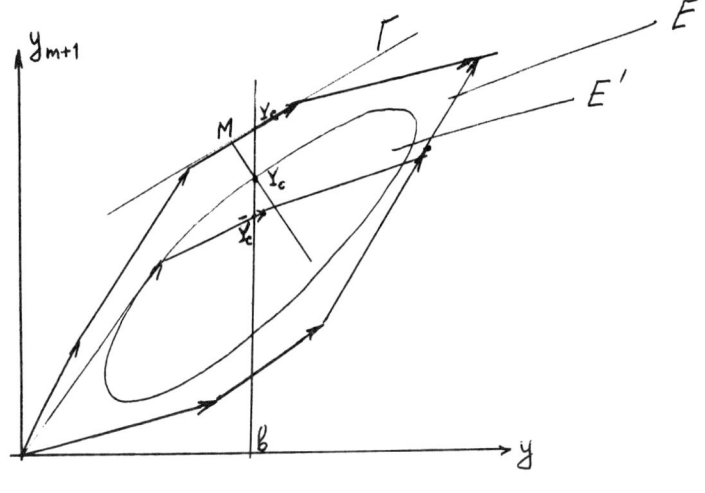

FIGURE 1.

continuous problem. However, such a method would be computationally unsatisfactory and it may be replaced by a substantially cheaper procedure. In fact, it is sufficient to consider the perpendicular MY_0 (see FIGURE 1), erected from Y_0 to the critical facet Γ (Γ is a hyperplane generated by the facet of E intersecting with L at Y_c) and then select the variables whose vectors of coefficients have projections on this perpendicular less than the distance MY_0. Such a criterion is, to a certain extent, weaker than the solution of reduced continuous problems; however, it is much more efficient computationally.

One of the most important problems is the estimation of the core size. In a degenerate case, the core may be rather large. However, some preliminary computations indicated that the core is sufficiently small for almost all cases. At the present time, the authors conjecture that if the coefficients of a problem are bounded by a polynomial of n, then the core size is expressed as $\log^{C(m)} N$, where $C(m)$ is a constant depending on m, and m is the number of constraints. Since the core size is supposed to be small, the ultimate solution may be obtained by any enumeration scheme.

Special attention should be paid to the approach where $[Y_0]$ is considered a true optimum by definition. In this approach, all the points outside the stability region are neglected. Integer points inside the stability region provide many solutions to (**10**), i.e., the value $[c_0]$ of the functional is provided by a large set of values of variables x_j. The latter can be of great importance in many practical situations. Thus, Y_0 is a point of stable optimum. The development of the concept of stability and its connection to optimization problems seems to have good prospects. A more detailed exposition of these considerations may be found in Reference 7.

COMMENTS ON SOME RESULTS

The approach to IP described above has been developed from considerations based on number theory techniques applied to the solvability of linear equations. At the same time, there are some details that are well known in the branch-and-bound scheme, which is one of the main schemes for solving IP. The concept of the core enables us to clarify the structure of this scheme and to understand some reasons for its successes and failures. It indicates which strategy would be better in the branch-and-bound methods.

One of the principal elements of branching is the choice of a variable, which is fixed with the values 0 and 1. This generates two subproblems, which are two branches of the search tree. If one of the branches leads to a more feasible solution than the bound of another, then the latter branch is cut off and the enumeration is reduced. The same is true for all stages of the process.

The faster branches are cut off and the faster algorithm proceeds. Let us try to explain why the branch-and-bound scheme often works successfully, despite the fact that the choice of branches is based on some heuristic ideas and conducted very often at random.

Which branch, in fact, would be cut off? In terms of the considerations above, it must be the branch that determines a reduced problem with, possibly, a small bound. The bound's increment could be characterized by the sum of the projections on the perpendicular to the critical facet of the vectors corresponding to those variables whose values have been fixed "wrong," i.e., values opposite those of a continuous solution of the initial problem. Obviously, if there are large projections, then such a branch would be cut off.

As discussed above, all the variables are partitioned into two groups, one of which is a core characterized by small projections, and the other of which is composed of unessential variables, characterized by large projections. The branches, where the core variables are not fixed, give solutions, one of which is given by the point Y_0 (see FIGURE 1). As was pointed out, the class of problems in question is assumed to have a small core, and so most variables would have large projections. Therefore, even for a random choice of the branching variable, we would get bounds enabling us to cut off the branch considered. This is provided by many heuristic rules. For instance, Reference 9, which considers the knapsack problem, suggests as a branching variable that with maximal c_j/a_j. This criterion is based on the fact that the vector (a_j, c_j) is the steepest vector to the critical facet. According to our considerations, we should take into account not only the angle, but also the length of this vector.

Let us point out some dangers encountered in the implementation of the branch-and-bound scheme. The choice of branching variables is often conducted, as has been mentioned, without appropriate justifications. Let there be k branching variables from the core. Obviously, no such branches could be cut off and there are 2^k branches altogether. If now we try a "good" variable (with a large projection), then the corresponding subbranch will be cut off in all 2^k branches, while, had it been considered in due time, this variable would have been excluded from further processes. For instance, if, in the knapsack problem, we take the steepest vector (a_j, c_j) (as recommended in Reference 9), then, for large numbers of short vectors, the algorithm could not reach the solution.

From our point of view, the best directions for further development are given in References 10–12. However, it should be noted that they all deal only with the knapsack problem.

Let us consider, briefly, Reference 10, in which the main ideas of References 11 and 12 are summarized.

Balas and Zemei are closest to the ideas presented in this article.[10] However, they have not developed the concept of the stability region; therefore, they cannot find the point Y_0 on its boundary. That is why they propose an iterative process leading to a solution. In fact, the more feasible the available solution, the more accurate the core approximation. This idea is used in Reference 10, where a few vectors, having minimal angles with critical facets, are selected as an initial core approximation. The core is then refined iteratively. If the core has too many elements, then a heuristic is applied to obtain an estimation of the optimum (i.e., an analogue of the point Y_0). The next step, according to Reference 10, consists of some reasonable procedures to refine the core at the new stage. The authors make use of the projections of the coefficients' vectors, as suggested in Reference 12. The refined core enables us to get an accurate optimum estimation and to approach the boundary of the stability region.

It worth noting that the method of Reference 10 invented for the knapsack problem would be difficult to generalize to a case with many constraints. It is not clear how to search for a feasible solution of linear simultaneous equations with 0,1 variables using the method suggested there. At this point, the possibility of finding Y_0, which we are able to obtain by means of the approach above, becomes very important.

REFERENCES

1. FREIMAN, G. A. 1980. An analytical method of analysis of linear Boolean equations. Ann. N.Y. Acad. Sci. **337**: 97.
2. BUZYTSKY, P. L. & G. A. FREIMAN. 1980. On the possibilities of solving combinatorial problems by analytic methods. Ann. N.Y. Acad. Sci. **337**: 87.
3. BERSTEIN, A. A., P. L. BUZYTSKY & G. A. FREIMAN. 1979. An application of analytical methods to combinatorial problems. 10th Int. Symp. Math. Programming, Canada.
4. BUZYTSKY, P. L. & G. A. FREIMAN. 1980. Analytical methods in integer programming. Preprint, CEMI.
5. BUZYTSKY, P. L. 1981. An effective formula for the number of solutions of linear Boolean equations. In press.

6. BUZYTSKY, P. L. & G. A. FREIMAN. 1981. An effective formula for the number of solutions of two Boolean equations' system. In press.
7. FREIMAN, G. A. 1981. On a new method in integer programming. In press.
8. BUZYTSKY, P. L. & G. A. FREIMAN. 1981. On the branch-and-bound method in integer programming. In press.
9. KOLESAR, P. J. 1967. A branch and bound algorithm for the knapsack problem. Manage. Sci. **13**: (9).
10. BALAS, E. & E. ZEMEL. 1979. An algorithm for large zero-one knapsack problems. Manage. Sci. Res. Rep. N408(R). Carnegie-Mellon University.
11. INGARGIOLA, G. P. & J. F. KORSH. 1973. Reduction algorithm for zero-one single knapsack problems. Manage. Sci. **20**(4).
12. DEMBO, R. S. & P. L. HAMMER. 1975. A reduction algorithm for knapsack problems. Research Report CORR 75-6. University of Waterloo.

TRENDS IN THE DEVELOPMENT OF COMPUTER APPLICATIONS*

Daniel D. McCracken

7 Sherwood Avenue
Ossining, New York 10562

INTRODUCTION

The bulk of my career has been devoted to writing books that teach beginners how to get results with a computer. My books on Fortran, Cobol, and other computer languages have been used by students in universities and in business to get started learning how to get useful work from a computer. I have also had a small role in promoting more effective use of traditional programming languages through the cluster of ideas known as structured programming. In some situations, it still makes sense to approach a new computer application by writing a conventional program, and, when that is the case, it is certainty advisable to use the most modern programming methods available.

But it is becoming more and more common for the user to take any one of several altogether different approaches in which no programming is done at all, at least not in the sense of Fortran, Cobol, or Basic programming. Instead, the user turns to one or more of the following classes of application development methods.

APPLICATIONS PACKAGES

In a wide variety of areas of computer use, there exist programs, which may be bought or leased, that will handle most aspects of the proposed work. In the business use of computers, there are packages that handle all the common tasks: general ledger, project management, inventory control, cash management, etc. In fact, in each of these and many other examples, the user has a choice of as many as dozens of competing packages. The subindustry that supplies software products in this area exceeds $1 billion annually in sales and is growing about as rapidly as any part of the computer industry.

Packages also exist for many types of engineering work. The chemical engineer wishing to study the transient flow of compressible gases in a proposed piping network can obtain any one of several packages that does the job. Or, if the problem involves steady-state flow, or incompressible fluids, there are other packages to choose from. An electrical engineer can pick from among several packages that analyze the performance of electrical networks containing any of the common circuit elements (including a range of nonlinear devices), perhaps looking at worst case analyses to determine tolerances on the specifications for critical elements. An optical designer

---

*This paper is a revision of one that had been scheduled to be presented at the Fourth International Conference on Collective Phenomena in Moscow in 1980; it was not presented because the USSR rejected, without explanation, the author's application for a visa to attend the meeting.

can use a package that incorporates more distilled knowledge about lenses than he probably commands himself. Supplying packages of this sort generates about $300 million annually for their vendors; this subindustry is growing at about 40% per year.

It commonly happens that no existing package does exactly what the user wants, and that small-to-moderate modifications must be made. The net effort is still usually far less than that of programming the application anew, however, and the project lead time is also dramatically shorter. Furthermore, successful packages are the end-product of a long development process in which a great deal of knowledge about the subject has been incorporated into the program.

Knowledge-Based Systems

In certain areas, the amount of factual data and lore about the field is so large that the program takes on the character of an intelligent assistant. The package named MACSYMA, for example, puts at the fingertips of the researcher man-decades of the expertise of some of the world's top experts in symbolic manipulation—the differentiation and integration of expressions in symbolic form, for example. The point here is not the tired comparison that a computer can do in minutes what would take a man a year or whatever, but that the program has so much expertise built into it that it can do things that only a few hundred people in the entire world can do at all. There are, perhaps not surprisingly, fewer examples of systems of this type, but others are to be found in mass spectrometer and x-ray diffraction analysis, certain aspects of internal medicine, and the analysis of seismic and other data related to oil exploration.

Very High Level Languages

There will always be applications that, for one reason or another, will not be amenable to treatment by any existing package. Perhaps the job is too new for a package to have been developed, or there are too few users to support the development cost of a package (which can take years and cost millions), or the job is simply too small to need the power of a big package.

In such cases we turn to "programming" of a type that is so different from what most people associate with computer programming that a new term is needed. Sometimes the phrase "nonprocedural language" is used, to contrast with the way in which conventional programming boils down to giving the computer a highly detailed sequence of operations for it to carry out. The current phrase is that we would rather tell a computer what to do than how to do it. However, it isn't very satisfying to define something in terms of what it is not, so I am still hoping for a better term. "Very high level language" is not as descriptive as one could hope, but it at least conveys the idea of a progression in the development of ways to tell a computer what we want done. In their early days, Fortran and Cobol were giant steps beyond the agonizing detail in which a computer must be instructed in its own language. As noted, they still have their uses, but today we need another giant step.

And if there are, at present, no giant steps, there are at least some fairly large ones.

The basis of perhaps a dozen systems now exceeding $100 million in annual revenues, one of these steps is the combination of a database management system with a very high level inquiry language. A database management system is a complex piece of computer software that provides a way to organize the relationships among data elements, enter data in forms specified in defining the database, and allow easy retrieval of the data presented in a form easily used by people. There are many dozens of competing database management systems available, but not all of them are intended for the type of application I have in mind here, where the emphasis is on the easy production of reports that may not even have been anticipated when the application was set up. Some representative systems are Ramis (Mathematica), NOMAD (National CSS), FOCUS (Information Builders), Query By Example (IBM), and Info (Henco). All of these and others make possible a fairly wide range of applications, from one-shot jobs that can be done in an hour to very large on-line databases with hundreds of simultaneous users. The common traits are: easy definition of what each data element looks like and what kinds of values are legitimate for it, easy specification of relationships among data elements, and very simple ways to say what kinds of reports are to be produced by various kinds of analyses of the data.

As I say, a number of such systems are already in extensive use; the oldest are approaching a decade of user experience. Their use is growing rapidly—as is their number—and new systems are announced almost monthly.

Summary

The entire field of computer applications development is currently in a state of feverish activity. There are simply not enough programmers to do all the computer work that people want done, using conventional programming methods, and there probably never will be. Newer methods are under intensive development, with most approaches having the effect of trading some increase in the use of computing hardware (which is rapidly getting cheaper) for a decrease in the use of the time and skills of computer programmers and analysts (for which demand far exceeds supply).

The demand for new computer applications is so large—backlogs for getting new projects programmed typically run to two to three years or longer—that it is not clear when these new methods will catch up. The growing availability and acceptance of alternatives to conventional programming, however, does offer some hope for the future.

DOOB-MEYER DECOMPOSITIONS FOR TWO-PARAMETER STOCHASTIC PROCESSES

Ely Merzbach

Department of Mathematics
Bar Ilan University
52100, Ramat-Gan, Israel

In this note, we generalize the well-known Doob-Meyer decomposition of a submartingale as the sum of a martingale and an increasing process, for processes with partially ordered parameter sets. Here the parameter set will be the positive quadrant of the plane $\mathbf{R}_+^2$ and the partial order induced by the Cartesian coordinates. In this case, several notions of martingales are naturally introduced,[1,2] and we obtain different decompositions. The techniques used here are principally the Doleans measure associated with a martingale and the notion of predictable projection for measurable and bounded processes.[3,4]

Besides being of interest for its own sake, the Doob-Meyer decomposition makes it possible to define the stochastic integral with respect to processes belonging to a broader class of processes than martingales and, above all, in the "multi-parameter" case, the decomposition makes it possible to stop any process of this class at an arbitrary point.

The notation and definitions of this note will follow those of Reference 1. For two points $z = (s, t)$ and $z' = (s', t')$ in the positive quadrant of the plane $\mathbf{R}_+^2$, $z < z'$ means $s \leq s'$ and $t \leq t'$, and $z \ll z'$ means $s < s'$ and $t < t'$.

If $z \ll z'$, $(z, z']$ will denote the rectangle $(s, s'] \times (t, t']$. Let $(\Omega, \mathcal{F}, P)$ be a complete probability space with a right-continuous filtration $\{\mathcal{F}_z, z \in \mathbf{R}_+^2\}$ satisfying the following property (the (F4) property of Reference 1): Let $\mathcal{F}^1_{(s,t)} = \mathcal{F}_{(s,\infty)}$ and $\mathcal{F}^2_{(s,t)} = \mathcal{F}_{(\infty,t)}$. Then, for each z, $\mathcal{F}^1_z$ and $\mathcal{F}^2_z$ are conditionally independent, given $\mathcal{F}_z$. A set in the product space $\mathbf{R}_+^2 \times \Omega$ is called measurable if it belongs to the σ-algebra $\mathcal{B}(\mathbf{R}_+^2) \otimes \mathcal{F}$.

Consider the measurable space $(\mathbf{R}_+^2 \times \Omega, \mathcal{B}(\mathbf{R}_+^2) \otimes \mathcal{F})$. Observe that finite unions of the following "rectangles" $((z, z'] \times G)$, $G \in \mathcal{F}_z$ (respectively, $G \in \mathcal{F}^i_z$, $i = 1, 2$), constitute an algebra. A set is called predictable (i-predictable, $i = 1, 2$) if it belongs to the σ-algebra generated by these "rectangles." This σ-algebra will be denoted by $\mathcal{P}$ (respectively, $\mathcal{P}^i$, $i = 1, 2$). A process will be called predictable (respectively i-predictable) if it is $\mathcal{P}$ (respectively, $\mathcal{P}^i$) measurable. Finally, recall the definition of an increasing process: A process $A = \{A_z, z \in \mathbf{R}_+^2\}$ is an increasing process if A vanishes on the axes, is right continuous, adapted, $\sup_z EA_z < \infty$ and if every rectangle $[z, z']$, $z \ll z'$, satisfies

$$A(z, z'] = A_{z'} + A_z - A_{s,t'} - A_{s',t} \geq 0.$$

A real σ-additive measure μ on $(\mathbf{R}_+^2 \times \Omega, \mathcal{B}(\mathbf{R}_+^2) \otimes \mathcal{F})$ such that P-evanescent subsets of $\mathbf{R}_+^2 \times \Omega$ are μ-null is called a *stochastic measure*. If A is an increasing process (bounded variation process), then A induces a (signed) stochastic measure μ_A, which is defined on all the measurable sets in the following way: If X is a positive measurable process, then $\mu_A(X) = E[\int X_z dA_z]$. Conversely, if μ is a stochastic measure defined on all the measurable sets, then there exists a unique increasing, but not necessarily adapted, process A such that $\mu = \mu_A$.

Definition 1. Let $X = \{X_z, z \in \mathbf{R}_+^2\}$ be an integrable process.

   a. X is called a strong martingale if X is adapted and if $\forall z \ll z'$, so we have $E[X(z, z']/\mathcal{F}_z^1 \vee \mathcal{F}_z^2] = 0$.

   b. X is called a martingale if $\forall z < z'$, so we have $E[X_{z'}/\mathcal{F}_z] = X_z$.

   c. X is called a i-martingale if X is $\mathcal{F}^i$-adapted and if $\forall z \ll z'$, so we have $E[X(z, z']/\mathcal{F}_z^i] = 0$ $(i = 1, 2)$.

   d. X is called a weak martingale if X is adapted and if $\forall z \ll z'$, so we have $E[X(z, z']/\mathcal{F}_z] = 0$.

Remark 1. In Reference 1, the following implications are proved: (1) A strong martingale that vanishes on the axes is a martingale. (2) A process is a martingale if and only if it is a 1-martingale and a 2-martingale. (3) An adapted i-martingale is a weak martingale.

Remark 2. Let μ be the Doleans measure associated with a process X vanishing on the axes and adapted. Then

X is a strong martingale $\Longleftrightarrow$ μ vanishes on $\mathcal{P}^1 \vee \mathcal{P}^2$.
X is a i-martingale $\Longleftrightarrow$ μ vanishes on $\mathcal{P}^i$ $(i = 1, 2)$.
X is a weak martingale $\Longleftrightarrow$ μ vanishes on $\mathcal{P}$.[7]

Recall now some facts about the notion of predictable projection. (Proofs can be found in Reference 4.)

THEOREM 1. Let X be a bounded and measurable process. There exists a unique predictable (i-predictable) process $^\pi X$ ($^{\pi_i}X$) called the predictable (i-predictable) projection of X, such that $E[\int X dA] = E[\int ^\pi X dA]$ $(= E[\int ^{\pi_i} X dA])$ for all predictable (i-predictable) and increasing processes A. Moreover, we have the following properties.

   a. $^\pi X = {}^{\pi_1}({}^{\pi_2}X) = {}^{\pi_2}({}^{\pi_1}X)$.
   b. $\forall z \in \mathbf{R}_+^2$, $^\pi X_z = E[X_z / \mathcal{F}_{z-}]$ $(\mathcal{F}_{z-} = \vee_{z' \ll z} \mathcal{F}_{z'})$.
   c. If X has a martingale property, then X has the same property with respect to the filtration $\{\mathcal{F}_{z-}\}$.

THEOREM 2. Let A be an increasing process. There exists a unique predictable (i-predictable) increasing process A^π (A^{π_i}) called the dual predictable (i-predictable) projection of A, such that $E[\int X dA] = E[\int X dA^\pi]$ $(= E[\int X dA^{\pi_i}])$ for all predictable processes X. Moreover, we have the following properties.

   a. $A^\pi = (A^{\pi_1})^{\pi_2} = (A^{\pi_2})^{\pi_1}$.
   b. $E[\int ^\pi X dA] = E[\int X dA^\pi] = E[\int ^\pi X dA^\pi]$.
   c. $A - A^\pi$ is a weak martingale and $A - A^{\pi_i}$ is a i-martingale.

Definition 2. We say that a process belongs to class D if $\{X_Z, Z$ being a bounded stopping point$\}$ is uniformly integrable.

A stopping point Z is a random point, such that $\forall z \in \mathbf{R}_+^2$, $\{\omega : Z(\omega) < z\} \in \mathcal{F}_z$.

The main result is the following:

THEOREM 3.

a. Let X be a positive, right-continuous, weak submartingale ($\geq$ instead of $=$ in Definition 1) process belonging to class D. Then there exists a unique predictable increasing process A such that $X - A$ is a weak martingale (same for $i = 1, 2$).

b. Let X be a submartingale and a weak submartingale belonging to class D for each parameter. Then X can be decomposed in the following way: $X = M + M_1 + M_2 + A$, where M is a martingale, M_i is a i-martingale ($i = 1, 2$) increasing in the other parameter, and A is increasing.

The idea of the proof of THEOREM 3a is to show that μ_X is a stochastic measure and to use THEOREM 2.[3] The second part of this theorem was proved by Cairoli.

Hereafter, assume that $M = \{M_z, z \in \mathbf{R}_+^2\}$ is a square integrable martingale. It is easy to see that, in this case, the process M^2 is a submartingale and a weak submartingale. We denote by $\langle M \rangle$ the unique predictable increasing process such that $M^2 - \langle M \rangle$ is a weak martingale. $\langle M \rangle$ is also the limit (in a weak sense) of the quadratic variation of M.[1]

THEOREM 4. If M is a strong martingale, then $M^2 - \langle M \rangle$ is a martingale.

Example. Let the process $B = \{B_z, z \in \mathbf{R}_+^2\}$ be the Brownian sheet. Then the process $B_z^2 - s \cdot t$ (where $z = (s, t)$) is a martingale. (for the converse of this result, see Reference 5).

Some indications of martingales that are not strong martingales, M, but such that $M^2 - \langle M \rangle$ are martingales are given by Zakai.[5]

Let us conclude with the following remark: It is easy to see that M is a martingale if and only if, for each increasing path $\mathcal{O}: [0,1] \to \mathbf{R}_+^2$, the process $\{M_{\mathcal{O}(t)}\}_t$ is a one-parameter martingale. Then, by the classical Doob-Meyer decomposition, there exists an increasing process $A^{\mathcal{O}}$ such that $M_{\mathcal{O}(t)}^2 - A_t^{\mathcal{O}}$ is a one-parameter martingale. If $A^{\mathcal{O}}$ does not depend on $\mathcal{O}$, we say that M is of path-independent variation (p.i.v., see Reference 1). We have the following: If M is a p.i.v. martingale, then there exists a unique predictable process A, increasing for the partial-order ($z < z' \Longrightarrow A_z \leq A_{z'}$), such that $M^2 - A$ is a martingale, but we don't know if A is increasing.

REFERENCES

1. CAIROLI, R. & J. B. WALSH. 1975. Stochastic integrals in the plane. Acta Math. **134:** 111–83.
2. WONG, E. M. ZAKAI. 1976. Weak martingales and stochastic integrals in the plane. Ann. Proba. **4:** 570–87.
3. MERZBACH E. 1980. Processus stochastiques à indices partiellement ordonnés. Rapport interne, Ecole Polytechnique.
4. MERZBACH, E. & M. ZAKAI. Predictable and dual predictable projections of two-parameter stochastic processes. In press.
5. ZAKAI M. 1980. Some classes of two-parameter martingales. Ann. Proba. In press.
6. BRENNAN, M. D. 1979. Planar semi-martingales. J. Multivar. Anal. **9:** 465–86.
7. MERZBACH E. 1980. Une remarque sur les martingales à deux indices et leurs mesures de Doléans associées. C.R. Acad. Sci. Ser. A **290:** 435–38.
8. STOICA L. 1978. On two-parameter semi-martingales. Z. Whar. Verw. Geb. **45:** 257–68.

MAGNETIC PROPERTIES OF RELATIVISTIC FERMI GAS

E. M. Chudnovsky

Kharkov, USSR

The expression for the magnetic susceptibility of a nonrelativistic degenerate electron gas is well known in the theory of metals. It consists of two parts corresponding to two different physical effects. The first of them is connected with the behavior of spin moments in the external magnetic field (Pauli paramagnetism). The second is the result of the circular motion of charges in the magnetic field (Landau diamagnetism). For nonrelativistic fermions, the paramagnetic and diamagnetic susceptibilities can be independently calculated.

This does not occur in the relativistic case. Magnetic properties of the relativistic Fermi gas are of interest because degenerate relativistic electrons and nucleons form white dwarfs and neutron stars. As we know, nucleons have anomalous magnetic moments that also contribute to the magnetic susceptibility. In this connection, we consider a Lagrangian that includes minimum and dipole interactions between the fermion ψ and the external electromagnetic field A_μ:[1]

$$\mathcal{L}_{int} = e\bar{x}\left\{q\gamma_\mu A_\mu + i\frac{\kappa}{8M}[\gamma_\mu, \gamma_\nu]F_{\mu\nu}\right\}\psi, \qquad (1)$$

where $F_{\mu\tau} = \partial_m A_\nu - \partial_\nu A_m$, $-e$ is the charge of the electron and γ_μ are Dirac matrices (see TABLE 1).

In a constant homogeneous magnetic field H, the energy levels for the Lagrangian (1) can be obtained exactly:[2]

$$\epsilon^2 = p_H^2 + \left\{\frac{\kappa e}{2M}\sigma H + [M^2 + |q|eH(2l+1) + q e\sigma H]^{1/2}\right\}^2, \qquad (2)$$

where p_H is the momentum projection on the magnetic field, and $l = 0, 1, 2 \ldots$ and $\sigma = \pm 1$ are the orbital and spin quantum numbers, respectively.

The thermodynamic potential density of the fermion gas has the form

$$\Omega = -T\sum_i \ln\{1 + \exp[(\mu - \epsilon_i)/T]\}, \qquad (3)$$

where ϵ_i are particle energy levels (2), μ is the chemical potential, and T is temperature.

In a plane normal to the magnetic field, fermions move in closed orbits, their area being quantized in the momentum space

$$S_\rho = \pi(2l+1)|q|eH. \qquad (4)$$

Taking this into account, considering T to be small compared to the Fermi energy, and using the Euler-McLoren summation formula

$$\sum_{l=0}^{n} f(l) = \int_0^n f(l)\, dl + \frac{1}{2}[f(0) + f(n)]$$

$$+ \sum_{m=1}^{\infty} \frac{B_{2m}}{(2m)!} [f^{(2m-1)}(n) - f^{(2m-1)}(0)], \quad (5)$$

one can obtain (to an accuracy of H^2),

$$\Omega = \Omega_0 - \frac{1}{2} \chi H^2, \quad (6)$$

where

$$\chi = \frac{e^2}{4\pi^2} \left\{ \frac{\kappa^2 v_F}{2(1 - v_F^2)} + \left(\frac{\kappa^2}{4} + \kappa q + \frac{1}{3} q^2 \right) \ln \frac{1 + v_F}{1 - v_F} \right\} \quad (7)$$

is the magnetic susceptibility, v_F is the Fermi velocity, and Ω_0 does not depend on the magnetic field.

TABLE 1

Particle	q	κ
Electron	-1	0
Proton	$+1$	$+1.79$
Neutron	0	-1.91

In the nonrelativistic approximation ($v_f \ll 1$), the expression for susceptibility may be represented as a sum of paramagnetic and diamagnetic terms:

$$\chi = \frac{e^2 v_F}{4\pi^2} \left[(\kappa + q)^2 - \frac{1}{3} q^2 \right]. \quad (8)$$

For $\kappa = 0$ and $q = -1$, these terms coincide with well-known expressions for the electron gas.[3,4]

The well-known electrodynamic condition of the matter stability with respect to the spontaneous magnetic field produced due to the spin and orbital moments ordering has the form

$$\chi < 1. \quad (9)$$

The stability of the relativistic electron gas with respect to spontaneous magnetization was studied by Canuto and Chiu.[5] Their analysis led them to a wrong conclusion on the stability of the electron gas at arbitrary density. As can be seen from our calculation (8), the stability of the electron gas ($\kappa = 0$ and $q = -1$) is lost in the ultrarelativistic limit $v_F \to 1$.

In connection with our results, we also note that the nucleon density phase transition into the state with $M \to 0$ ($v_F \to 1$) studied by Lee and Wick[6] can lead to the spontaneous magnetization of a neutron star.

References

1. GELL-MAN, M. 1956. Nuovo Cimento Suppl. **2:** 848.
2. TSAI, W. & A. YILDIZ. 1971. Phys. Rev. D **4:** 3643.
3. PAULI, W. 1977. Z. Phys. **41:** 81.
4. LANDAU, L. D. 1930. Z. Phys. **64:** 629.
5. CANUTO, V. & H. Y. CHIU. 1968. Phys. Rev. **173:** 1229.
6. LEE, T. D. & G. C. WICK. 1974. Phys. Rev. D **9:** 2291.

ENTROPY AND IRREVERSIBILITY

Oliver Penrose

Faculty of Mathematics
The Open University
Milton Keynes MK7 6AA, England

Introduction

Even in equilibrium statistical mechanics, there are a variety of ways of defining entropy, and the various ways are not necessarily equivalent—the Gibbs paradox is an example of the difficulties one can get into through failing to take this into account. In nonequilibrium statistical mechanics, an even greater variety is possible. In the first place, the definition of entropy depends on what macroscopic description of the system is chosen; and, in the second place, a variety of possible definitions are possible for any given macroscopic description. In a paper by Goldstein and Penrose,[1] which will be referred to in the following as GP, a particular method of defining nonequilibrium entropy for a given macroscopic description was put forward. It is the purpose of the present paper to review this definition of entropy and to discuss its physical significance.

Coarse Graining

Let Ω represent the phase space of our system and μ some standard measure on it. We shall require later that μ be invariant under time shifts, and so the Liouville measure is an appropriate choice, when it applies. An ensemble or probability distribution over Ω (assumed to be absolutely continuous with respect to μ) can be represented by a phase-space density function $\rho(\omega)$, where ω is a general phase-space point. The *fine-grained entropy* of this ensemble is given by the Gibbs formula

$$h(\rho) = -k \int \eta(\rho(\omega)) d\mu(\omega), \qquad (1)$$

where k is Boltzmann's constant and η is the function defined by

$$\eta(x) = \begin{cases} x \log x & \text{if } x > 0 \\ 0 & \text{if } x = 0 \end{cases}. \qquad (2)$$

We assume that $h(\rho)$ is finite, and, in particular, that

$$h(\rho) > -\infty. \qquad (3)$$

When the system evolves in time, ρ changes, but, as Gibbs noted, $h(\rho)$ does not. If we want entropy to increase with time, it follows that (1) is not satisfactory as a definition of entropy for systems out of equilibrium. A possible way round this difficulty, also suggested by Gibbs, is to replace ρ in (1) by a *coarse-grained* phase-space density, which is supposed to contain the same information about experimentally measurable quantities as ρ, but exclude irrelevant information about quantities that cannot be measured. To construct a coarse-grained phase-space density, we first partition phase space into a number of nonoverlapping "cells" P_1, $P_2, \ldots P_n$, each of which is, for the time being, assumed to have finite measure. The interpretation is that, at any given time, we can determine by experiment which cell the phase point ω representing our system is in, but not where it is in that cell. We may call this collection of cells "the partition $\mathcal{P}$." We then define the coarse-grained phase-space density $\rho_\mathcal{P}$ to be constant within each cell and to give the same probabilities as ρ does for the various cells. In symbols, these conditions are

$$\rho_\mathcal{P}(\omega_1) = \rho_\mathcal{P}(\omega_2), \tag{4a}$$

if ω_1 and ω_2 are in the same set P_i from the partition $\mathcal{P}$, and

$$\int_P \rho_\mathcal{P}(\omega) d\mu(\omega) = \int_P \rho(\omega) d\mu(\omega), \tag{4b}$$

if P is any union of sets from the partition $\mathcal{P}$. Then the coarse-grained entropy of ρ with respect to the partition $\mathcal{P}$, which we shall denote by $h(\rho,\mathcal{P})$, is defined by

$$\begin{aligned} h(\rho, \mathcal{P}) &= h(\rho_\mathcal{P}) \\ &= -k \int \eta(\rho_\mathcal{P}(\omega)) \, d\mu(\omega). \end{aligned} \tag{5}$$

If the sets comprising the partition $\mathcal{P}$ are countable, this definition is equivalent to

$$h(\eta,\mathcal{P}) = k \sum_i p(P_i) \log\left(\frac{\mu(P_i)}{p(P_i)}\right), \tag{6}$$

where

$$p(P_i) = \int_{P_i} \rho(\omega) d\mu(\omega) \tag{7}$$

is the probability that the phase point ω is in P_i.

We shall need to generalize this idea of coarse-graining to cases where the sets comprising the partition $\mathcal{P}$ do not have positive measure; for example, if Ω were the (x, y) plane, the elements of the partition could consist of lines $x = $ const. It is still possible to satisfy (4), however, provided that we require that the sets P in (4b), which are unions of sets from the partition, be measurable. The collection of all measurable unions of sets from the partition $\mathcal{P}$ forms an algebra, which we shall also denote by $\mathcal{P}$, and the phase-space density $\rho_\mathcal{P}$ defined in (4) is called the *conditional expectation* of ρ with respect to this algebra, often written $E(\rho|\mathcal{P})$,

$$\rho_\mathcal{P} = E(\rho|\mathcal{P}). \tag{8}$$

It is shown in GP that the entropy formula (5) satisfies the following:

Property 1. $h(\rho,\mathcal{A})$ is a nonincreasing function of $\mathcal{A}$. This means that if two algebras $\mathcal{A}_1$ and $\mathcal{A}_2$ satisfy

$$\mathcal{A}_1 \subset \mathcal{A}_2$$

(which is equivalent to saying that every element of the partition $\mathcal{A}_1$ is a union of elements of the partition $\mathcal{A}_2$), then

$$h(\rho, \mathcal{A}_1) \geq h(\rho, \mathcal{A}_2).$$

Property 2. $h(\rho, \mathcal{A})$ is a continuous function of $\mathcal{A}$. This means that if $\mathcal{A}_1, \mathcal{A}_2, \ldots$ is an increasing or decreasing sequence of algebras with limit $\mathcal{A}$ (i.e., if $\mathcal{A}_n \uparrow \mathcal{A}$ or $\mathcal{A}_n \downarrow \mathcal{A}$), then

$$\lim_{n \to \infty} h(\rho, \mathcal{A}_n) = h(\rho, \mathcal{A}).$$

TIME EVOLUTION

To describe the time evolution of our system, we introduce a family of functions ϕ_t ($t \in \mathcal{R}$) such that if the system has phase point ω at any given time t_1, then it has phase point $\phi_t(\omega)$ at time $t_1 + t$. The functions ϕ_t form a one-parameter group: they satisfy

$$\phi_{t+s} = \phi_t \circ \phi_s \tag{9}$$

We are requiring that the measure μ be invariant under time shifts; that is, if A is any measurable set, then the set $\phi_t A$, consisting of the images under ϕ_t of all the points ω in A, is also measurable and satisfies

$$\mu(\phi_t A) = \mu(A). \tag{10}$$

It follows that if the phase-space density of our ensemble at some given time, say time 0, is ρ, then the phase-space density at time t is ρ_t, defined by

$$\rho_t(\omega) = \rho(\phi_{-t}\omega). \tag{11}$$

The time evolution of the coarse-grained entropy can be calculated from the formula

$$h(\rho_t, \mathcal{A}) = h(\rho, \phi_{-t}\mathcal{A}), \tag{12}$$

where $\phi_{-t}\mathcal{A}$ is the partition (or algebra) consisting of sets having the form $\phi_{-t}A$, where A is a set in $\mathcal{A}$. We would like to ensure that $h(\rho_t, \mathcal{A})$ does not increase with the time t; by (12), this requirement is equivalent to

$$h(\rho, \phi_{-t}\mathcal{A}) \geq h(\rho, \phi_{-s}\mathcal{A}) \quad \text{if} \quad t > s \tag{13}$$

and, by Property 1, it can be achieved, for arbitrary phase-space densities ρ, by making

$$\phi_{-t}\mathcal{A} \subset \phi_{-s}\mathcal{A} \quad \text{for all} \quad t > s. \tag{14}$$

Because of (9), this is equivalent to

$$\mathcal{A} \subset \phi_t \mathcal{A} \quad \text{if} \quad t > 0. \tag{15}$$

That is, the image of the partition $\mathcal{A}$ under any positive time shift must be a finer partition than $\mathcal{A}$. If the partition $\mathcal{A}$ is finite, this condition can only be satisfied by trivial choices of $\mathcal{A}$. However, the following construction gives a possible method of constructing a nontrivial partition satisfying (15). We start with any partition $\mathcal{P}$, which may or may not be the partition of observations defined above, and we can construct a new partition $\mathcal{P}^-$ defined by

$$\mathcal{P}^- = \bigvee_{t=0}^{\infty} \phi_{-t}\mathcal{P}, \qquad (16)$$

which means the coarsest partition that is finer than all the partitions $\mathcal{P}, \phi_{-1}\mathcal{P}, \phi_{-2}\mathcal{P}, \ldots$; in other words, the algebra $\mathcal{P}^-$ is the smallest one containing all the algebras $\mathcal{P}, \phi_{-1}\mathcal{P}, \phi_{-2}\mathcal{P}, \ldots$. It is shown in GP that there are cases where $\mathcal{P}^-$ is not the "trivial" partition whose elements are the individual points of Ω.

Equation (5), with (15), gives us a method of defining an entropy that cannot decrease with time, but, if the definition is to correspond with our normal ideas about entropy, there must be some phase-space density for which the entropy so defined actually increases.

It was shown in GP that, provided the system is ergodic and has a finite positive Kolmogorov-Sinai entropy, there is a *finite* partition $\mathcal{P}$ such that the entropy constructed in this way, $h(\rho, \mathcal{P}^-)$, has the properties

$$\lim_{t \to \infty} h(\rho_t, \mathcal{P}^-) = k \log \mu(\Omega) \qquad (17)$$

and

$$\lim_{t \to -\infty} h(\rho_t, \mathcal{P}^-) = k\, h(\rho), \qquad (18)$$

so that, unless ρ is a constant, the entropy does increase at some time. But this existence proof does not guarantee that any given partition $\mathcal{P}$ will lead to an entropy that can increase with time. As we shall see, for systems of physical interest, such as the hard-sphere gas, or even the Lorentz gas, it is not easy to show that the natural choice of $\mathcal{P}$ does lead to an entropy $h(\rho, \mathcal{P}^-)$ that can increase with time.

The Hard-Sphere Gas

As an illustration of the use of the entropy definition (5), let us consider a system consisting of N hard spheres of diameter d, enclosed in a container (say a cube) of unit volume. We shall be interested in the limiting properties in the Boltzmann-Grad limit

$$\left.\begin{array}{c} N \to \infty, \quad d \to 0 \\ Nd^2 \to \text{const} \end{array}\right\}. \qquad (19)$$

Lanford shows that, for suitable initial probability distributions at time 0, the one-particle distribution function would, in this limit, satisfy Boltzmann's kinetic equation for times t satisfying $0 < t < t_0$, where t_0 is about $1/5$ of the mean free time.[2] The behavior for larger times is not covered by Lanford's theorem, but it is plausible that some such theorem may be provable for all positive values of t. We shall show here how such a theorem could also enable us to relate the definition (5) of entropy to the more conventional one based on the formula for Boltzmann's "H."

Boltzmann's formula is

$$H(f) = \int_{\Omega_1} f(\omega_1) \log f(\omega_1) d\omega_1, \tag{20}$$

where f is the one-particle distribution function normalized so that $\int_{\Omega_1} f(\omega_1) d\omega_1 = 1$, Ω_1 is the phase space of a single particle, ω_1 stands for a general point in Ω_1, and $d\omega_1$ is the Liouville measure in Ω_1.

The one-particle distribution function f is related to the phase-space density ρ by

$$f(\omega_1) = \int \cdots \int \rho(\omega) \, d\omega_2 \cdots d\omega_N. \tag{21}$$

For the particular case of independent particles, we have

$$\rho(\omega) = \prod_{i=1}^{N} f(\omega_i), \tag{22}$$

so that the fine-grained entropy then satisfies

$$h(\rho) = -NkH(f). \tag{23}$$

In general, (22) does not hold, and so neither does (23), but what we may hope to do is show that a suitably defined coarse-grained entropy, $h(\rho, \mathcal{A})$ satisfies an approximate relation of an analogous form:

$$h(\rho, \mathcal{A}) \simeq -NkH(f). \tag{24}$$

The law of nondecrease (13) for coarse-grained entropy would then correspond to Boltzmann's result that if f varies with time in accordance with Boltzmann's kinetic equation, then $H(f)$ is a nonincreasing function of time.

To obtain a result in the form of (24), we must specify the partition. A natural choice is this. We divide the one-particle phase-space into a countable set of cells Δ_1, Δ_2, ...; for example, these cells might be hypercubes. Then, for each phase point ω, and each integer i, we denote by $N_i(\omega)$ the number of particles in the cell Δ_i. Then we can define a mapping ψ from the phase-space Ω to the space l_1 of infinite sequences $(x_1, x_2, \ldots)$ by

$$\psi(\omega) = \left(\frac{N_1(\omega)}{N}, \frac{N_2(\omega)}{N}, \ldots \right). \tag{25}$$

To define our partition $\mathcal{A}$, we use the construction described at the end of the last section, taking

$$\mathcal{A} = \mathcal{P}^- = \bigvee_{t=0}^{\infty} \phi_{-t} \mathcal{P}, \tag{26}$$

where $\mathcal{P}$ is a partition of Ω describing a model of the act of observation. In the present case, we may suppose that the most detailed observation possible at any instant is to measure how many particles are in each of the cells Δ_i of one-particle phase-space: that is, we measure the numbers $N_1(\omega), N_2(\omega), \ldots$. This is equivalent to determining the point $\psi(\omega)$ in the "observation space" l_1.

If this point could be determined with perfect accuracy (i.e., if we could count the precise number of particles in each cell Δ_i), then the sets of observationally indistinguishable points in Ω, which are the sets comprising the partition $\mathcal{P}$, would be sets of the form $\psi^{-1}(\mathbf{x})$ where $\mathbf{x}(=x_1,x_2,\ldots)$ is a point in l_1. However, since we are interested in the limit of large N, it is somewhat more realistic to assume that the point $\psi(\omega)$ in l_1 can be measured only with finite accuracy. This leads us to consider also the case where the partition $\mathcal{P}$ consists of sets in Ω having the form $\psi^{-1}(M)$, where M is a set in the space l_1. Indeed, any partition $\mathcal{M}$ of the space l_1 into nonoverlapping sets M will generate a partition of the space Ω into nonoverlapping sets $\psi^{-1}(M)$; this last partition is a candidate for the partition $\mathcal{P}$ in (26).

It is not at all easy to evaluate the entropy $h(\rho,\mathcal{A})$, and only some partial results will be given here. First of all, there is a connection between $h(\rho,\mathcal{A})$ and $h(\rho,\mathcal{P})$. Equation 26 shows that $\mathcal{P}$ is a subalgebra of $\mathcal{A}$, and hence, by Property 1 above, that

$$h(\rho,\mathcal{A}) \leq h(\rho,\mathcal{P}). \tag{27}$$

Further, it is shown in section 7 of GP that if the successive observational states (elements of $\mathcal{P}$) in which the system finds itself form a Markov chain, then

$$h(\rho,\mathcal{A}) = h(\rho,\mathcal{P}). \tag{28}$$

Lanford's work suggests that, in a suitable limit, the one-particle distribution function f satisfies Boltzmann's kinetic equation, which is deterministic and a fortiori Markovian. This suggests that, in a suitable limit (which would presumably combine the Boltzmann-Grad limit $N \to \infty$ with making the partition Δ, and perhaps also $\mathcal{M}$, infinitely fine), the successive observational states P may indeed be deterministic, in which case (28) would hold in some limiting or approximate sense:

$$h(\rho,\mathcal{A}) \simeq h(\rho,\mathcal{P}). \tag{29}$$

Both the firm result (27) and the conjecture (29) indicate that we can obtain information about the nondecreasing coarse-grained entropy $h(\rho,\mathcal{A})$ by studying the simpler coarse-grained entropy $h(\rho,\mathcal{P})$.

First of all, we consider the measures of the individual elements in the partition $\mathcal{P}$. This is easiest in the case where $\mathcal{M}$ is the partition of l_1 into its individual points. Let $\mathbf{x}(=x_1,x_2,\ldots)$ be any point in l_1 (that is, a sequence satisfying $\Sigma_i |x_i| < \infty$). In order for the inverse image $\psi^{-1}(\mathbf{x})$ to be nonempty, we must have

$$\sum_i x_i = 1 \tag{30}$$

$$Nx_i = \text{integer} \geq 0, \quad i = 1, 2, 3, \ldots$$

The inverse image is then a region in Ω whose Liouville measure is given by

$$\mu(\psi^{-1}(\mathbf{x})) = N! \prod_{i=1}^{\infty} \frac{\mu_1(\Delta_i)^{N_i}}{N_i!}, \tag{31}$$

where $N_i = Nx_i$ and μ_1 is the Liouville measure in Ω_1. Conditions 30 ensure that only a finite number of factors in the product are different from 1. From (31), it follows, by

Stirling's approximation for factorials, that, for large N,

$$\frac{1}{N} \log \mu(\psi^{-1}(\mathbf{x})) \simeq \sum_i x_i \log \left(\frac{\mu_1(\Delta_i)}{x_i}\right)$$
$$= -\sum_i \mu_1(\Delta_i) \, \eta\left(\frac{x_i}{\mu_1(\Delta_i)}\right) \qquad (32)$$
$$= -H(f_\mathbf{x}),$$

where H is defined in (20) and $f_\mathbf{x}$ is the function on Ω_1 defined by

$$f_\mathbf{x}(\omega_1) = x_i/\mu_1(\Delta_i) \quad \text{if} \quad \omega_1 \in \Delta_i, \quad i = 1, 2, \ldots.$$

So, in this case, there is a simple relation between the measures of the sets forming the partition $\mathcal{P}$ on the one hand and Boltzmann's H on the other (also mentioned by Lanford,[2] pp. 81–82).

In the more complicated case where the elements of the partition $\mathcal{M}$ are regions in l_1 rather than single points, we may expect a similar result to hold, provided that the logarithm of the number of points in each region M of l_1 that satisfies (30) increases more slowly than N for large N. One way of satisfying this condition is to use regions $M_\mathbf{x}$ having the form

$$M_\mathbf{x} = \{(\xi_1, \xi_2, \ldots) : |\xi_i - x_i| < \delta x_i, \quad i = 1, 2, \ldots\},$$

where the numbers δx_i are chosen so that the number of them greater than $1/N$ increases more slowly than $N/\log N$ as N increases. For example, we could require $\delta x_n < \text{const} \cdot n^{-1-\epsilon}$, where $\epsilon > 0$. Provided that the regions M are chosen in such a way, we would again expect to find, for large N,

$$\frac{1}{N} \log \mu(\psi^{-1}(M_\mathbf{x})) \simeq -H(f_\mathbf{x}), \qquad (33)$$

provided that $\Sigma x_i = 1$.

Now we can consider the coarse-grained entropy $h(\rho, \mathcal{P})$. Of course, this depends very much on the phase-space density ρ. We are interested here in the type of phase-space densities considered by Lanford. These have the property that, in the Boltzmann-Grad limit, the one- and two-particle distributions are related by

$$\lim_{N \to \infty} f_2(\omega_1, \omega_2) = f_1(\omega_1) f_1(\omega_2). \qquad (34)$$

From this it follows, as noted by Lanford (page 98 of his article),[2] that the occupation number fractions $N_i(\omega)/N$ converge in probability to the numbers $\int_{\Delta_i} f_1(\omega_1) d\omega_1$. Consequently, the probability distribution over the various elements of the partition $\mathcal{M}$ in l_1 will, in the limit, be concentrated on the element of that partition containing the point

$$\mathbf{x}(f_1) = (x_1, x_2, \ldots) \quad \text{with} \quad x_i = \int_{\Delta_i} f_1(\omega_1) d\omega_1. \qquad (35)$$

The probability distribution over the partition $\mathcal{P}$ in Ω will therefore become concentrated on the element $\psi^{-1}(\mathbf{x}(f_1))$. In accordance with (6), we may therefore expect to find that, in the Boltzmann-Grad limit, $h(\rho,\mathcal{P})$ is asymptotically equivalent to $k \log \mu(\psi^{-1}(\mathbf{x}(f_1)))$. That is, we expect to find

$$\lim_{N\to\infty} \frac{1}{N} h(\rho, \mathcal{P}) = \lim_{N\to\infty} \frac{1}{N} k \log \mu(\psi^{-1}(\mathbf{x}(f_1))) \tag{36}$$

$$= -kH(\tilde{f}_1),$$

where $\tilde{f}_1$ is the "coarse-grained" version of f_1, defined by

$$\tilde{f}_1(\omega_1) = \int_{\Delta_i} f(\omega_1)d\omega_1/\mu(\Delta_i), \qquad \omega \in \Delta_i. \tag{37}$$

If we now take a further limit where the partition $\{\Delta\}$ of Ω_1 becomes infinitely fine, the result will be

$$\lim_{\{\Delta\}} \lim_{N\to\infty} \frac{1}{N} h(\rho,\mathcal{P}) = -kH(f_1). \tag{38}$$

Combined with (29), this gives a formula of the same type as (24), which we would like to establish.

Clearly, these remarks are very far from being a rigorous argument. All they do is make it plausible that, for the hard-sphere system in the Boltzmann-Grad limit, the definition of entropy proposed in GP is equivalent to the one suggested by Boltzmann and, hence, that the nonincrease property of Boltzmann's H can be regarded as a special case of the nonincrease property of the coarse-grained entropy $h(\rho,\mathcal{A})$, with a suitable choice of $\mathcal{A}$.

Entropy and Irreversibility

Having looked at the example of the hard-sphere gas in some detail, we can now draw some general conclusions about the GP definition of entropy in (5).

The virtue of the definition is that it gives us a way of ensuring that the nondecrease property is satisfied, and in a way such that the entropy so defined approaches the equilibrium entropy as $t \to +\infty$ and the fine-grained entropy as $t \to -\infty$. However, this nondecrease property arises not from any deep physical principle but from the mathematical condition, (15), which we require the algebra $\mathcal{A}$ to satisfy. If we had used, instead, an algebra $\mathcal{B}$, say, satisfying the time inverse of (15),

$$\mathcal{B} \subset \phi_t \mathcal{B} \qquad \text{if} \quad t < 0, \tag{39}$$

then we would have obtained an "entropy" that decreased with time instead of increasing. The "entropy" so defined would not have been a serious candidate for consideration as the physical entropy, whereas, as the discussion above shows, it is plausible that $h(\rho,\mathcal{A})$, with $\mathcal{A}$ satisfying the original condition, (15), can, for suitable $\mathcal{A}$, correspond well with our normal ideas about the physical entropy.

Since irreversibility is built into our definition of $h(\rho, \mathcal{A})$, one is led to ask, What is it that makes $h(\rho,\mathcal{A})$ suitable for use in an entropy definition whereas the time inverse $h(\rho,\mathcal{B})$ is not? I think the answer must lie in the nature of the phase-space densities, ρ, that are appropriate for describing a physical system undergoing an irreversible process. If we use the construction (**16**) to obtain $\mathcal{A}$ and $\mathcal{B}$ from a time-symmetric algebra, $\mathcal{P}$, that models the observations at a particular time, then we have

$$\left.\begin{aligned} \mathcal{A} = \mathcal{P}^- = \bigvee_{t=0}^{\infty} \phi_{-t}\mathcal{P} \\ \mathcal{B} = \mathcal{P}^+ = \bigvee_{t=0}^{\infty} \phi_{t}\mathcal{P} \end{aligned}\right\} \tag{40}$$

Then the results of the previous section indicate that

$$h(\rho,\mathcal{A}) \simeq h(\rho,\mathcal{P}), \tag{41}$$

but

$$h(\rho,\mathcal{B}) \neq h(\rho,\mathcal{P}). \tag{42}$$

That is to say, the partition $\mathcal{A}$ "fits" the phase-space density ρ in a way that $\mathcal{B}$ does not.

It appears to be a fundamental principle of nonequilibrium statistical mechanics that some restriction must be placed on the phase-space densities that appropriately describe a nonequilibrium ensemble.[3,4] It is clear that this restriction must be unsymmetrical under time reversal, but not at all clear how it should be formulated. But the above discussion gives us a hint that a possible formulation of this unsymmetrical condition may be given by the approximate equation, (**41**), with $\mathcal{A}$ defined as in (**40**). The fact that the condition on ρ is only approximate is not necessarily a disadvantage: there may (except in special limiting cases such as the Boltzmann-Grad limit) be no precise condition that a good phase-space density must satisfy and it may be, in general, only at equilibrium that the two sides of (**41**) can be precisely equal. Indeed, the difference between the two sides of (**41**) may give us some measure of how appropriate the phase-space density ρ is as a representation of physically possible processes in a system out of equilibrium.

References

1. GOLDSTEIN, S. & O. PENROSE. 1981. An H-theorem for dynamical systems. J. Stat. Phys. **24**: 325–43.
2. LANFORD, O. E. 1975. Time evolution of large classical systems. *In* Dynamical Systems, Theory and Applications, Lecture Notes in Physics, Vol. 38. J. Moser, Ed.: 1–111. Springer-Verlag. New York.
3. PENROSE, O. 1980. Foundations of statistical mechanics. Rep. Prog. Phys. **42**: 1937–2006.
4. LEBOWITZ, J. L. 1981. Microscopic dynamics and macroscopic laws. This volume.

MICROSCOPIC DYNAMICS AND MACROSCOPIC LAWS*

Joel L. Lebowitz†

Institute for Advanced Study
Princeton, New Jersey 08540

Time present and time past
Are both perhaps present in time future,
And time future contained in time past.
If all time is eternally present
All time is unredeemable.

T. S. Eliot
Four Quartets, Burnt Norton. 1935.

INTRODUCTION

These lines from one of my favorite poems show that scientists are not the only ones who care about the nature of time. The poet describes how, despite our experience of time as ever flowing and continually changing, with the past inherently different from the future—"time's arrow"—we also feel the simultaneous coexistence of past and future, as in a trajectory.

The problem of time's arrow is one with which philosophers and scientists have often grappled.[1] Indeed, it is hard to believe that there is anything left to say on this subject that has not been said better before (not to mention later). Yet every time I try to think through the relation between microscopic laws and the observed irreversible behavior of macroscopic objects I get quite confused. Can one really explain the latter *solely* on the basis of the former? If not, then what else has to be added? This note is an attempt to clarify some of these questions.

Let me begin by saying that I start from the assumption that the deterministic, reversible microscopic laws (classical or quantum) apply to the evolution of macroscopic systems, e.g., to the evolution of the local density in an initially nonuniform fluid. Similarly, I believe that irreversibility is not due to approximations, i.e., the diffusion equation is made irreversible by more than just some small missing term. There remains, therefore, the question of how to derive the diffusion equation or the Boltzmann equation from microscopic dynamics. I deliberately said derive rather than reconcile because I do not believe that there is any contradiction, but there certainly is a need for a convincing mathematical derivation. A derivation would at least help dispel some of the confusion surrounding the subject. It may, however, not convinc-

---

*This research was supported, in part, by a grant from the National Science Foundation, no. PHY 78-15920.

†Permanent address: Departments of Mathematics and Physics, Rutgers University, New Brunswick, N.J. 08903.

ingly answer all questions; for instance, Why are the assumptions about initial states the right ones?

To illustrate what I mean, let me consider a somewhat idealized version of a typical time—asymmetric macroscopic event. It will contain parts that may not be easy or even possible to achieve in practice—I will consider an isolated system, use a classical description, etc. The relevant question, however, is not simply, What idealization can actually be carried out practically? but rather, What is the right idealization in order to understand rationally and be able to predict what will be observed at a particular level of precision under given circumstances? (Misunderstanding this leads to statements like the one I read some years ago in a respectable popular magazine that Galileo was wrong and Aristotle right in the latter's assertion that heavy objects fall faster than light ones under the action of gravity; just drop a feather and a penny together.) I will assume from now on, and this may be wrong, that quantum mechanics does not play an essential role in the qualitative understanding of the irreversible behavior of macroscopic systems.

Example

The situation I wish to consider is illustrated in FIGURE 1. There we have a box Λ, 10 cm on each side, divided into two equal parts connected by a channel. There are, altogether, $N \simeq 10^{21}$ atoms in the box and the macroscopic experiment or observation consists of determining the numbers N_1 and N_2 of particles in the left and right parts, Λ_1 and Λ_2, $N_1 + N_2 = N$. Consider first just the left side of the figure. The sequence is a qualitative representation of a macroscopic experiment carried out as follows: We start with a plug in the hole of the box and fill the left side Λ_1 with a gas, say helium, at room temperature and 0.01 atmospheric pressure. We then wait for a few minutes and then, at time $t = t_1$, the beginning of our observations, we remove the plug. The configurations x_j are my imagined ones at the observations times t_j, $j = 1, 2, 3, 4$. No surprise here as long as I tell you that $t_1 < t_2 < t_3 < t_4$. The system goes from a highly nonuniform density to a uniform one, which is precisely the kind of behavior we are used to seeing in such experiments. It is, I am sure you will agree, consistent with Hamilton's equations of motion for a system of particles interacting with a Lennard-Jones or hard sphere pair potentials. It is also true, and consistent with the microscopic equations, that the evolution of the density, kinetic energy, and other similar quantities are described *for the times observed* very accurately by an irreversible kinetic equation, e.g., the Boltzmann equation. (If you don't believe this, try a computer experiment.)

What *is* inconsistent is to say that the behavior of the system will be accurately described by the Boltzmann equation in *all* imaginable situations. Thus, if we imagine that, at time t_2, we somehow put this system in the microscopic state y_2 obtained from x_2 by reversing all velocities, $y_2 = \overline{x}_2$, then, for the time interval (t_2, t_3), we will not observe that the density becomes more uniform, as is predicted by the Boltzmann equation. What should happen to such a system is shown in the right side of FIGURE 1, including, on top, the microscopic state at t_1 that would give y_2 at t_2 without further intervention. The appropriate question then is, Why are we typically, or essentially always, in a situation where Boltzmann's equation (or other irreversible equations) makes the right predictions? This has to do with what is going to happen (or is likely to happen) when we carry out certain types of experiments or observations on a *macroscopic* system prepared explicitly or implicitly in a certain way.

In terms of the example given above, the question is, Why is it that, when we observe the density of a gas in the position shown at t_2 in the second row of pictures, we can quite safely predict that it will follow the course on the left rather than the one on

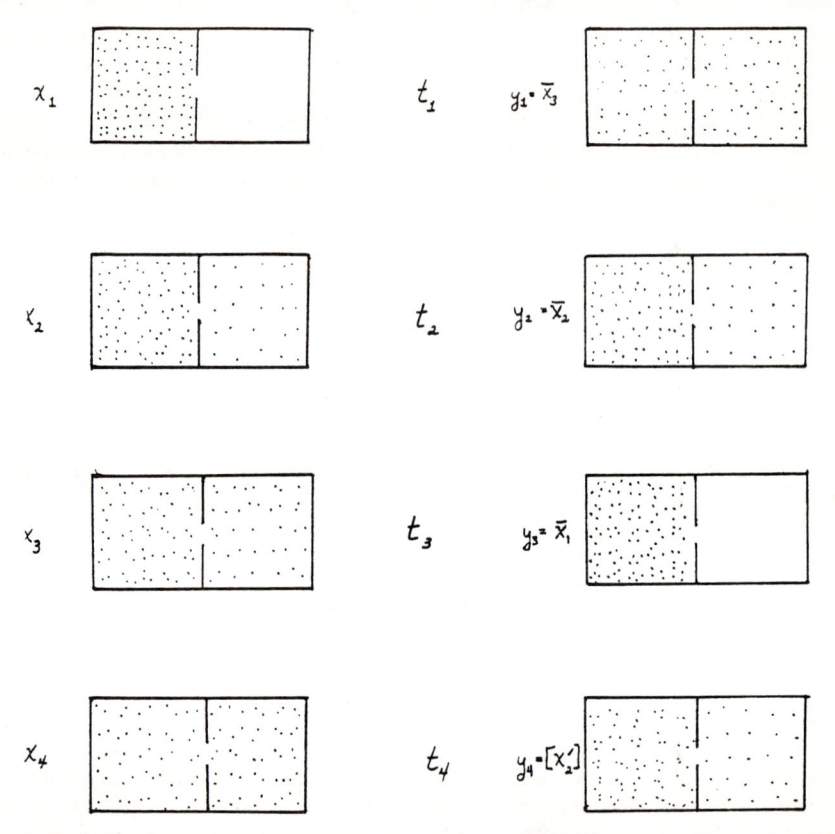

FIGURE 1. Left: the typically observed evolution of an isolated microscopic system. Right: an untypical evolution.

the right? The apparent answer is that I gave you a prescription for constructing the pictures on the left, but I don't know how to construct state y_2 or y_1. There are, of course, situations, like Hahn's spin echo experiments,[2] in which a spin state "similar" to our y_2 is produced. The spin reversal is, however, considered to be something special. Also, the "isolation" of the spin system is relatively low and interactions with other degrees of freedom in the system, which have not been reversed, mean that the effect is limited and of short duration. This has permitted us to accept this occasional "antiuniformization" behavior in spin systems without modifying our predictions about the course of events in general macroscopic systems such as our example following the observation at time t_2. Unfortunately, or fortunately, no one has succeeded in reversing all velocities in a macroscopic system and all our experience corresponds to the sequence on the left rather than the one on the right. We explain this behavior intuitively by saying that the amount of "phase space volume" consistent with a macroscopic observation of N_1 and N_2 and energy E increases in the left sequence. But, since each trajectory on the left has a counterpart trajectory on the right, the preparation of the system at t_1 must somehow be relevant in assigning appropriate probabilities to different configurations and this is where we need more understanding.

To repeat then: we know how to construct certain kinds of states of macroscopic systems (and not others). The time evolution of "observables" in these states, isolated *after* their construction, is described, over appropriate time periods, by time-asymmetric (irreversible) equations. What is still required then is: (1) a criterion that characterizes the kind of states for which the kinetic equations hold, and (2) a convincing explanation of why we cannot, in general, construct the other kind of states. The first question is a precise mathematical one and one of the points of this note is to describe its answer in a low density limiting case. This will also furnish a concrete example of a derivation of an irreversible macroscopic law from microscopic dynamics. No attempt will be made to answer the second question. I don't even know whether it can be answered on the basis of dynamics alone or requires a separate principle—valid for our universe. Some general points, relevant to both questions, concerning our assumption that we can focus our attention on an isolated system, will now be discussed.

SYSTEM ISOLATION

The fact that we have not, so far, taken outside perturbations into account is certainly a neglect in our analysis. I do not, however, feel that this is really very relevant to the example considered, i.e., I do not believe that an isolated macroscopic system would behave differently for the *times considered*. Certainly neither the Boltzmann nor other kinetic equations include any terms due to cosmic rays or other outside interactions. It seems, therefore, inappropriate to invoke such outside perturbations in justifying the derivation of irreversible kinetic equations. This does not necessarily mean, however, that the "outside universe" does not play a role, perhaps even a central role, in determining the arrow of time. Many people certainly believe that they do, c.f. Reference 1c, but that concerns the second point and I shall try to see how far we can go without involving any "cosmological principles."

It should be noted, however, that even very small interactions with the outside world are able to "greatly affect" the unstable microscopic trajectory of a macroscopic system.[1] Again, however, the question is, Just how different would the macroscopic behavior be if there were absolutely no outside interference?—my stated assumption is that it would not be very different for the times observed.

Poincare's famous theorem about isolated systems should, however, be mentioned here.[3] The theorem states that if we surround the trajectory of an isolated

Hamiltonian system by a tube of "diameter" d, then—since the energy surface has finite area—any point on the trajectory will be inside the tube infinitely often, no matter how small δ. This means that, if we *continue* to observe the system in FIGURE 1, we will see it returning again and again to configurations "close" to x_1, to x_2, etc. Boltzmann's equation, or, indeed, any equation predicting an approach to uniformity without reversal, cannot, therefore, even be an approximately valid description of an isolated system for *all* times. These "recurrence times" are, however, likely to increase very rapidly with the size of the system—being longer, probably, than the age of the universe for $N \simeq 10^{21}$. As Boltzmann is supposed to have told Zermelo, who raised this objection to Boltzmann's equation, "You should live so long."[4]

This extrapolation to *arbitrary times* is, therefore, invalid for isolated systems. But since no experiment lasts for a very long time, this need not worry us. It is just that if we want to rigorously derive kinetic equations, then it appears easier to consider situations in which they would be formally valid for all times. Then the idealization of an isolated finite system is not appropriate. On the other hand, when we consider ensembles—as we shall soon do—this is not a problem even for an isolated finite system if it is at least mixing. For such systems, ensembles can and do approach a uniform (coarse grained) state.[5] Just how important such ergodic properties are for real systems is not clear at present.

It is certainly true, however, that good ergodic properties reflect strong dynamical instabilities of trajectories. Thus, any external perturbations are likely to completely destroy this recurrence for dynamically unstable systems. Even more important, perhaps, is the fact that, due to the dynamical instability of a system's trajectory, any reversal of velocities that is not absolutely precise might make the subsequent motion behave very differently from the exactly reversed one. Doing a nonexact velocity reversal might then not have much observable effect on a real system. I shall come back to this point at the end.

STATISTICAL MECHANICS

The above discussion is clearly far from conclusive and, while it would be possible to go on talking about it, I am not sure that it would clarify things further. More likely it would just add to the confusion. I will, therefore, go on to a more concrete analysis of this problem. The central question is, How can we describe, in a way amenable (at least in principle) to quantitative study, the observed "typical" behavior of macroscopic systems? The answer to this is our beloved statistical mechanics: the study of dynamics combined with probability. Here we replace the study of the microscopic trajectory of a macroscopic system (prepared initially by some macroscopic means) by the study of the time evolution of an initial ensemble. The question is now simpler: How do we characterize the appropriate ensembles? This question is discussed in detail, although not entirely resolved, in a recent article by Oliver Penrose,[6] which I recommend to you highly. His article also contains a discussion of various approaches to the problem of irreversibility and a very extensive list of references.

Instead of considering the general problem, however, I would like to discuss in more detail a further, more drastic idealization of the above example for which one can actually prove some results. This involves consideration of a well-defined limit first introduced by Harold Grad.[7] It is the appropriate idealization of a dilute classical gas for which the Boltzmann equation ought to hold exactly and, therefore, might perhaps be proven rigorously. Grad's program was carried out brilliantly by Oscar Lanford, with certain limitations.[8]

I will now use the Lanford Theorem to make the above discussion more precise and, therefore, hopefully, more clear. My remarks are taken from a joint paper with van Beijern, Lanford, and Spohn.[9] In that paper, the Lanford theorem was extended to all times for certain initial states corresponding to the motion of a test particle in a dilute gas in equilibrium. In the considerations here, however, I deal with the original theorem, as explicated in Kiang's thesis,[10] so that, in order to be strictly applicable, the observation times t_j would have to be much sooner than those indicated in FIGURE 1. (Truth is a necessary but not sufficient condition for a mathematical proof.)

LANFORD'S THEOREM

We consider a system of hard spheres of diameter ϵ and unit mass inside a box Λ. The spheres are elastically reflected among themselves and at the boundary of Λ. Let the state of the system be specified by the absolutely continuous distribution functions $\{\rho_n^\epsilon | n > 0\}$. These satisfy the BBGKY equation for hard spheres:‡

$$\frac{\partial}{\partial t} \rho_n^\epsilon (x_1, \ldots x_n, t) = H_n^\epsilon \rho_n^\epsilon (x_1, \ldots x_n, t)$$

$$+ \epsilon^2 \sum_{j=1}^n \int_{R^3} dp_{n+1} \int_{S^2} d\omega \, \omega \cdot (p_{n+1} - p_j) \rho_{n+1}^\epsilon (x_1, \ldots x_n, q_j + \epsilon\omega, p_{n+1}, t).$$

(1)

Here

$$x_i = (q_i, p_i) \in \Lambda \times R^3,$$

where ω is a unit vector in R^3 and $d\omega$ is the surface measure of the unit sphere S^2 in three dimensions. H_n^ϵ describes the evolution of n hard spheres of diameter ϵ inside Λ. The solutions of the BBGKY hierarchy are denoted by

$$\rho_n^\epsilon (x_1, \ldots x_n, t) = (V_t^\epsilon \rho^\epsilon)_n (x_1, \ldots x_n) \tag{2}$$

for the initial vector of distribution functions $\rho^\epsilon = (\rho_1^\epsilon, \rho_2^\epsilon, \ldots)$.

We want to study the low density Boltzmann-Grad limit of the solutions of the BBGKY hierarchy. This limit is obtained by letting the fraction of volume occupied by the particles, $\rho\epsilon^3$, with ρ the average density, go to zero while keeping the mean free path of the hard spheres, $\simeq 1/\epsilon^2\rho$, constant. This requires that, as $\epsilon \to 0$, the density be increased as ϵ^{-2}. Therefore, for each value of the diameter ϵ, one chooses an initial state with distribution functions ρ_n^ϵ such that $\rho_n^\epsilon \simeq \epsilon^{-2n}$. With this in mind, we define the rescaled distribution functions:

$$r_n^\epsilon(x_1, \ldots x_n) = \epsilon^{2n} \rho_n^\epsilon(x_1, \ldots x_n). \tag{3}$$

Regarding the sequence $\{r_n^\epsilon | n \geq 0\}$ as the vector r^ϵ, one can write (1) compactly as

---

‡The derivation of (1) requires some careful analysis; see, for example, References 8 and 10.

$$\frac{d}{dt} r^{\epsilon}(t) = H^{\epsilon} r^{\epsilon}(t) + C^{\epsilon} r^{\epsilon}(t), \tag{4}$$

where H^{ϵ} is a diagonal matrix with entries H_n^{ϵ} and C^{ϵ} is a matrix with entries $C_{n,n+1}^{\epsilon}$ and zero otherwise.

For $t > 0$, the time evolution of $r_n^{\epsilon}(t)$ is determined by backward streaming. Therefore, it seems natural to replace, for a collision, the phase point $(x_1, \ldots q_j, p_j, \ldots q_j + \epsilon\omega, p_{n+1})$ with outgoing momenta by the phase point $(x_1, \ldots q_j, p'_j, \ldots q_j + \epsilon\omega, p'_{n+1})$ with incoming momenta. (These are just two different representations of the same phase point.) This leads to

$$\frac{\partial}{\partial t} r_n^{\epsilon}(x_1, \ldots x_n, t) = H_n^{\epsilon} r_n^{\epsilon}(x_1, \ldots x_n, t)$$

$$+ \sum_{j=1}^{n} \int_{+} dp_{n+1} \, d\omega \, \omega \cdot (p_j - p_{n+1}) \tag{5}$$

$$\times \{r_{n+1}^{\epsilon}(x_1, \ldots q_j, p'_j, \ldots q_j - \epsilon\omega, p'_{n+1}, t)$$

$$- r_{n+1}^{\epsilon}(x_1, \ldots q_j, p_j, \ldots q_j + \epsilon\omega, p_{n+1}, t)\},$$

where $\int_{+}$ indicates that the integration over ω is restricted to the upper hemisphere $\omega \cdot (p_j - p_{n+1}) \geq 0$.

Formally, the limiting form of (5), which the distribution functions $r(t) = \lim_{\epsilon \to 0} r^{\epsilon}(t)$ might satisfy for $t \geq 0$, is obtained by simply setting $\epsilon = 0$ in (5):

$$\frac{\partial}{\partial t} r_n(x_1, \ldots x_n, t) = \sum_{j=1}^{n} p_j \frac{\partial}{\partial q_j} r_n(x_1, \ldots x_1, t)$$

$$+ \sum_{j=1}^{n} \int_{+} dp_{n+1} \, d\omega \, \omega \cdot (p_j - p_{n+1}) \tag{6}$$

$$\times \{r_{n+1}(x_1, \ldots q_j, p'_j, \ldots q_j, p'_{n+1}, t)$$

$$- r_{n+1}(x_1, \ldots q_j, p_j, \ldots q_j, p_{n+1}, t)\}$$

(Implicitly, the free motion $-\sum_{j=1}^{n} p_j \partial/\partial q_j$ includes the specular reflection at the boundaries of Λ.)

For $t < 0$, the time evolution of $r_n^{\epsilon}(t)$ is determined by forward streaming. In that case, for a collision, the phase point with incoming momenta should be replaced by the phase point with outgoing momenta. The formal limit of the resulting equation is then again (6), but with the sign of the collision term reversed.

Equation 6 for $t \geq 0$ (and with the sign of the collision term reversed for $t \leq 0$) is called the Boltzmann hierarchy, which can be written in the form

$$\frac{d}{dt} r(t) = H r(t) + C r(t). \tag{7}$$

Let $r^{\epsilon}(t) \equiv V^{\epsilon}(t) r^{\epsilon}(0)$ and $r(t) \equiv V(t) r(0)$ be the solutions of (4) and (7), respectively, as defined (for example) by the Dyson series with $r(0) = \lim_{\epsilon \to 0} r^{\epsilon}(0)$.

To prove that $r^\epsilon(t)$ converges to $r(t)$, for $t \neq 0$ as $\epsilon \to 0$, we need two conditions.

First, the initial distributions $r^\epsilon(0)$ have to be uniformly bounded in ϵ. This guarantees the uniform convergence of the Dyson series solution for some interval $|t| < t_0$. If h_β denotes the normalized Maxwellian at inverse temperature β, then a suitable choice for this bound is as follows:

Condition 1. There exist a pair (z, β) such that

$$r_n^\epsilon(x_1, \ldots x_n) \leq M z^n \prod_{j=1}^{n} h_\beta(p_j) \tag{8}$$

for all $\epsilon < \epsilon_0$ with a positive constant M independent of ϵ.

Second, $r_n^\epsilon(0)$ has to converge to $r_n(0)$ in such a way that the Dyson series for $r^\epsilon(t)$ converges term by term to the series for $r(t)$. For the initial phase point $x^{(n)} = (x_1, \ldots x_n) \in (\Lambda \times R^3)^n$, let $q_j(t, x^{(n)})$, $j = 1, \ldots n$, be the position of the jth point particle at time t under the free motion. Then

$$\Gamma_n(t) = \{x^{(n)} = x_1, \ldots x_n \in (\Lambda \times R^3)^n | q_i(s, x^{(n)}) \neq q_j(s, x^{(n)})\} \tag{9}$$

for $i \neq j = 1, \ldots n$ and $-t \leq s \leq 0$ if $t \geq 0$, $0 \leq s \leq -t$ if $t \leq 0\}$.

In words, $\Gamma_n(t)$ is the restriction of the n-particle phase space to the set of phase points that, under free backward streaming over a time t, if t is positive (or free forward streaming over a time $|t|$, if t is negative), do not lead to a collision between any pair of particles regarded as point particles. By this restriction only a set of Lebesgue measure zero is excluded from $(\Lambda \times R^3)^n$.

Note that (1) $\Gamma_n(t)$ depends only on the free motion, (2) $\Gamma_n(t) \subset \Gamma_n(t')$ for $t' = \alpha t$, $\alpha \leq 1$, (3) $\Gamma_n(t) \neq \Gamma_n(-t)$, and (4) $x^{(n)} \in \Gamma_n(t)$ is equivalent to $\bar{x}^{(n)} \in \Gamma_n(-t)$, where $\bar{x}^{(n)} \equiv R x^{(n)}$ is the phase point obtained from $x^{(n)}$ under the reversal $p_j \to -p_j$. In particular, $\Gamma_n(t)$ is not invariant under reversal of velocities.

The suitable choice of convergence is then:

Condition 2. There exists a continuous function r_n on $(\Lambda \times R^3)^n$ such that

$$\lim_{\epsilon \to 0} \epsilon^{2n} \rho_n^\epsilon = \lim_{\epsilon \to 0} r_n^\epsilon = r_n \tag{10}$$

uniformly on all compact sets of $\Gamma_n(s)$ for some $s > 0$.

THEOREM (*Lanford*). Let $\{\rho_n^\epsilon | n \geq 0\}$ be a sequence of initial distribution functions of a fluid of hard spheres of diameter ϵ inside a region Λ and let the sequence $\{r_n^\epsilon | n \geq 0\}$ of rescaled distribution functions satisfy Conditions 1 and 2. Let $r_n^\epsilon(t)$ be the solution of the BBGKY hierarchy with initial conditions r_n^ϵ and let $r_n(t)$ be the solution of the Boltzmann hierarchy with initial conditions r_n.

Then there exists a $t_0(z, \beta) > 0$ such that, for $0 \leq t \leq t_0(z, \beta)$, the Dyson series for (4) and (7) converge and such that $r_n(t)$ satisfies a bound of the form of Condition 1 with $z' > z$ and $\beta' < \beta$. Furthermore,

$$\lim_{\epsilon \to 0} r_n^\epsilon(t) = r_n(t) \tag{11}$$

uniformly *on compact sets of* $\Gamma_n(s + t)$.

For $-t_0(x, \beta) \le t \le 0$, (11) holds, provided that, in Condition 2, $s \le 0$ and that, in the Boltzmann hierarchy, the collision term $C_{n,n+1}$ is replaced by $-C_{n,n+1}$.

Remark. An interesting property of the Boltzmann hierarchy is the well-known "propagation of chaos": if the initial conditions of the Boltzmann hierarchy factorize,

$$r_n(x_1, \ldots x_n) = \prod_{j=1}^{n} f(x_j), \tag{12}$$

then the solutions with this initial condition stay factorized,

$$r_n(x_1, \ldots x_n, t) = \prod_{j=1}^{n} f(x_j, t), \tag{13}$$

where $f(x, t)$ is the solution of the Boltzmann equation

$$\frac{\partial}{\partial t} f(q, p, t) = -p \frac{\partial}{\partial q} f(q, p, t) + \int_{+} dp_1 d\omega \, \omega \cdot (p - p_1) \tag{14}$$

$$\times \{f(q, p', t) f(q, p'_1, t) - f(q, p, t) f(q, p_1 t)\},$$

with initial condition $f(q, p)$.

Note, however, that, even when (12) is not satisfied, the solutions $r_n(t)$ of the Boltzmann hierarchy are *not* reversible while the $r_n^\epsilon(t)$ are—so what has happened?

The answer lies in the fact that the set on which the $r_n^\epsilon(t)$ converge to $r_n(t)$ gets smaller and smaller as t increases, $\Gamma_n(s + t_2) < \Gamma_n(s + t_1)$ for $t_2 > t_1 \ge 0$ and $\Gamma(t) \ne \Gamma(-t)$.

When I get confused at this point (this happens at least four out of five times), it sometimes helps me to consider the following picture (suggested by H. Spohn):

Let the phase space of the system $\bigcup_n (\Lambda \times R^3)^n$ be represented schematically by the (u,v) plane. Let the complement of $\Gamma(z) = \bigcup_n \Gamma_n(z)$ correspond to the line segment $\Gamma^c(z) = \{u,v \{ 0 \le u = v \le z\}$ for $z > 0$ and to the set $\{u,v \mid 0 \le -u = v \le |z|\}$ for $z < 0$. Then, for the case where Condition 2 is satisfied on $\Gamma_n(s)$, $s \ge 0$, the convergence at $t \ge 0$ holds on the set $\Gamma(s+t)$ but

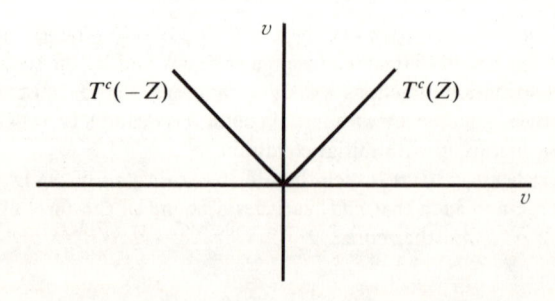

may not hold on $\Gamma^c(s+t)$. If we now reverse all velocities at t, then the convergence of the new r_n^ϵ at t will be on $\Gamma(-s-t)$, but need no longer hold on $\Gamma(z)$ for any $z \ge 0$.

Therefore, since Condition 2 is no longer satisfied for any $s > 0$, the Lanford theorem need not hold for any time $t_2 + \tau$, $\tau > 0$.

Armed with these mathematical weapons, let us now return to our previous example.

Example Revisited

Let us now consider our example (FIGURE 1) from the point of view of ensembles. The initial state, at $t = 0$, corresponds to a canonical Gibbs state of N hard spheres of diameter ϵ, all in the left half box Λ_1. It is clear that, since the initial state is invariant to reversal of velocities, its distribution functions, $\rho^\epsilon = (\rho_1^\epsilon, \rho_2^\epsilon, \ldots)$, satisfy the equality

$$V_t^\epsilon \rho^\epsilon = RV_{-t}^\epsilon \rho^\epsilon, \tag{15}$$

where

$$(R\rho)_n(q_1, p_1, \ldots q_n, p_n) = \rho_n(q_1, -p_1, \ldots q_n, -p_n). \tag{16}$$

Furthermore,

$$V_t^\epsilon(RV_t^\epsilon \rho^\epsilon) = \rho^\epsilon, \tag{17}$$

while

$$V_t^\epsilon(V_t^\epsilon \rho^\epsilon) = V_{2t}^\epsilon \rho^\epsilon. \tag{18}$$

Equation 17 states that if we reverse all velocities at time t, then the system, after another time interval t, will return to its initial state in which all the particles are in Λ_1.

Consider now the sequence of initial states with distribution functions ρ^ϵ in which, as $\epsilon \to 0$, the number of particles inside Λ_1 increases with fixed $N\epsilon^2 = z$. Then

$$\lim_{\epsilon \to 0} \epsilon^{2n} \rho_n^\epsilon(x_1, \ldots x_n) = \lim_{\epsilon \to 0} r_n^\epsilon(x_1, \ldots x_n) = r_n(x_1, \ldots x_n)$$
$$= \prod_{j=1}^n \{x_{\Lambda_1}(q_j) z h_\beta(p_j)\} \tag{19}$$

on $\Gamma_n(0)$, where X_{Λ_1} is the characteristic function of the set Λ_1, and, since Conditions 1 and 2 are satisfied, by Lanford's theorem,

$$\lim_{\epsilon \to 0} \epsilon^{2n} (V_t^\epsilon \rho^\epsilon)_n(x_1, \ldots x_n) = (V_t r)_n(x_1, \ldots x_n) = \prod_{j=1}^n \{f(x_j, p_j, t)\} \tag{20}$$

on $\Gamma_n(t)$ for $|t| < t_0(z, \beta)$, where $f(x, t)$ is the solution of the Boltzmann equation with initial conditions $f(q, p) = X_{\Lambda_1}(q) z h_\beta(p)$.

Let us now reverse the velocities at time t, $0 < t < t_0/2$, and let us consider $RV_t^\epsilon \cdot \rho^\epsilon$ as a new initial state. Clearly,

$$V_t(RV_t r) \neq r = \lim_{\epsilon \to 0} V_t^\epsilon(RV_t^\epsilon r^\epsilon), \tag{21}$$

according to (17), so the limiting r does not have the time reversibility of r^ϵ. Indeed, the Boltzmann H-function decreases up to t, remaims unchanged by R, and continues to decrease as $RV_t r$ is evolved for a time interval t.

At first sight, this seems to contradict Lanford's theorem, which appears to assert that the right side of (21) should, indeed, equal the left side. There is, however, no such contradiction for, while

$$\lim_{\epsilon \to 0} \epsilon^{2n} (V_t^\epsilon \rho^\epsilon)_n = (V_t r)_n \quad \text{on } \Gamma_n(t), \tag{22}$$

$$\lim_{\epsilon \to 0} \epsilon^{2n} (RV_t^\epsilon \rho^\epsilon)_n = (RV_t r)_n \quad \text{on } \Gamma_n(-t) \neq \Gamma_n(t+s), \quad \forall s \geq 0. \tag{23}$$

Therefore, continuing in the same time direction as before the reversal of velocities, $RV_t^\epsilon \rho^\epsilon$ no longer satisfies the Condition 2 of Lanford's theorem. The theorem asserts nothing about the convergence of $\epsilon^{2n}(V_t^\epsilon(RV_t^\epsilon\rho^\epsilon))_n$ as $\epsilon \to 0$. (Of course, by (17), we can say something about this limit. The point is that we cannot conclude from Lanford's Theorem that the limit is $(V_t (RV_t r))_n$, since Condition 2 is violated.) For the theorem to be applicable with the initial condition at time t, one has to consider either $V_t^\epsilon(V_t^\epsilon \rho^\epsilon)$ or $V_{-t}^\epsilon(RV_t^\epsilon\rho^\epsilon)$. In both cases, the system evolves further toward equilibrium, e.g., the third picture on the left or the first picture on the right of FIGURE 1.

CONCLUDING REMARKS

1. Lanford's theorem deals with correlations that are absolutely continuous with respect to Lebesgue measure; see (8) and (10). This singling out of Lebesgue measure, while "intuitively" very reasonable, cannot be justified on mathematical grounds alone. It may, however, not be really essential—there may be other physical conditions that rule out "bad" initial configurations of macroscopic systems, whether prepared in the laboratory or found in nature.

The essential physical ingredient in Lanford's derivation of the Boltzmann equation is that there are no recollisions in the Boltzmann-Grad limit—real or virtual—and thus no correlations build up on $\Gamma(t)$ for initially uncorrelated states. This justifies, then, under precise conditions, the usual derivation of the Boltzmann equation in which one assumes no correlations prior to collisions.[6] For small ϵ, one would expect that (for a macroscopic system $\Lambda \to \infty$, $N \to \infty$) $r_n^\epsilon(t)$ would be "close" to $r_n(t)$ for fixed t and n, the difference becoming larger as n or t increases.

The limitation of Lanford's theorem to $t < t_0$, t_0 about a mean free time appears to be entirely technical. It is Condition 1 (not Condition 2) that fails for $t > t_0$. This prevents a proof of the convergence of the Dyson series for either $r_n^\epsilon(t)$ or $r_n(t)$ for longer times. Indeed, there is no proof at present of the existence of solutions of the Boltzmann equation—for general initial conditions—at longer times. It seems entirely reasonable to expect that the theorem should hold for a large class of initial conditions for all times. Indeed, this is the case for the Lorentz gas where one derives the linear Boltzmann equation in the Boltzmann-Grad limit.[13] In any case, t_0 is large enough for the entropy to change significantly.

2. The irreversible Boltzmann hierarchy is consistent with the reversible BBGKY hierarchy, since the approximation by the Boltzmann hierarchy is valid only for a

particular class of initial states. Condition 2 excludes initial states such as the one just constructed by reversal of velocities. While the question of how to generally characterize good initial states for systems other than very dilute gases remains, this example does illustrate what form such an answer might take, i.e., condition 2. This is time asymmetric in just the right way[6]—the property is preserved under *forward* time evolution, V_t^r, but is not invariant under R.

More precisely, if Condition 2 is satisfied for some $s > 0$ (say 1 hour) then, if the state evolves for some t, $t > 0$, Condition 2 is still satisfied on the smaller set $\Gamma(s + t)$. However, if we do a reflection at t, then Condition 2 is no longer satisfied for forward times. The same statements are true for $s < 0$, $s < t < 0$. The initial state considered in the example is (1) symmetric under R and (2) such that Condition 2 is satisfied with $s = 0$. We can, therefore, derive either the forward or backward Boltzmann equation (both leading to identical uniformization of the density as $|t|$ increases), but we cannot go first in one direction and then "backtrack" by using R. It is this restriction which permits one to derive irreversible equations from reversible dynamics and symmetric initial conditions. Which way we actually go physically is determined by the fact that we first prepare the system, i.e. select the state, and then observe it.

3. I sometimes wonder how our conception of the "arrow of time" would change if a method was found for actually reversing velocities in a fluid (or change appropriate quantum phases). I don't mean reverse all the velocities in the universe, but just in systems like that of our example. Just something modest that would permit the right side of FIGURE 1 to represent the result of an actual experiment in which the system was isolated beginning with t_1. Would this be just like the spin echo experiment,[2] which is now almost forgotten, or would this substantially change our concept of time's arrow? Put differently, Do the laws of nature, as we understand them at present, exclude the possibility of ever observing the right side of FIGURE 1 in a real life experiment? The fact that trajectories are unstable may be relevant here—precluding sufficiently exact reversal of velocities to give a macroscopically observable effect—like Maxwell's demon, it would cost more in entropy than it would gain.

This instability of trajectories relates to the good ergodic properties that physical systems are believed to possess. So, while I have played down the role of ergodicity in providing the key to understanding irreversible macroscopic behavior, I do believe that macroscopic systems do generally have good ergodic properties. Their absence would, I think, lead to effects that have not been observed. Note, however, that Lanford's theorem never uses the ergodic properties of hard spheres—it is equally valid for cubes, with an appropriate modification of the collision kernel.

4. To emphasize the importance of the macroscopic size of the system in the observation of irreversible behavior, note that the example in FIGURE 1 would not make much sense if there were only three particles in the system. Systems with few degrees of freedom can certainly exhibit instabilities in their trajectories. They can even be Bernoulli, like the baker's transformation and the point particle moving among fixed convex scatterers (Sinai's billiard) are. In this case, certain types of initial ensembles will behave irreversibly, but a single trajectory will not exhibit irreversible behavior. For a macroscopic system, however, a simple trajectory can give observational results that we would call irreversible, as in the above example.

The "stochastic-type" behavior of trajectories of nonlinear dynamical systems with a few degrees of freedom can thus only serve as one ingredient in the derivation of kinetic equations describing the time evolution of real macroscopic variables. This is so despite the fact that the study of the consequences of good ergodic properties on the behavior of measures absolutely continuous to a given stationary measure may be directly relevant to the behavior of ensembles for macroscopic systems. The situation here is similar to, but much less well understood than, the situation in equilibrium. While the use of equilibrium Gibbs ensembles is formally similar for systems of few or many particles, the relation between ensemble averages and observations is quite different in the two cases. Only for macroscopic size systems can these be expected to (approximately) coincide. It should, in fact, be strongly emphasized that Lanford's theorem, like equilibrium results in the thermodynamic limit, are valid not only for ensemble averages but are true for almost all "systems," i.e., they occur with a probability of one in the ensembles considered.[8b]

5. Consider again the second row in FIGURE 1. Suppose we measure, at time t_2, the numbers N_1 and N_2, and also the total energy E of the system. We then want to predict the future behavior of this system without knowing anything else about its past or future history. Being statistical mechanicians, we would then construct an ensemble to represent the initial state at t_2 and use Liouville's equation (which is equivalent to the Hamiltonian equations of motion) for the time evolution of this ensemble. It would seem appropriate to use an initial ensemble that is symmetric under velocity reversal. It is also reasonable that the ensemble be a smooth function on the energy surface E.

One such ensemble is $\mu(dx) = \chi(x|N_1, N_2)dx$, where dx is the Liouville measure projected on the energy surface S_E, $H(x) = E$ on S_E, and $\chi(x|N_1, N_2)$ is the characteristic function of the set in which there are N_1 particles on the left side and N_2 particles on the right side, i.e., $\chi(x|N_1, N_2)$ is 1 or 0 depending on whether the phase point is consistent with the observation. This is the so-called generalized microcanonical ensemble for the use of which (or of its relatives, the generalized canonical or grand canonical ensembles) many "justifications" have been given.[6] In my opinion, none of the arguments is entirely convincing on logical grounds alone—but then perhaps neither are the arguments for equilibrium Gibbs ensembles. The important question is, How well will this predict the outcome of measurements? It seems clear on the basis of phase space volume arguments that the predictions would favor the left sequence over the right one, at least in a qualitative way. I, furthermore, think that the prescription would also work quantitatively for simple macroscopic systems, e.g., an inert fluid, after some "short" transient time necessary for the system to establish its own quasi-steady state consistent with the macroscopic constraints. In our example, this would presumably be something close to a product state with a one particle distribution given by the Chapman-Enskog solution of the Boltzmann Equation. What the appropriate quasi-steady state ensemble looks like for a more general system, even just a dense gas or an anharmonic crystal, is an open question. It is the big question in nonequilibrium statistical mechanics at the present time.[12] I offer two bottles of champagne and a bouquet of flowers to whomever shows any good progress on this problem.

6. We note, finally, that the Boltzmann hierarchy (7) does not have any flow underlying it in the phase space. Thus, unlike the BBGKY hierarchy (4), where the evolution of the states having these correlations can be implemented via the evolution of the phase points, there is no point transformation in the phase space that yields the

time evolution of the correlation functions given by (7). This is yet another manifestation of the "loss of information" resulting from the use of the Boltzmann-Grad limit—a loss necessary to make dissipative macroscopic laws consistent with reversible microscopic dynamics.

Acknowledgments

It is a great pleasure to thank P. G. Bergmann, S. Goldstein, O. Lanford, O. Penrose, I. Prigogine, H. Spohn, and the late P. Resibois for many useful discussions and arguments about the subject of irreversibility. Many thanks also to the organizers of this very pleasant meeting, particularly L. Reichel.

References

1a. GOLD, T. & D. I SCHUMACHER, Eds. 1967. The Nature of Time. Cornell University Press. Ithaca, N.Y.
1b. WHITROW, G. 1973. The Nature of Time. Holt, Rinehart and Winston. New York.
1c. LAYZER, D. 1975. Sci. Am. **233**: 56.
1d. KRYLOV, N. S. 1979. Works on the Foundation of Statistical Mechanics (A. B. Mugdal, Ya. G. Sinai, and Yu. L. Zeeman, translators). Princeton University Press. Princeton, N.J. (See the precis of the book written by A. B. Mugdal and V. A. Fock.)
1e. PRIGOGINE, I. 1980. From Being to Becoming: Time and Complexity in the Physical Sciences. W. H. Freeman. San Francisco.
1f. SAKHAROV, A. D. 1980. Zh. Eksp. Teor. Fiz. **79**: 669. (English translation: 1981. JETP **52**: 49.)
2. HAHN, E. L. 1950. Phys. Rev. **80**: 580; HARTMAN, S. & E. L. HAHN. 1962. Phys. Rev. **128**: 2024.
3. EHRENFEST, P. & T. EHRENFEST. 1911. Encykl. Mat. Wiss. IV 2.II. Taubner. Leipzig. (English translation: The Conceptual Approach in Mechanics. 1959. Cornell University Press. Ithaca, N. Y.
4. For a history of the subject, see BRUSH, S. G. 1976. The Kind of Motion We Call Heat, Vols 1 & 2. North Holland Publishing Co. Amsterdam.
5. LEBOWITZ, J. L. & O. PENROSE. 1973. Phys. Today. February, p. 23.
6. PENROSE, O. 1979. Rep. Prog. Phys. **42**: 1937.
7. GRAD, H. 1958. Principles of the Kinetic Theory of Gases. In Handbuch der Physik, Vol. 12, S. Flugge, Ed. Springer-Verlag. Berlin.
8a. LANFORD, O. E., III. 1975. In Dynamical Systems, Theory and Applications, Lecture Notes in Physics, Vol. 38. J. Moser, Ed. Springer-Verlag. Berlin.
8b. LANFORD, O. E., III. 1981. Physica A **106**: 70.
9. VAN BEYERN, H., O. E. LANFORD III, J. L. LEBOWITZ & H. SPOHN. 1980. J. Stat. Phys. **22**: 237.
10. KIANG, F. 1975. Ph. D. Thesis. University of California. Berkeley, Calif.
11. LEBOWITZ, J. L. 1978. Prog. Theor. Phys. Suppl. **64**: 35.
12. SPOHN, H. 1978. Commun. Math. Phys. **60**: 277.